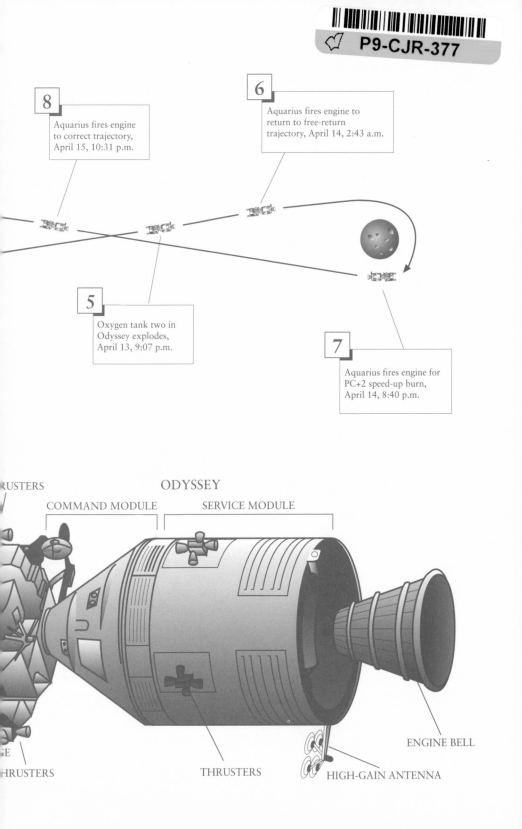

8
Aquarius fires engine to correct trajectory, April 15, 10:31 p.m.

6
Aquarius fires engine to return to free-return trajectory, April 14, 2:43 a.m.

5
Oxygen tank two in Odyssey explodes, April 13, 9:07 p.m.

7
Aquarius fires engine for PC+2 speed-up burn, April 14, 8:40 p.m.

ODYSSEY

RUSTERS

COMMAND MODULE   SERVICE MODULE

E
HRUSTERS   THRUSTERS

ENGINE BELL

HIGH-GAIN ANTENNA

APOLLO 13

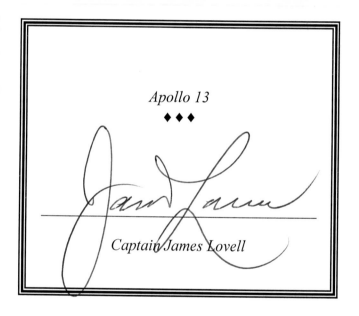

Apollo 13

♦ ♦ ♦

Captain James Lovell

Jim Lovell & Jeffrey Kluger

APOLLO 13

HOUGHTON MIFFLIN COMPANY

BOSTON • NEW YORK

Library of Congress Cataloging-in-Publication
Data is available.

ISBN 0-618-05665-3

Printed in the United States of America

MP 10 9 8 7 6 5 4

Book design by Robert Overholtzer

Endpaper art by Tech-Graphics

This book was previously published as *Lost Moon.*

*Frontispiece: Official Apollo 13 emblem,
courtesy of NASA*

This true adventure is dedicated to those earthbound astronauts: my wife, Marilyn, and my children, Barbara, Jay, Susan, and Jeffrey, who shared with me the fears and anxieties of four days in April, 1970.

— JIM LOVELL

With love to my family — nuclear and extended, past and present — for providing an always stable orbit.

— JEFFREY KLUGER

PREFACE

THE MEN of Apollo 13 had a lot of plans for the afternoon of April 21, 1970. If the schedule ran the way it was supposed to run — and there was no reason to assume it wouldn't — April 21 was the day they would splash down in the Pacific Ocean and climb aboard the deck of the helicopter carrier Iwo Jima, having just completed humanity's third and most ambitious landing on the moon. Unlike Apollos 11 and 12, which had been sent to the glass-smooth plains of the Sea of Tranquillity and the Ocean of Storms, Apollo 13 would be heading for a high-wire touchdown in the moon's Fra Mauro highlands. Negotiating a terrain that treacherous would require some sublimely good piloting, the kind that would prove not only the soundness of the translunar ships but the skill of the men who had been tapped to fly them. Pull off a mission like that, and the day you return to the familiar waters of the South Pacific ought to be a historic one indeed.

While the Apollo 13 crew — commander Jim Lovell and his rookie crewmates Jack Swigert and Fred Haise — indeed made history in the days leading up to April 21, it wasn't the kind they had expected. The mission they flew didn't involve pinpoint landings and memorable words and great, bounding bunny hops in the moon's otherworldly gravity. Rather, it involved an exploding oxygen tank and a mortally wounded spacecraft and a crew marooned 200,000 miles from Earth with no right to expect they could patch their ship together again, much less turn it around and nurse it back home.

Nonetheless, the Apollo 13 astronauts did come home, hitting the ocean not on April 21 but on April 17 — a four-day difference that represented the time it would have taken them to descend to the lunar surface, visit awhile, and take off again. Lovell, Swigert, and Haise could not forget how unlikely their safe return home was. But they could also not forget that those four days were something that dark fortune and flawed hardware would now forever deny them.

For a crew that had been assigned a snakebit spaceship like the flawed Apollo 13, there was always the possibility of finagling another lunar trip. It was the machines that had failed on this mission, after all, not the men. But Lovell, Swigert, and Haise instinctively knew they were not returning to space any time soon. The Apollo 13 astronauts had barely escaped with their necks during their week in space, and everyone aboard the recovery carrier realized it. The fact that they did escape was the longest of piloting long shots, and a space agency surviving on public goodwill was not about to push things now, packing the same three men into the same sort of ship, flinging them back into the void, and inviting fate to have a go at them again.

If the astronauts could not rewrite their personal story, however, they could still tell the one they had. Apollos 11 and 12 might have landed on the moon a year earlier; Apollos 8 and 10 might have orbited it before them. But no other crew — no Americans, no Soviets, not a soul who had ridden a rocket into space before — had ever come so heartstoppingly close to cashing it in out there and yet somehow managed to make it home. Refract the failed Apollo 13 mission through the right prism, and it started to look an awful lot like a success.

When the Apollo 13 crewmen were plucked from the ocean and flown to the carrier on April 17, they didn't have much opportunity to talk. But as the afternoon wore on and the welcoming ceremonies wound down, and Lovell, Swigert, and Haise stood with a handful of carrier officers, saying their final thanks in the crisp blue jumpsuits they'd donned in the helicopter to replace the wilted white flight suits they'd been wearing for the last six days, Lovell had a moment to pull Swigert aside.

"We ought to write it up, Jack," the commander said to the command module pilot. Swigert looked at him perplexed. "The story; *this* story," Lovell said. "We ought to write it."

The idea of telling the tale of Apollo 13 was always a natural one — or at least it seemed that way to the men who had been involved in the mission. The problem was, who in the world would read it? By 1970, the United States had been flying in space for nine years, and while Americans had developed a fondness for their spacecraft and their astronauts, they had come to expect the missions to turn out one way: successfully. Oh sure, there had been that nasty bit of business back in 1961 when Gus Grissom's Mercury 7 capsule sank in the Atlantic the moment after it splashed down. There had been that hair-raising time five years later when Neil Armstrong and Dave Scott's Gemini 8 suddenly began pinwheeling in Earth orbit, requiring the crew to slam on the brakes and return home just ten hours into a flight that was supposed to have lasted five days. But while Gus and Neil and Dave might have known the kind of deadly mess they were in, the public — to NASA's relief — never fully grasped it, figuring that as long as the boys got into space and came down all right, any minor scrapes they had before their feet hit the carrier deck didn't amount to much.

It was only in 1967, when the hard-luck Grissom, along with crewmates Ed White and Roger Chaffee, were killed in a launch-pad fire aboard the Apollo 1 spacecraft, that the taxpayers picking up the check for the missions realized the mortal price tag traveling in space could carry — and they didn't much like it. Americans might be willing to continue funding NASA's increasingly risky cosmic expeditions, but give them too many flag-covered coffins or too many crepe-draped widows, and they just might drop the hammer on the whole operation.

For that reason as much as any other, the moment Apollo 13 returned from space, NASA did everything it could to get the nation's mind off the near-catastrophic mission. Lovell, Swigert, and Haise got their requisite parades; Congress held its requisite hearings into the accident, determining which pieces of hardware and which pre-launch errors had led to the explosion. But after that, the space agency moved on to other things, gearing up as quickly as it could for Apollo 14 and pointing proudly to all those other magnificent Apollos that had come before. Even the Apollo 13 spacecraft itself was soon forgotten. After the command module sizzled into the Pacific, it was hauled onto the carrier deck and then shipped to California for a post-flight physical at the plant where it had been manufactured. Engineers descended on the ship, stripping its innards, testing and retesting its on-board systems to determine how well they had survived their journey and what could be done to improve them for future missions. Afterward, the gutted spacecraft was shipped to Florida where it sat, largely overlooked, as part of an out-of-the-way Cape Canaveral display. After a few years of such internal exile, it was banished even further, expatriated to France where it would be kept in an aviation museum outside Paris.

NASA, of course, was not always so skittish about the idea of risk. The American space program was born not of ambition or passion or celestial wanderlust, but rather of something closer to fear — the fear of being second best. In 1957, the Soviet Union rocked the West with the announcement that it had placed the 184-pound Sputnik satellite into orbit around the Earth. Putting a satellite that weighed 184 pounds in orbit required globe-spanning missiles that packed a propulsive punch of more than 50,000 pounds. And a nation with that kind of ballistic muscle was a fearsome thing. The United States had fallen behind the Soviet Union in both the technology game and the propaganda game; if it was going to close the gap, it was going to have to ramp up fast.

In order to stake their country's claim in space in the short time they had been given, NASA's engineers had to develop a whole new type of engineering ethos. After 1961, when President Kennedy outrageously promised to put Americans on the moon by 1970, the space agency knew the old way of doing business had to change. The cautious customs of the aeronautical inventor would continue to be observed, but this natural reserve would be enlivened by a new willingness to gamble. So no one's ever tried building a 36-story rocket able to accelerate a manned spacecraft to 25,000 miles per hour, fling it out of a circular Earth orbit, and send it on a screaming beeline to the moon? Then it's about time someone did. So no one's ever contemplated how to build a bungalow-sized, four-legged ship so flimsy it can't even support its own weight on Earth, but so strong it can take off and land under the moon's reduced gravity? Then maybe NASA was the place to do it. There is a thin line between arrogance and confidence, between hubris and true skill, and the engineers and astronauts of NASA spent more than a decade surefootedly straddling it.

Much of that changed, however, after Apollo 11, America's first manned lunar landing, in the summer of 1969. The moment NASA's astronauts set foot on the moon's soil, the twelve-year space race was won and the Soviets grudgingly sued for peace. Other Apollos would still be allowed to travel back to the moon for their victory laps, but the public was beginning to have its doubts. Risking the lives of astronauts now, when the moon was already tattooed with boot-prints, was a little like sending soldiers out to fight the day after an armistice has been signed. The near-death experience of Apollo 13 only confirmed the belief that further exploration of deep space was lethal folly. Even before the liftoffs of Apollos 14 through 17, in 1971 and 1972, Apollos 18 through 20, which had been poised to make their own lunar excursions, were ordered to stand down. The geologists and chemists may have seen the moon as a scientific horn of plenty, but the engineers and the public saw it as little more than a goal that had to be reached. Once it *was* reached, there was little purpose in repeating the feat again and again and again.

For NASA, the decades that followed were in many respects wilderness years. After the Apollo program came Skylab, an upper stage of a Saturn 5 rocket that had been refitted as a makeshift space station. In May 1973, the first Skylab crew flew up to the station in a leftover Apollo spacecraft and spent 28 days circling the planet, conducting experiments, and learning to live in orbit. In July 1973, the second crew flew up to the station in another leftover Apollo spacecraft and spent 59 days circling the planet, conducting experiments, and learning to live in orbit. In November of 1973, the third Skylab crew spent 84 days in orbit doing much the same thing. The science the astronauts brought home was good, solid stuff, but for pure exploratory octane, the missions came up mighty short.

NASA's horizons receded even further in 1981 when the space shuttle began to fly. By any measure, the shuttle was a fantastic piece of engineering — a bona fide skyliner compared to the small, wingless pods of the Mercury, Gemini, and Apollo eras. But as the name of the new ship implied, its mission was a modest one. The shuttle was principally a delivery truck — a huge, multi-billion-dollar delivery truck, to be sure, one that would put such glamorous payloads as the Hubble Space Telescope into orbit; but it was a truck nonetheless. For a time, a dusting of glamour clung to the new ships, particularly during the early flights when senior astronauts like John Young and Ken Mattingly, who had earned their stripes flying the old translunar route, took the ships out for their shakedown cruises. But once the legends surrendered the stick and the new crews took over, the public began to view the shuttle flights with little more interest than they would bus departures.

But spacecraft aren't buses, a lesson an indifferent America learned anew on January 28, 1986, when the shuttle Challenger took off for what was to be a routine mission with a routine, seven-member crew — and never came home again. For the first time since Apollo 13, blameless astronauts were betrayed by their hardware — in this case a flawed seal in a solid-fuel booster that failed catastrophically before the ship could even reach orbit. And for the first time ever, the lives of American astronauts were claimed in flight.

As with the Apollo 13 mission, hearings were convened, charges were hurled, and blame — some fair, some unfair — was assigned. NASA, most people concluded, had simply pushed its ships too hard, launching imperfect equipment in poor weather with little more on its mind than improving the shuttle fleet's turnaround time. Cargo dispatchers make those kinds of mistakes; space explorers shouldn't.

Disastrous as the loss of Challenger was — far more disastrous than the near loss of Apollo 13 — this time few people spoke of canceling the space program. NASA, for better or worse, had shackled itself to its shuttles. Three more of the infernal machines were idling in their hangars, manufacturers were poised to build the government a fourth one, and paying customers were waiting in line to get their satellites aboard. With few other American launchers available to carry commercial payloads — and no others to carry human beings — NASA was either going to space aboard the shuttle or not going at all.

But if the space agency's spacecraft were safe for now, its reputation was another matter. Nothing threatens a federal program like doubts about the competence of the people who run it, and the public now had doubts indeed. The engineers and flight planners who had once been the symbol of America's Cold War preeminence — coolly cutting cards with the Soviets for dominance of the very cosmos, and ultimately drawing the high hand — now seemed to have gone fumble-fingered. In the wake of Challenger, NASA was being thought of as little more than another government agency — slow, mistake-prone, bureaucratically arthritic, the Department of Labor with niftier machines.

But NASA was not, and never could be, just another government agency. The business of exploring space may have appeared to have become dull and clumsy, stripped of the crapshoot exuberance that made it the grand adventure it once was, but it was still the business of exploring space. Put enough engineers at enough drafting tables in enough NASA facilities, and someone is almost certain to dream up something extraordinary. Quietly, in the late 1980s, extraordinary things indeed started happening. Even as the shuttles were plying their cautious trade routes from Canaveral to orbit and back to Earth again, other ships — unmanned ships — were accomplishing much more, much farther away.

Within NASA, the manned and unmanned space programs had long operated side by side, but the unmanned program had always been thought of as sort of a poor institutional relation. Little robot ships with names like Ranger and Voyager are no match for charismatic astronauts with names like Buzz and Gordo, and the machines were forever getting lost in the glamorous glare given off by the men. But with the Buzzes and Gordos long since gone, and the manned ships that still flew no longer making headlines, other space achievements slowly started making news. In 1989, the Voyager 2 spacecraft, which had been speeding through the deep freeze of deep space since 1977, reconnoitered Neptune, completing an improbable four-world tour during which it had planet-hopped from Jupiter to Saturn to Uranus. That same year, the Magellan probe took off for Venus, equipped with a high-risk flight plan and high-resolution radar that would allow it to peer through the opaque Venusian air and photograph the surface of that world for the first time. Six months later, the Galileo probe set out on a decade-long mission to Jupiter during which it would photograph the planet's spangle of moons and fire a probe into the flank of the giant world itself.

Even as those missions were flying, other probes targeted for other destinations were being built on workbenches throughout the NASA labs. There was the Cassini mission, launched in 1997 and now en route to Saturn, which will slalom through the moons of the ringed planet and launch a probe down to the surface of the huge Saturnian moon Titan. There was the Stardust spacecraft, which will fly through the tail of a comet and bring home a sample of its pixie-dust ice. There was the Pathfinder spacecraft, which on July 4, 1997, bounced improbably down on Mars in a swaddling of air bags and released an even more improbable robot car that began toddling across the surface, sniffing rocks and sampling soil on a world where no other machine had ever budged so much as an inch from its landing spot. And there was the Hubble Space Telescope which, from the close-to-home perch of near-Earth orbit, has now spent more than a decade peering deeper into space than human or machine eyes had ever seen before.

And as the rockets went up and the data came back and the photos of all the blue and red and orange-black worlds began streaming home to Earth, a small, tectonic shift occurred in the popular consciousness. Space — despite the explosive death of Challenger, despite the plodding work of the shuttle fleet, despite the decades that had elapsed since the last human crew did anything more dramatic than orbit and orbit and orbit the Earth — was slowly becoming thrilling again.

It was in this newly adventurous environment in the 1990s that the story of Apollo 13 reemerged. If our machines could achieve greatness in deep space, there was no reason humans couldn't once again do the same — but only if we were willing to assume equally great risks. In a business as perilous as space travel, the measure of success — and heroism and good, dramatic yarn-spinning — is not in avoiding technological breakdowns but in how engineers and astronauts handle those breakdowns when they do occur. People will die during space missions; it's the nature of the enterprise and something the space community itself level-headedly accepts. But when you find yourself in space, eyeball to eyeball with death, and through imagination, resourcefulness, and flat-out fine flying see to it that death is the one who blinks, you've achieved something truly extraordinary. Measured by that yardstick, the mission of Apollo 13 deserved to be thought of as far more than the forgotten child of NASA's lunar program; it might well have been thought of as its favorite son.

That was the point we hoped to make and the tale we hoped to tell when we set about writing *Lost Moon* (now *Apollo 13*) in 1992. In a popular environment that was intolerant of the fallibility of human beings and their machines and was content to limit space exploration to paddling about in the familiar harbor of near-Earth orbit, the tale of Apollo 13 would not have had much appeal. But by the final decade of the millennium — and the fourth decade of humanity's travels in space — that had already begun to change. The people of the world intoxicated themselves once before on their spaceships and their spacemen, and a generation after the last Apollo astronaut returned from the moon, they appeared ready to do it again.

The truck driver who brought Apollo 13 home to Hutchison, Kansas, in 1995 knew just what it was he was carrying in his hitch. The big wooden crate that had been loaded aboard earlier in the day looked like any other wooden crate — large enough to hold a car or a bedroom's worth of furniture. But cars and furniture aren't welcomed at the dock when they come off the boat with the kind of ceremony this crate received. Nor do they have a government in Paris and a government in Washington keeping an eye on their welfare as they make the transatlantic crossing from France to Houston and then begin the overland journey to the American plains. Nor, finally, would a more ordinary crate have a team of restorers waiting for it at the Kansas Cosmosphere and Space Center, anxious to crack the lid and return the machine inside to its once-pristine condition.

Since 1976, Apollo 13 had been out of the country on indefinite loan. Not until 1995 — when the name of the mission had once again become part of the popular lexicon and the story of its six days in space had become part of its lore — did the government at last call the banished spacecraft home. Max Ary, the founder and chief of the Kansas museum, had spent two decades collecting more than 80,000 bits and bolts from the eviscerated ship, planning for the day it might return.

On the final leg of its final journey, the Apollo spacecraft — jouncing over the 600 statute miles from Houston to Hutchison — moved a lot more slowly than it had when it covered the 500,000 miles from the Earth to the moon and back. This time, however, the ship that had suffered so grievously in space was well protected against injury: its great dish of a heat shield rested in a molded bed of Styrofoam; its scorched skin was wrapped in a layer of plastic; the big crate was hammered together around it all.

When the truck reached the museum, Ary was waiting to meet it. The driver hopped down from his cab, walked across the parking lot to meet him, and handed him a clipboard. Ary scanned the yellow tissue sheet that served as the shipment's invoice and allowed himself a smile.

"Contents," the top line read, "one spacecraft."

Ary signed the sheet with a flourish. Apollo 13 was at last home.

Jim Lovell
Jeffrey Kluger
April 2000

APOLLO 13

PROLOGUE

Monday, April 13, 1970; 10:00 P.M. Houston time

NOBODY KNEW how the stories about the poison pills got started. Most people had heard them; most people even believed them. The press and the public certainly did; even some people at the Agency did. A new person would show up for his first day on the job at NASA, meet his first crewman, and as soon as he got back to his desk would turn to the guy sitting next to him and want to know: Have you heard about the poison pills?

Stories about poison pills always made Jim Lovell laugh. Poison pills! Forget about it! There just weren't any situations in which you'd ever really consider making, well, an early exit. And even if there were, you had lots of easier ways to do it than poison pills. The command module did have a crank for the cabin vent, after all. One turn of the handle and the 5 pounds per square inch of cozy capsule pressure would instantly be exposed to zero pounds per square inch of nasty space pressure. As the atmosphere inside rushed out and the vacuum outside rushed in, whatever air was left in your lungs would explode out in an angry rush, your blood would instantly — and literally — boil, your brain and body tissues would scream for oxygen, and your traumatized system would simply shut up shop. The whole thing would be over in just a few seconds. It was no slower, really, than some ridiculous poison pill, and it was a hell of a lot more respectable.

Of course, it wasn't as if Lovell or anyone else had spent so much as an instant thinking about the damage the cabin vent could do. None of the crew members in the twenty-two ships that had gone up previously had ever found themselves at the point where they might even remotely consider using it. Lovell himself had been aboard three of those ships, and the only time he'd had to let the air out of the cockpit was when he was supposed to: at the end of the flight, when the spacecraft was bobbing around in the ocean and the parachutes were floating about on the water and the frogmen were closing in on the dye marker and the rescue cage was being lowered from the chopper and the band was tuning up on the carrier and he was rehearsing the little speech he might have to make before going in for a medical check, a debriefing, and a shower.

Until today, it had looked as if this trip were going to be just as routine as all the others. Actually it was, until *tonight,* if you were going by Houston time — although out here, close to 200,000 miles away from home and five sixths of the way to the moon, the time in south Texas seemed pointless. But whatever time it was, this trip into the nasty vacuum had turned suddenly sour. At the moment, there was almost too much going on inside the cabin for Lovell and his two crewmates to keep track of, but the things that were claiming most of their attention were their oxygen, which was almost gone, their power, which was almost spent, and their main engine, which was probably — though not certainly — dead.

It was a bad position to be in, exactly the kind of position that the press and the public and the new people at the Agency would dream up when they were in the mood to ask about the poison pills. For their part, Lovell and his crewmates weren't thinking about pills or vents or anything of the sort. They were thinking about fixing the power, fixing the oxygen, fixing whatever else ailed the ship. Whether they actually could was open to question; no ship had ever before gotten quite so sick quite so far from home. The people in Houston felt bad about all this, and came on the line to say so.

"Apollo 13, we've got lots and lots of people working on this," said a voice from Mission Control. "We'll get you some dope as soon as we have it, and you'll be the first one to know."

"Oh," Lovell replied, his voice revealing more irritation than he intended it to. "Thank you."

Behind Lovell's annoyance was the fact that, according to everyone's calculations, Houston had only about an hour and fifty-four minutes to come up with their good ideas. That was all the time that was left in the cabin's oxygen tanks. After that, the crew would slowly start rebreathing their own carbon dioxide, gasping and sweating, wide-eyed, as they strangled to death on their own waste gas in an enclosure the size of a large automobile. If that should happen, the all-at-once unpiloted craft would continue toward the moon, whipping around its far side and heading toward Earth at speeds reaching 25,000 miles per hour. Alas, it wouldn't be aimed precisely for Earth, but instead would miss the home planet by about 40,000 miles, entering a huge, absurd, egg-shaped orbit that would send it 240,000 miles back into space, then back toward the Earth then back into space, then back toward the Earth then back into space, and on and on in a hideously pointless, hideously permanent circuit that could easily outlast the very species that launched it. With Lovell and his crewmates sealed inside, the zipping ship would remain visible to humans for millennia, enduring indefinitely as a twinkling, mocking monument to the technology of the twentieth century.

It was enough to make people start talking about the poison pills.

Monday, April 13, 11:30 P.M. eastern time

Jules Bergman buttoned his gray blazer, tightened his blue and black rep tie, and looked into the camera while the last ten seconds to airtime were counted down. The bustle around him began to quiet as it always did in the instants before a broadcast. Bergman would be getting only about a minute or so of live air for this report, and, as always with such emergency dispatches, he knew he'd have to pack a lot of information into that short sweep of the clock.

The atmosphere in the studio had been electric from the moment Bergman arrived. Nobody on the space beat expected to be here so late tonight, but when the wire services started sending news from Houston and the ABC correspondents started phoning in with their disjointed scraps of data, people seemed to stream in from everywhere. A novice might have been impressed by the alacrity with which the network's titanic news machine could pick itself up and get itself running, but Bergman was not a novice. Just why a major news organization would even have considered shutting down its cameras and punching out for the night when a ship of astronauts was 200,000 miles from home was a complete mystery.

Bergman had been covering manned space flight since Alan Shepard's popgun suborbital in 1961, and he had long since learned that the surest way to get your head handed to you in the space-travel business was to take it for granted that a smooth flight was going to stay smooth. Like no other newsman before him, Bergman had made it his business to learn the ways of space flight, getting himself spun in centrifuges, lofted in zero-g aircraft, and set adrift in splashdown rafts, all in the service of better understanding the tightrope the pilots walked so he could better explain it to the public that paid the bills.

The problem was, these days the public didn't seem to want the explanations. This wasn't Shepard's Freedom 7 or Glenn's Friendship 7 anymore, and it certainly wasn't Neil Armstrong, Michael Collins, and Buzz Aldrin's Apollo 11 — the magnificent mission that just nine months earlier had made the first landing on the surface of the moon. This was Apollo 13, the third such landing, and by the spring of 1970, both the network and the country it reported to were bored.

What ABC was showing instead of the late-breaking moon news was *The Dick Cavett Show.* Cavett would be talking to Susannah York, James Whitmore, and a few members of the world-champion New York Mets tonight, but for the first minutes of this evening's show, he at least had viewers *thinking* about the moon.

"It's a great day in New York today," Cavett bantered with his band and his audience before his guests were introduced. "It's girl-watching weather. And speaking of girl watching, did you know our first bachelor astronaut is on his way to the moon? It's Swigert, right? He's the kind of guy who they say has a girl in every port. Well, that may be, but I think he's kind of foolishly optimistic taking nylons and Hershey bars to the moon." The audience laughed along. "And did you read that three million fewer viewers watched this space shot than the last one? Colonel Borman was here the other day, and he admitted that space shots are sort of losing their glamour. But in fairness, the problem might have been partly that it was a nice day and a lot of people were out, and partly that a lot of people thought the launch was a summer rerun." The audience laughed along with that, too.

While Cavett talked, Jules Bergman's director over in the ABC news studio finished counting down from ten to one, and all at once the TV image of the talk-show host was replaced by the bright red words "Apollo 13" and the bright blue words "Special Report." A second later, Bergman's face replaced the text.

"The Apollo 13 spacecraft has suffered a major electrical failure," he began, "leaving the astronauts in no immediate danger, but ruling out any chance of a lunar landing. Seconds after inspecting the Aquarius lunar module, Jim Lovell and Fred Haise crawled back into the command module and then reported hearing a loud bang followed by a power loss in two of their three fuel cells. They also reported seeing fuel, apparently oxygen and nitrogen, leaking from the spacecraft and also reported that gauges for those gases were reading zero. Mission Control ordered the astronauts to power down the spacecraft, cutting electrical usage while troubleshooters looked for solutions for the problems. Without all three fuel cells, the problem becomes getting enough power to fire the spacecraft's onboard engine to get them back to Earth. Another problem still to be determined is an apparent loss of breathing oxygen in the command module. Mission Control confirms the seriousness of the problem. Repeating, the Apollo 13 astronauts are in no immediate danger, but the flight itself is in danger of being aborted."

As quickly as he appeared, Bergman vanished from the screen, replaced once again by the happy Dick Cavett. In the news studio, the bustle returned the instant the camera was shut off. The experienced space hands had reason to be less than satisfied with the news they had just broadcast. The astronauts were "in no immediate danger"? That was the line NASA was putting out? How you could be in no immediate danger when you were close to a quarter of a million miles from home and down to your last molecules of oxygen was unclear, but it was more than likely that the prognosis from the Agency would change soon. NASA officials were always reluctant to use the word "emergency" when they could get away with "glitch," but when they were staring down the barrel of a full-bore crisis, they usually fessed up. Already, the New York studio was back on the phone to correspondent David Snell in Houston, to get the latest Agency line. Already, consultants from North American Rockwell, formerly North American Aviation, the builder of the Apollo spacecraft, were being asked to come to the studio to explain the problem on the air.

Across the studio, the phones began ringing with the correspondents' latest reports from Houston, and the news crew snatched up the receivers, listened to the dispatches, and passed the updates on to Bergman. Only minutes after going on the air with his guardedly optimistic report, the newsman could see that the prognosis had indeed changed — and it was not for the better. The Apollo 13 command module, NASA's updated statement now admitted, was completely without air or power; the astronauts, it now appeared, would have to abandon ship and make their way into the lunar module; and their lives, the Agency now conceded, were indeed endangered.

Near Bergman, the director prepped his cameramen to go back on the air. Clearly, there would be no more Dick Cavett tonight.

1

January 27, 1967

JIM LOVELL was having dinner at the White House when his friend Ed White burned to death.

Actually, it wasn't really dinner Lovell was having, just finger sandwiches, orange juice, and unmemorable wine laid out on linen-covered tables in the Green Room. But being that the sun had already gone down and no other time was formally set aside for chow that day, this was as close to dinner as Lovell was going to get.

Actually, too, Ed White didn't really burn to death. The fumes claimed him long before the flames ever could have. By most estimates, it was only fifteen seconds before he — along with his commander, Gus Grissom, and his junior crewmate, Roger Chaffee — succumbed to the poisons they were drawing into their lungs. In the end, it might have been for the best. Nobody knew exactly how hot it got inside the cockpit, but with a flame-feeding atmosphere of 100 percent pure oxygen, it was a good bet that the thermometer climbed above 1,400 degrees. At that temperature, copper glows, aluminum melts, and zinc can burst into flame. Gus Grissom, Ed White, and Roger Chaffee — fragile agglomerations of skin and hair and flesh and bone — wouldn't have stood a chance.

Jim Lovell had no way of knowing what was happening to the three men at the moment it was happening. He would know soon, but at the moment he didn't. At the moment, Lovell was concerned with the job in front of him, and that job was to circulate and socialize and shake some hands. There were dozens of dignitaries gathering around to scarf up the White House's snacks and drinks, and it was Lovell's business to say hello to as many of them as possible. The guest pass Lovell had been sent in the mail was very specific about this part of the job:

"Green and Blue Rooms for individual pic with ambassadors and handshake," it said. It didn't say "You're invited here for the food"; it didn't say "You're invited here for the fun." It said, in so many words, "You're invited here — if you must know — to work the crowd."

Lovell was not unaccustomed to this kind of evening, of course, and the candor of the invitation was no surprise. This was just more of what he and the other members of the astronaut corps called their "time in the barrel": those occasions when some chief of state or chamber of commerce needed a showpiece spaceman to round out a reception and NASA would dispatch a crewman or two to attend the party, pose for pictures with the host, and generally spread goodwill. All of the astronauts were good at this drill, but Lovell was especially good. At 5 feet 11 inches and 170 pounds, with mainstream midwestern looks, he projected an almost archetypal astronaut image, perfect for the VIP who wanted just the right photo to complete his office wall. This evening, there would be fewer opportunities for such photos than most. The invitation called for the event to begin promptly at 5:14 P.M. — it actually said 5:14 — and conclude no later than 6:45. What the White House was hoping to achieve with an extra 60 seconds at the front end of the evening was unclear, but all Lovell and the other four crewmen here tonight had to do was work the crowd for the 91 minutes they'd be on call, then they'd be free to go enjoy Washington.

Truth be told, if Lovell had to put in an hour and a half or so in the barrel, there were worse places to do it than the White House. Lyndon Johnson, who was always at his best at nibble-and-gab sessions like these, was here, and Lovell looked forward to saying hello to the president. The two had met once before, just a month or so earlier when Lovell and his copilot Buzz Aldrin were invited down to the ranch for a medal and a speech after their Gemini 12 spacecraft splashed down in the Atlantic, ending the triumphantly successful ten-flight run of the tiny craft.

In the deepest part of his deepest heart Lovell had felt that a medal might not actually be warranted. It wasn't politic to say so, but he had thought so. It wasn't as if the flight hadn't been a huge accomplishment; it was. It wasn't as if it hadn't achieved all the goals the mission planners had set out for it, and more; it had. But the nine previous flights had achieved most of their goals too, and if it weren't for the astronautical expertise accumulated on Geminis 3 through 11, Gemini 12 would never have been possible. Johnson, however, had a taste for high drama, and as this final Gemini flight unfolded — as Lovell docked his two-man spacecraft with an unmanned Agena spacecraft as effortlessly as if he were pulling a Pontiac into a parking space; as Buzz climbed outside and rode the back of the Agena like a dickey bird on a rhino's back — the president became more and more pleased with his multi-billion-dollar space program. No sooner had Lovell and Aldrin plopped back into the ocean than Johnson called out the photographers and proclamation writers and had the heroes down for a ceremony and a little south Texas hospitality.

After that, Lovell had a soft spot for the president and counted himself among Johnson's most enthusiastic admirers. But even if there were no chief executive here today, this reception would be one worth attending. The purpose of the evening was to celebrate the signing of the much debated, prosaically named "Treaty on Principles Governing the Activities of States in the Exploration and Use of Outer Space." As treaties went, Lovell knew this was not a truly big deal; it wasn't Versailles, it wasn't Appomattox, it wasn't a nuclear test ban. It was one of those treaties that come about because, as the diplomats say, "something should be put on paper."

That something had to do with space — specifically, the boundaries that define space. Since the first proto-nation drew the first line in the soil of the first populated savanna, countries had been steadily, greedily extending their boundaries. It began with a circle around the campfire, then it was a zone from the campfire to the coast, then it was from the coast out to a three-mile line in the ocean. In the last ten years, since the dawn of the space age, the three miles had changed to two hundred miles, the direction had changed from out to *up,* and the world's nations had been fretting about if and how the lines would continue to be drawn in this most exotic of new frontiers.

The pact being signed today by more than five dozen countries would see to it that there *were* no lines. Among its provisions were guarantees that outer space would remain forever nonmilitarized, that no country would declare any Earth orbital zone as its own, and that land claims would never be made on the moon, Mars, or any other place humanity's rockets might one day reach. More important to Lovell and the other astronauts here tonight, however, was article 5 of the document — the safe return of space travelers clause. This provision guaranteed that any astronauts or cosmonauts who veered off course and splashed down in a hostile ocean or thumped down in a hostile wheat field would not be scooped up and carted off by security forces of the violated country. Rather, they would be treated as "envoys of mankind," to be "safely and promptly returned to the state of registry of their space vehicle."

In picking the astronaut delegation tonight, NASA had chosen carefully. In addition to Lovell, who had flown twice in the Gemini program, was Neil Armstrong, a veteran NASA test pilot, whose sole Gemini flight, Gemini 8, had ended in near disaster ten months earlier when one of his thrusters suddenly went south on him, causing his ship to begin spinning at a stomach-knotting 500 r.p.m. and forcing flight controllers to abort the mission and bring him down in the first ocean or sea or duck pond they could find. Also on hand was Scott Carpenter, whose Mercury flight had gone almost as haywire five years earlier when he spent too much time in his final orbit monkeying around with some onboard astronomy experiment, aligned his retrorockets improperly, and splashed down in the Atlantic 250 miles from his recovery crew. While the Navy scrambled this way and that, the second American to orbit the Earth found himself bobbing in his life raft, nibbling his ration crackers, and scanning the horizon for a ship he fervently hoped would be flying the Stars and Stripes.

Both Armstrong and Carpenter could have used the protection of the treaty during their flights, and this no doubt was on NASA's mind when it sent them here tonight. The other two members of the delegation, Gordon Cooper and Dick Gordon, were tougher to explain, though it was likely that NASA had simply spun its dial and picked the first two names that came up.

Lovell got a brief hello from Johnson almost as soon as the reception began — *just* a brief hello, nothing like the presidential fawning of only a month earlier — and wandered to the buffet table to get a sandwich and survey the minefield of milling dignitaries. The room would be a big one to work. Kurt Waldheim was here from Austria; Ambassador Patrick Dean had come from Great Britain; Anatoly Dobrynin strolled over from the Soviet embassy; Dean Rusk, Averell

Harriman, and Arthur Goldberg were on hand for the United States. The presence of so many geopolitical giants was catnip for legislators from Capitol Hill as well. Senate minority leader Everett Dirksen, Senator Albert Gore Sr. of Tennessee, and Senators Eugene McCarthy and Walter Mondale of Minnesota were here, as well as other Washington heavyweights who had wangled invitations of their own.

Lovell was about to wade into the crowd when he noticed Dobrynin standing to his right. The Soviet ambassador had a solid reputation among the astronauts who had met him before. He was said to be an accomplished student of both the American and Soviet space programs, a good-natured fellow who spoke first-rate English, and a man who, on the whole, did not at all fit the image of a representative of the socialist superstate. Lovell extended his hand.

"Mr. Ambassador?" he said. "Jim Lovell."

The ambassador broke into a grin. "Ah, Jim Lovell," Dobrynin said. "Nice to meet you. You're the, uh . . ."

The expectant trail-off at the end of Dobrynin's sentence, of course, was Lovell's cue to say "astronaut," after which Dobrynin would nod vigorously and smile easily, as if to say, "Yes, yes, I know what you are, I just forgot the English word." Lovell suspected he could just as easily answer, "shortstop" or "sculptor" or "professional wrestler" and Dobrynin would react the same way.

"Astronaut, Mr. Ambassador."

Dobrynin responded immediately. "Yes, you're the one who just came back. A wonderful flight, a real accomplishment."

Lovell smiled, impressed. "Well, we're working hard to keep up with you folks."

"Maybe one day we won't have to compete so much," Dobrynin said. "Maybe this treaty is the first step toward working peacefully together."

"We certainly hope so. It would be nice if all humanity could explore the moon one day."

"I don't know if I'll get there," the diplomat said. "But I wouldn't be at all surprised if you do."

"That's what I'm working toward," Lovell said.

"The best of luck." With that, the ambassador shook Lovell's hand and moved off into the throng to find other people to charm.

Lovell turned in the other direction and spied Hubert Humphrey, deep in conversation with Carpenter and Gordon. As he approached, he could hear the vice president's characteristically nasal voice going on in its characteristically engaging way.

"This is a landmark treaty, just a landmark," Humphrey was saying as Lovell reached them. "Everybody wins, even countries that don't have a space program, because now the superpowers won't militarize the areas beyond Earth."

"The astronauts have always thought that was a great idea," Carpenter said, echoing the NASA party line, but one he wholeheartedly believed. "For a long time there's been a lot of camaraderie between American crewmen and Russian crewmen. We've always thought the peaceful exploration of space is bigger than any one country."

"Much bigger," Humphrey agreed.

"What the astronauts are most worried about," Lovell said after introducing himself, "is the safety question. It would be nice to know that we could fly over any country, even a hostile country, and be guaranteed a cordial reception if we have to abort."

"That's one of the biggest objectives of this treaty," the vice president answered. "The safety of the astronauts."

The crewmen made party talk with Humphrey for another minute or two, just long enough to register on the administration that NASA's goodwill ambassadors were doing their jobs, just short enough to give other guests a chance at the vice president. The three men were about to disperse to greet more visitors when Lovell became suddenly troubled. The mention of crew safety had brought a forgotten worry to mind.

"What time did they start the countdown at the Cape today?" Lovell asked Gordon as they walked off.

"Early afternoon," Gordon answered.

Lovell looked at his watch: it was shortly after six. "So they should be finished soon," he said. "Good."

The test Lovell was concerned about was no small matter. Today NASA had scheduled a full-scale dress rehearsal of the countdown for the first mission of the Apollo spacecraft, set to begin three weeks from now. If things had gone as planned, at this moment the three-man crew would be zipped into their pressure suits, strapped to their seats, and locked behind their command module's hatch, sealed in a 16-pound-per-square-inch atmosphere of pure oxygen. Lovell himself had gone through such tests countless times in preparation for his Gemini 12 flight, his two-week Gemini 7 flight, and the two other

Gemini missions on which he had served as part of the backup crew. There was nothing inherently dangerous about a countdown test, yet if you asked anyone at the Agency, they'd tell you they couldn't wait until this one was over.

The worry wasn't the crew, of course. The commander, Gus Grissom, had flown in space in both the Mercury and Gemini programs and had run through these counterfeit countdowns dozens of times. The pilot, Ed White, had flown in Gemini too, and had also had more than his share of pad training. Even the junior pilot, Roger Chaffee, who had never been in space, was rigorously tutored in the art of flight rehearsal. No, the worry in this exercise was the ship.

The Apollo spacecraft, by even the most charitable estimations, was turning out to be an Edsel. Actually, among the astronauts it was thought of as worse than an Edsel. An Edsel is a clunker, but an essentially harmless clunker. Apollo was downright dangerous. Earlier in the development and testing of the craft, the nozzle of the ship's giant engine — the one that would have to function perfectly to place the moonship in lunar orbit and blast it on its way home again — shattered like a teacup when engineers tried to fire it. During a splashdown test, the heat shield of the craft had split open, causing the command module to sink like a $35 million anvil to the bottom of a factory test pool. The environmental control system had already logged 200 individual failures; the spacecraft as a whole had accumulated roughly 20,000. During one checkout run at the manufacturing plant, a disgusted Gus Grissom walked away from the command module after leaving a lemon perched atop it.

Yesterday afternoon, so the whispers went, all of this finally reached a head. For much of the day, Wally Schirra — a veteran of Mercury and Gemini, and commander of the backup crew that would replace Grissom, White, and Chaffee if anything happened to them — ran through an identical countdown test with his crew, Walt Cunningham and Donn Eisele. When the trio climbed out of the ship, sweaty and fatigued after six long hours, Schirra made it clear that he was not pleased with what he had seen.

"I don't know, Gus," Schirra said when he met later with Grissom and Apollo program manager Joe Shea in the crew's quarters at the Cape, "there's nothing wrong with this ship that I can point to, but it just makes me uncomfortable. Something about it doesn't ring right."

Saying that a craft of any kind didn't "ring" was one of the most worrisome reports one test pilot could offer another. The term conjured up the image of a subtly cracked bell that looks more or less O.K. on the surface but emits a flat clack instead of a resonant gong when struck by its clapper. Better that the craft should go to pieces when you try to fly it — that its engine nozzle should drop off, say, or its thrusters break away; at least you'd know what to fix. But a ship that doesn't ring right could get you in a thousand insidious ways. "If you have any problem," Schirra told his colleague, "I'd get out."

Grissom was almost certainly disturbed by the report, but he reacted to Schirra's warning with surprising nonchalance. "I'll keep an eye on it," he said.

The problem, as many people knew, was that Gus had "go fever": he was itching to fly this spacecraft. Sure there were glitches in the ship, but that's what test pilots were for, to find the glitches and work them out. And even if there was a problem, just getting out — as Schirra had suggested — wouldn't be so easy. The Apollo's hatch was a three-layer sandwich assembly designed less to permit easy escape than to maintain the integrity of the craft. The inner cover was equipped with a sealed drive, a rack-drive bar, and six latches that clamped onto the module's walls. The next cover was even more complicated, equipped with bell cranks, rollers, push-pull rods, an over-center lock, and twenty-two latches. Before liftoff, the entire craft was also surrounded by a form-fitting "boost protective cover," a layer of armor that would shield it from the aerodynamic stresses of powered ascent. The cover was meant to pop off well before the spacecraft reached orbit, but until then, it provided one more layer between the crew inside and a rescue team outside. Under the best conditions, astronauts and rescue crews working together could remove all three hatches in about ninety seconds. Under adverse conditions, it could take much longer.

Standing in the White House Green Room, Lovell glanced at his watch. In half an hour or so the test would be over. He'd be relieved to get word that his friends were out of that ship.

On Florida's Atlantic coast, a thousand miles south, the countdown at Cape Kennedy was not going well. From the time the crew members were strapped into their seats, at about one in the afternoon, the Apollo spacecraft had begun fulfilling its critics' worst expectations.

When Grissom first plugged his suit hose into the command module's oxygen supply, he reported a "sour smell" flowing into his helmet. The odor soon dissipated and the environmental control team promised they'd look into it. Shortly afterward, and throughout the day, the astronauts found nettlesome problems with the air-to-ground communications system as well. Chaffee's transmissions were more or less clear, White's were spotty at best, and Grissom's hissed and crackled like a cheap walkie-talkie in an electrical storm.

"How do you expect us to talk to you from the moon," the commander snapped through the static, "when we can't even communicate from the pad to the blockhouse?" The technicians promised they'd look into this too.

At 6:20 Florida time, the countdown reached T minus 10 minutes, and the clock was stopped temporarily while the engineers fiddled with the communications problem and a few other glitches. As in any real launch, this ersatz one was being monitored at both the Cape and the Manned Spacecraft Center in Houston. The protocol called for the Florida team to run the show from countdown through liftoff through the moment the booster's engine bells cleared the tower; then they would hand the baton to Houston.

Helping to run the show in Florida were Chuck Gay, the chief spacecraft test conductor, and Deke Slayton, one of the original seven Mercury astronauts. Before ever getting a chance to fly in space, Slayton had been grounded because of an irregular heartbeat, but he had managed to make lemonade out of that particular lemon, getting himself appointed director of Flight Crew Operations — in essence, chief astronaut — while quietly and insistently lobbying for a return to flight status. So much an astronaut at heart was Slayton that earlier today, when the communications from the ship first started to go to hell, he had offered to climb into the spacecraft, fold himself into the lower equipment bay at the astronauts' feet, and remain there for the countdown to see if he couldn't dope out the static problem himself. The test directors vetoed the idea, however, and Slayton instead found himself seated at a console next to Stu Roosa, the capsule communicator, or Capcom. In Houston, the overseer today — as on most days — was Chris Kraft, deputy director of the Manned Spacecraft Center and the man who had served as flight director on all six Mercury flights and all ten Geminis.

Kraft, Slayton, Roosa, and Gay were eager to get this exercise over with. For more than half a day, the crew had been flat on their backs under the weight of their own bodies and their bulky pressure suits, in couches designed not for the oppressive load of a one-g environment but the friendly float of weightless space. In a few more minutes, they could get the countdown rolling again, complete their simulated blastoff, and then get those men out of there.

But this was not to be. The first sign that something was amiss came moments before the clock was set to start running again, at 6:31 P.M., when technicians watching the video monitor of the command module noticed a sudden movement through the hatch window, a shadow moving rapidly across the screen. Controllers accustomed to the deliberate movements of well-drilled crewmen plodding through a familiar countdown snapped their heads to the screen. Anyone who didn't have a monitor directly in front of him or who was out on the scaffold-like gantry surrounding the Apollo ship and its 224-foot booster would have noticed nothing. A moment later, a voice crackled down from the tip of the rocket.

"Fire in the spacecraft!" It was Roger Chaffee, the rookie, calling out.

On the gantry, James Gleaves, a mechanical technician monitoring communications through his headset, turned with a start and began running toward the White Room, which led from the uppermost level of the gantry to the spacecraft. In the blockhouse, Gary Propst, a communications control technician, looked instantly to his top-left monitor, the one connected to a camera in the White Room, and thought — *thought* — he could make out a bright glow of some kind through the hatch's porthole. At the Cape's Capcom console, Deke Slayton and Stu Roosa, who had been reviewing flight plans, looked at their monitor and believed they saw something that looked like a flame playing about the seam of the hatch.

At a nearby console, assistant test supervisor William Schick, who was responsible for keeping a log of every significant event in the course of the countdown, looked immediately at his flight clock and then dutifully recorded: "1831: Fire in the cockpit."

On the communications line, those same words echoed down from the spacecraft. "Fire in the cockpit!" shouted Ed White through his balky radio. The flight surgeon glanced at his console and saw that White's heartbeat had spiked dramatically. Environmental control officers looked at their readouts and noticed that spacecraft motion detectors were picking up furious movement inside the craft. On the gantry, Gleaves

16

heard a sudden *whoosh* coming from the command module, as if Grissom were opening the O_2 vent to dump the spacecraft's atmosphere — precisely what you'd want to do if you were trying to choke off a fire. Nearby, systems technician Bruce Davis saw flames shoot from the side of the ship near the umbilical cord that connected the ship to the ground systems. An instant later fire began dancing along the umbilical itself. At his blockhouse monitor, Propst could see flames behind the porthole; through them, he could also see a pair of arms — from the position, they had to be White's — reaching toward the console to fumble with something.

"We're on fire! Get us out of here!" Chaffee shouted, his voice clear on the ship's one perfect channel. From the left of Propst's screen, a second pair of arms — they had to be Grissom's — appeared in the porthole. Donald Babbitt, the pad leader, whose desk was just twelve feet away from the spacecraft, on the top level of the gantry — level 8 — shouted to Gleaves, "Get them out of there!" As Gleaves dashed toward the hatch, Babbitt turned to grab his pad-to-blockhouse communications box. Just then, a huge burst of smoke erupted from the side of the craft. Beneath it, a duct that was supposed to vent steam now sent out tongues of flame.

From the blockhouse Gay, the test director, called up to the astronauts in disciplined tones: "Crew egress." There was no answer. "Crew, can you egress at this time?"

"Blow the hatch!" Propst screamed to no one in particular. "Why don't they blow the hatch?"

Through the smoke on the gantry, someone shouted, "She's going to blow!"

"Clear the level," someone else ordered.

Davis turned and ran toward the southwest door of the gantry. Creed Journey, another technician, threw himself to the ground. Gleaves backed warily away from the ship. Babbitt stayed at his desk, intent on raising the blockhouse on his comm box. On the ground, the environmental control console recorded the cabin pressure at 29 pounds per square inch, twice sea level, and the temperature off the scale. At that moment, with a crack and a roar and a burst of hideous heat, the Apollo 1 spacecraft — America's flagship moonship — surrendered to the inferno inside it, splitting at the seam like an old treadless tire. Fourteen seconds had elapsed since Chaffee's first cry of distress.

A dozen feet away from the Apollo command module, Donald Babbitt felt the full force of the explosion. The pressure wave knocked him back on his heels and the blast of heat felt as if someone had flung open the door of a giant furnace. Sticky, molten globules shot from the ship, splattered his white lab coat and burned through to his shirt beneath. The papers on his desk charred and curled. Nearby, Gleaves felt himself slammed backward against an orange emergency escape door — an escape door that he only now discovered had been installed to open in, not out. Davis, turning away from the ship, felt a scorching breeze at his back.

At the Capcom station in the blockhouse, Stu Roosa frantically tried to raise the crew by radio while Deke Slayton collared the blockhouse medics. "Get out to the pad," he ordered them. "They're going to need you." In Houston, a helpless Chris Kraft saw and heard the chaos on the gantry and found himself in the utterly unfamiliar position of having no idea what was going on aboard one of his ships.

"Why can't they get them out of there?" he said to his controllers and technicians. "Why can't somebody get to them?"

At the assistant test supervisor's station, Schick wrote in his log: "1832: Pad leader ordered to help crew egress."

On gantry level 8, Babbitt picked himself up from his desk, ran to the elevator, and grabbed a communications technician. "Tell the test supervisor we're on fire!" he shouted. "I need firemen, ambulances, and equipment." Babbitt then ran back inside and grabbed Gleaves and systems technicians Jerry Hawkins and Stephen Clemmons. Wherever the ship had ruptured, it wasn't visible to the pad leader, which meant that the rip could provide no access to the men in the cockpit. This meant there was only one way to get to them. "Let's get that hatch off," he shouted to his assistants. "We've got to get them out of there."

The four men gathered fire extinguishers and dove into the black cloud vomiting from the spacecraft. Blindly firing the extinguishers, they beat back the flames just a bit, but the inky smoke and dense cloud of poisonous fumes proved a killing combination, and the men quickly retreated. Behind them, at a supply station, systems technician L. D. Reece found a cache of gas masks and handed them to the choking pad crew. Gleaves tried to remove the strip of tape that activated the mask and noticed with incongruous clarity that the tape was the same color as the surrounding mask and was thus nearly impossible to see with all the smoke. (*Remember to report that for next time. Yes, must remember to report that.)* Babbitt got his mask activated and in place, but found that it formed a vacuum around his face, causing the rubber to cling uncomfortably and making it impossible to breathe. Flinging the mask away and trying another one, he discovered that it worked only a little better.

Diving into the smoke, the pad crew wrestled with the hatch bolts only for as long as the heat and the fumes and their faulty gas masks would allow them to. Then they stumbled out again, gasping and hacking in the marginally cleaner air until they had enough breath for another try. On the gantry levels below, word had now spread that a flaming pandemonium was playing out above. At level 6, technician William Schneider heard the cries of fire from overhead and ran for the elevator to take him up to level 8. The car had just left, however, and Schneider headed for the stairs. On his way up, he found that the fire was now licking down to levels 6 and 7, reaching the spacecraft's service module. Seizing a fire extinguisher, he began somewhat futilely to spray carbon dioxide into the doors that led to the module's thrusters. Down on level 4, mechanical technician William Medcalf heard the cries of alarm and dove into another elevator to take him up to level 8. When he reached the White Room and opened the door, he was unprepared for the wall of heat and smoke and the tableau of choking men that greeted him. He took the staircase down to a lower level and returned with an armload of gas masks. When he arrived, he was greeted by the wide-eyed, soot-smeared Babbitt, who shouted, "Two firemen right now! I have a crew inside and I want them *out!*"

Medcalf radioed the alarm to the Cape's fire station, alerting them that trucks were needed at launch complex 34; the response came back that three units had already begun to roll. When Medcalf waded into the White Room, he nearly stumbled over the pad crew, who, having given up on their poor, porous masks, were now on all fours, crawling to and from the spacecraft just beneath the densest smoke, working the hatch bolts until they could take it no longer. Gleaves was almost unconscious, and Babbitt ordered him away from the command module. Hawkins and Clemmons were little better off. Babbitt glanced back into the room, spied two other, fresher technicians, and motioned them into the cloud.

It was another several minutes before the hatch was opened, and then only partway — barely a six-inch gap at the top. This was enough, however, to release a final blast of heat and smoke from the interior of the spacecraft and to reveal that the fire itself was at last out. With some more shoving and manipulating, Babbitt managed to pry the hatch loose and drop it down inside the cockpit, between the head of the astronauts' couches and the wall. Then he fell away from the ship, exhausted.

Systems technician Reece was the first to peer into the maw of the cremated Apollo. He poked his head nervously inside, and through the blackness saw a few caution lights winking on the instrument panel and a weak floodlight glowing on the commander's side. Apart from this he saw nothing — including the crew. But he heard something; Reece was certain he heard something. He leaned in and felt around on the center couch, where Ed White should have been, but he felt only burned fabric. He took off his mask and shouted into the void, "Is anyone there?" No response. "Is anyone there?"

Reece was pushed aside by Clemmons, Hawkins, and Medcalf, who were carrying flashlights. The three men played their lights around the interior of the cockpit, but their smoke-stung eyes could make out nothing but what appeared to be a blanket of ashes across the crew's couches. Medcalf backed away from the ship and bumped into Babbitt. He choked.

"There's nothing left inside," he told the pad leader.

Babbitt lunged to the spacecraft. More people crowded around the ship, and more light was trained on its interior. With his eyes slowly recovering, Babbitt saw that there was, most assuredly, something inside. Directly in front of him was Ed White — lying on his back with his arms over his head, reaching toward where the hatch had been. From the left Grissom was visible, turned slightly in the direction of White, reaching through his junior crewman's arms for the same absent hatch. Roger Chaffee was still lost in the gloom, and Babbitt guessed he was probably strapped in his couch. The emergency escape drill called for the commander and the pilot to handle the hatch while the junior crewman stayed in his seat. Chaffee was no doubt there, waiting patiently — now eternally — for his senior crewmates to finish their work.

From the back of the crowd, James Burch of the Cape Kennedy fire station pushed his way to the spacecraft. Burch had seen this kind of scene before. The other men here hadn't. The technicians, who made their living maintaining the best machines science could conceive, now made respectful room for the man who takes over when something in one of those machines goes disastrously wrong.

Burch crawled through the hatch and into the cockpit and, unknowingly, stopped atop White. He swept his light across the charred instrument panel and the spider web of singed wires dangling from it. Just beneath him, he noticed a boot. Not knowing if the crew was dead or alive, and not having the time to find out gingerly, he grabbed the boot and pulled hard. The still-hot mass of molded rubber and cloth came off in his hands, revealing White's foot. Burch then patted his hands farther up and felt ankle, shin, and knee. The uniform was partly burned away, but the skin underneath was unmolested. Burch tugged the skin this way and that to see if it would slip from the flesh — a consequence of traumatic burns that, he knew, could cause a victim to shed his outer dermis like a tropical gecko. This skin, however, was intact; indeed the entire body appeared intact. The fire had been exceedingly hot, but it had also been exceedingly fast. It was fumes that claimed this man, not flames. Burch pulled up on White's legs with as much force as he could, but the body budged only six inches or so and he let it fall back into its couch. The fireman backed away to the edge of the hatch and took another look around the cruel kiln of the cockpit. The two bodies flanking the one in the center looked the same as White's, and Burch knew that whatever life had been in this spacecraft just fourteen minutes earlier had certainly been snuffed out. He climbed out of the ship.

"They are all dead," Burch intoned quietly. "The fire is extinguished."

Over the next several hours, photographers and technicians arrived to record the scene, including the position of every switch in the cockpit, for the excruciatingly detailed investigation that would surely follow. It would be well past two in the morning, more than thirteen hours after the fatal countdown test began, before the crew of Apollo 1 would be lifted from their spacecraft and carried to an ambulance at the bottom of the gantry.

The White House celebration honoring the signing of the space treaty ended, as scheduled, at precisely 6:45 P.M. The affair wound down, like all White House affairs, almost undetectably. The president vanished from the room without fanfare. The food and drink disappeared from the tables in much the same way. After that, the crowd began to flow slowly, uniformly, and without instruction toward the exits, as if rising air pressure at one end of the room had nudged all of the occupants toward the other. By shortly before seven, the quintet of astronauts who had been summoned here tonight stood on Pennsylvania Avenue, competing with tourists for the few available cabs that plied the boulevard at this time of day. Scott Carpenter claimed the first taxi and headed off to the airport for an engagement in another city. Lovell, Armstrong, Cooper, and Gordon, who had all flown here in NASA aircraft and were not due back in Houston until tomorrow, had already booked rooms at the Georgetown Inn, on Wisconsin Avenue.

Ever since 1962, when Wally Schirra came to town for a medal and a handshake from President Kennedy following his successful nine-hour Mercury flight, the inn had served as unofficial host to many a NASA dignitary visiting the capital. The place was out-of-the-way enough to offer the privacy the country's space pioneers craved, and new enough to offer the poshness they had come to enjoy. Collins Bird, the hotel's first and only owner, had his inn done up in a subdued colonial style, with four-poster beds, bent-cane rockers, and matching upholstery and drapes. The five guest floors were each distinguished by their own color schemes: above the lobby, the second floor was blue, the third gold, the fourth red, the fifth turquoise, the sixth black and white and gray. The astronauts tonight were housed on the turquoise floor — not Bird's first choice of colors for the Magellans of the late twentieth century, but the reservations had been made late and the management did the best it could.

Even before Lovell, Armstrong, Cooper, and Gordon returned tonight, Bird knew there had been trouble. Bob Gilruth, the director of the Manned Spacecraft Center and another guest at the White House this evening, arrived back at the hotel looking stunned and drawn; he walked blankly by the front desk where the owner was working. Gilruth had been on the phone to Houston and had been told what had happened on pad 34.

"Anything wrong, Mr. Gilruth?" Bird asked.

"We've got troubles, Collins," Gilruth answered flatly. "Real troubles."

"Anything we can do here?"

Gilruth said nothing and kept walking.

When the astronauts arrived and got to their rooms, they each found the red message light on their phones flashing. Lovell buzzed the front desk and was told simply that he was to call the Manned Spacecraft Center, and was to do so immediately. He phoned the number he was given and reached a voice he didn't recognize — some official or administrator or public affairs officer in the Apollo program office. Lovell could hear ringing phones and noisy voices in the background.

"The details are still sketchy," the man on the phone told him, "but there was a fire on pad 34 tonight. A bad fire. It is probable the crew did not survive."

"What do you mean 'probable'?" Lovell asked. "Did they survive or didn't they survive?"

The man paused. "It is probable the crew did not survive."

Lovell closed his eyes. "Does anyone know this yet?"

"People who have to know, know. It won't be long before the media gets word. When they do, they'll pounce on anyone connected to the Agency. It is strongly suggested that you four disappear until we get further word to you."

"What does 'disappear' involve?" Lovell asked.

"Don't leave the hotel tonight. In fact, don't leave your rooms. If you need anything, call the desk. If you want to eat, call room service. We don't want any loose cannons."

Lovell hung up, dazed. He had known Grissom, White, and Chaffee for years and had been friendly with them all, but it was White he had known best. Fifteen years earlier, when Lovell was an Annapolis midshipman attending the Army–Navy game in Philadelphia, he met a congenial West Point cadet, whose name he never quite caught, at a crowded hotel party. As was the tradition, the friendly foes would exchange makeshift gifts that would serve as mementos of the game and the subsequent celebration. With nothing else handy, Lovell removed one of his Navy cuff links and gave it to the West Pointer; the West Pointer reciprocated with an Army link, and the two young men parted.

More than a decade later, when Lovell had joined the astronaut corps, he told the story to fellow astronaut Ed White. White's jaw dropped. *He* was the West Pointer; he, like Lovell, had told the story numerous times over the intervening years; and he, like Lovell, still had the cuff link. The two astronauts became fast friends. Grissom was more of a stranger to Lovell, but the reputation of the veteran Mercury pilot was well known throughout the astronaut corps; like everyone who knew Grissom, Lovell had profound respect for the man's accomplishments and admiration for his flying skills. Chaffee was an unknown quantity. As a member of the third astronaut class, the junior pilot had little occasion to work with any of the men who flew in the Gemini program. However, NASA had tapped Chaffee for the first Apollo mission, and that said a lot. More important, Grissom had once referred to his Apollo apprentice as "a really great boy." And that said even more.

Lovell walked abstractedly into the hallway of the turquoise floor just as the other astronauts were emerging from their rooms. Gordon and Armstrong had also spoken to Houston. Cooper, as the senior member of the group and one of the original seven Mercury astronauts, got his call from Congressman Jerry Ford, the ranking Republican member of the House space committee.

"You heard?" Lovell asked.

The other three nodded.

"What the hell happened?"

"What happened?" Gordon said. "That spacecraft happened, that's what. They should've deep-sixed that thing long ago."

"Do the wives know?" Lovell asked.

"No one said yet," Cooper answered.

"Who's on hand to tell them?" Armstrong asked.

"Mike Collins is around," Lovell said. "Pete Conrad and Al Bean should be. Deke's at the Cape, but his wife is at home, near Gus's place." He paused. "Does it really make any difference who tells them?"

Down in the lobby, Collins Bird at last got word from Houston about the disaster at the Cape. NASA's unofficial host knew what the astronauts on the fifth floor would need tonight, and told his staff to open up room 503, a suite with a living room where the pilots could sit undisturbed and talk. Lovell and the others moved into the room, phoned the kitchen, and ordered dinner and, more important, Scotch. Tomorrow they would be expected to fly to Houston for postmortems and emergency meetings. Tonight, however, was their own, and they would do what flight jocks traditionally do when a member of their small, insular circle dies. They would talk about how and why it happened, and they would get drunk while doing it.

The conversation went on well into the early morning, with the astronauts moving from concern for the future of the program, to predictions about whether it would now be possible to get to the moon before the end of the decade, to resentment of NASA for pushing the program so hard just to make that artificial deadline, to rage at NASA for building that piece of crap spacecraft in the first place and refusing to listen to the astronauts when they told the Agency bosses they were going to have to spend the money to rebuild it right.

Inevitably, as the liquor went down and the sun began to come up, the talk turned to death, and the astronauts quietly agreed that while Grissom, White, and Chaffee had indeed met a hero's end, a fire on the pad in a locked-down, unfueled missile was not the way to go. If you've got to buy the farm, better to do it while riding a corkscrewing rocket up through the atmosphere, or steering a tumbling spacecraft down to Earth, or getting stuck in orbit with a dead retrorocket, or being marooned on the face of the moon. Perhaps it was disrespectful to admit it, especially tonight, but while no violent death was to be envied, the astronauts knew that an earthbound death was to be envied least of all.

Gus Grissom, Ed White, and Roger Chaffee were buried four days later, on January 31, 1967. Grissom and Chaffee would be interred with full military honors at Arlington National Cemetery. White, as was his wish, would be buried where his father planned one day to be buried, at their alma mater, West Point. The surviving members of Grissom's first class of astronauts and Chaffee's third class attended the Arlington ceremony, along with dozens of other dignitaries, including Lyndon Johnson. Jim Lovell and the rest of the second class of astronauts, along with Lady Bird Johnson and Hubert Humphrey, went to West Point. Lovell flew to the academy in a T-38 jet with Frank Borman, his commander on the Gemini 7 mission. After spending a fortnight together in the anchovy can of the Gemini capsule, Lovell and Borman never had trouble making conversation, but on this flight they were mostly quiet. Borman offered a reminiscence or two about the dead crewmen, Lovell told his cuff link story; otherwise they fell to reflective silence.

Of the two ceremonies that day, White's was decidedly the more low-key. The funeral, held inside the Old Cadet Chapel, was attended by nine hundred spectators. After the service, Lovell, Borman, Armstrong, Conrad, Aldrin, and Tom Stafford wrestled the coffin up to a bluff overlooking the frozen Hudson, where a few more words were spoken and White was lowered into the cement-hard ground.

At Arlington things were not nearly so understated. With the president in attendance, Phantom jets flying in formation overhead, and bands, buglers, riflery corps, and honor guards arrayed about the graves, Grissom and Chaffee were bid farewell like departed heads of state. Schirra, Slayton, Cooper, Carpenter, Alan Shepard, and John Glenn served as pallbearers for fellow Mercury veteran Grissom. Chaffee was borne to his grave by Navy men and members of his own astronaut class. President Johnson murmured sympathetic words to the grieving kin. As one of the men who had helped flog the space program to a breathless (reckless?) gallop in recent years, Johnson found his condolences only tepidly received. Chaffee's father barely acknowledged the president when they met at graveside, glancing at him briefly and nodding his head before looking away again. Grissom's parents did not meet the Texan's eyes at all.

The speeches, of course, praised the astronauts' achievements lavishly. Grissom was described as a "pioneer" and "one of the great heroes of the space age." Similar homage was heaped upon White at West Point. It was only in Chaffee's eulogy that the plaudits might have seemed a bit strained. The novice astronaut had yet flown no higher than an ordinary Navy plane could take the ordinary pilot, and so the odes to the departed explorer could not be to the wonderful things he'd done, but to the wonderful things he might have done.

At least one person at Arlington knew that Chaffee had already achieved more than many people ever would. Standing among the mourners, Wally Schirra thought back to that week in October 1962 when he visited the White House to receive his medal. The ceremony that day was noticeably more perfunctory than some of the earlier astronaut welcomes, not only because the novelty of the Mercury program had begun to wear off, but because President Kennedy had other things on his mind. Recently, surveillance flights had overflown Cuba and revealed the presence of silos and launchers and, significantly, ICBMs where vacant fields or sugar crops should have been. Though Schirra couldn't know it, at the same time he and his wife and daughter were standing in the Oval Office, another pilot in another surveillance plane was taking to the skies over Castro's angry island to gather a bit more evidence to send back to his president. The pilot of that plane was naval aviator Roger Chaffee.

Schirra said a silent farewell to the spaceman who never flew. A really great boy, all right.

2

December 21, 1968

FRANK BORMAN, Jim Lovell, and Bill Anders were awakened in the Kennedy Space Center crew quarters at just after three in the morning on the Saturday before Christmas. The sun was hours from rising, but fluorescent light leaking under the door illuminated the rooms just enough to remind the crew where they were.

As government barracks went, the place was not half bad. NASA spared little expense on accommodations for the men it was planning to fling into space, decorating the dorms with new carpeting, surprisingly stylish furniture, and reproductions of paintings in expensive frames. The facilities were also equipped with a conference room, a sauna, and a complete kitchen with a private chef. All the lavishness was less a case of Agency excess than it was a smart precaution. Flight planners knew that isolating a crew in the final days before a launch was the only way to keep them focused on the mission at hand and insulated from any errant bugs that might cause a launch-scrubbing cold or flu; but they also knew that quarantined men were not, on the whole, happy men, and unhappy men did not, on the whole, make good pilots. In order to keep the mood of the crews as bright as possible, therefore, the Agency decided that their quarters would be as plush as possible. On this day, of all days, that was more important than ever.

Lovell heard the knock on his door, opened one eye, and saw the face of Deke Slayton peering in from the hallway; he greeted the chief of the astronaut office with a grumble, a half wave, and a secret wish he'd go away. More than his two crewmates, Lovell was familiar with this morning-of-liftoff ritual. There would be the long hot shower, the last for eight days; the final medical checkup; the traditional steak and eggs breakfast with Slayton and the backup crew; the gladiatorial ceremony of suiting up in the bulky, inflatable pressure garment with the fishbowl helmet; the smiling, waving, stiff-legged walk to the air-conditioned van; the hushed drive to the launch pad; the rattly ride up the gantry elevator; the awkward climb into the cockpit; and finally the slamming and sealing of the spacecraft hatch.

Lovell had been through it all twice before, and NASA had been through it seventeen times before. So there was no real reason to feel that today would be different. But the fact was, today was very different. For the first time, after all of those ceremonial showers and suit-ups and breakfasts and blastoffs, the crew's destination was not near-Earth orbit. This was the day NASA planned to launch Apollo 8, and the destination was the moon.

It had been just under two years since the fire in the cockpit that killed Gus Grissom, Ed White, and Roger Chaffee, and the memories of that day were only now beginning to fade. Borman, Lovell, and Anders were not the first American crew to fly in space in those twenty-three months; the first had been Wally Schirra, Donn Eisele, and Walt Cunningham just eight weeks earlier, and that day, the reminders of the lost crew had been everywhere. Though Schirra, Eisele, and Cunningham were the first men ever to pilot an Apollo spacecraft, their mission was officially known as Apollo 7. There had been five unmanned Apollo flights previously, and these had been designated Apollos 2 through 6. Before the fire, Grissom, White, and Chaffee had informally claimed the Apollo 1 honorific for themselves, but NASA officials had not yet given their O.K. Two of the unmanned flights had actually preceded the dead crewmen's scheduled mission, and the best the astronauts could technically have hoped for was Apollo 3. After the fire, however, the sentiment within NASA changed, and the Agency decided to grant the astronauts' wishes posthumously, permanently retiring the Apollo 1 name.

Also contributing to the cloud hanging over the pre-launch ritual that morning eight weeks ago was the fact that Wally Schirra still did not completely trust the ship he would be commanding, and he didn't give a damn who knew it. In the days, indeed in the very hours, after the Apollo 1 fire, NASA did what most governmental agencies do when they're blind-sided by events: it appointed a commission to find out what went wrong and what could be done to set it right. The seven-man panel was made up of six ranking officials in NASA and the aerospace industry and one astronaut, Frank Borman. Borman and his colleagues, knowing that they could not hope to analyze every system and component in the ship by themselves, in turn appointed twenty-one other panels, each of which would autopsy a different portion of the Apollo spacecraft until the source of the fire could be found and fixed.

Of the twenty-one panels, the one with the most straightforward job was panel twenty, the group investigating in-flight fire emergency procedures. Among the members of that group were rookie astronauts Ron Evans and Jack Swigert and two-time orbital veteran Jim Lovell. While Borman and the NASA brass heading up the fire investigation became something close to media darlings, Lovell, Swigert, Evans, and the other men on the rest of the panels toiled in near obscurity.

This rankled some of the men in the astronaut corps. Who the hell was Borman to be tapped as the one astronaut among dozens to help get the Agency through its darkest hour? For Lovell, however, the obscurity was just fine. Conducting a postmortem on a mission that cost lives could be a grisly business, something that once experienced was not gladly repeated. This was not the first time NASA's astronaut corps had been struck by tragedy. The first time had been just over two years earlier, and it had been Lovell's job to sort through the mess.

It was October of 1964, and Lovell, an astronaut now for less than two years, was coming back from a day of goose hunting with Pete Conrad, a fellow member of the astronaut class of '62. Passing Ellington Air Force Base near the Manned Spacecraft Center in Houston, Lovell and Conrad noticed a crowd gathering around what appeared to be the smashed remains of a T-38 jet lying in a field just off the runway. Lovell slammed the car to a stop, and the two men ran up to the edge of the crowd and asked a bystander what happened.

"Fellow was making a routine flight," the witness answered, "just flying in a big circle and coming back to the runway. All of a sudden, at about 1,500 feet, the plane just started to pancake in. The guy tried to eject but it was way too late — he came out mostly horizontal and hit the ground before his chute could open all the way."

"Do you know who he was?" Lovell asked.

"Yeah," the man said. "Ted Freeman."

Lovell and Conrad exchanged stricken looks. Ted Freeman was a rookie astronaut who had come into the program about a year after they had. They didn't know him all that well, but they knew him by reputation, and he was thought to be a formidable competitor for the limited number of slots on the Gemini missions still to be announced. To date, no American astronaut had been lost in space, and now poor Freeman had augered in before even getting the chance to climb into a spacecraft.

Lovell pushed his way through the crowd, with Conrad half a step behind. During his time as a Navy flight instructor, Lovell, having studied aviation safety at the University of Southern California, was appointed squadron safety officer. The first rule of thumb he'd learned in training was that there was no better way to figure out what caused an aircraft to crash than to eyeball the remains. To the untutored observer, a pancaked plane is just a pancaked plane, but to someone who knows what to look for, the precise condition of the wreck can tell a lot about what brought it down.

What Lovell saw when he reached Freeman's T-38, however, only deepened the mystery surrounding the crash. With the exception of its flattened nose, the plane was not badly damaged. The front canopy — essentially a metal framework fitted with a bubble of Plexiglas — had popped free, as it should have when Freeman ejected. Found in the grass several hundred feet behind the plane, it too appeared to have weathered the crash reasonably well, though curiously, most of the Plexiglas was missing. Lovell noticed that the back seat of the T-38's cockpit, which had been empty during the flight, had a splatter of blood on it, and the back canopy, which was still secured to the aircraft, was also missing much of its Plexiglas.

When NASA officials arrived and began taking statements, Lovell and Conrad pointed out what they had discovered. Later that day, Deke Slayton contacted Lovell, thanked him for his assistance, and told him that, given his timely arrival at the crash site and his experience in aviation safety, he would be placed in charge of the full investigation still to come.

Lovell attacked the new assignment with relish, but there was not much to go on. A detailed examination of the plane revealed that the cause of the crash was engine failure; sometime before Freeman bailed out, the air-breathing jets on either side of the fuselage had quit on him, leaving him in a dead, silent glide. But what had caused the engines to quit in the first place? With the jet itself yielding no other information, Lovell sorely wished he could find the one part that still eluded examination: the Plexiglas missing from the two canopies. Since the transparent shards could have landed anywhere in a several-mile radius around the airfield, however, he knew his chances of finding them were not good.

There was one possible solution. When the engines of a T-38 shut down, Lovell knew, the generators that run the instrument panel quit too. This meant that at the precise instant generator power was lost, all of the navigation hardware would have frozen in place — including the TACAN tracker, the instrument that continuously records the plane's direction and distance in relation to the tracking station at the airfield. By reading this instrument, Lovell could, in theory, pinpoint the spot where the engines had quit. Below that, on the ground, the Plexiglas should be lying.

Lovell copied the data from the instruments, dug out a map of the area, and the TACAN pointed him to a field about four miles from the air base. Conrad volunteered to helicopter out to the site and conduct a search. Touching down in the tall Texas grass, the astronaut began wading around when, almost immediately, he spotted something glinting in the distance. On approaching he could see that the object was indeed a section of Ted Freeman's lost Plexiglas, shattered beyond recognition. In the grass only a few feet away were the remains of a badly lacerated Canada snow goose.

The conclusion was obvious: tearing along at 400 knots, Freeman must have been been hit broadside by the far slower goose, which smashed through the canopy, shattering the Plexiglas. The goose flew out the back of the plane, leaving blood on the rear seat, and the Plexiglas from both canopies flew in every direction, choking the intake vents of the engines and causing them to flame out. Freeman tried to swing the plane around to land on the closest runway he could find, but without engines his 400-knot speed quickly slowed and he began to sink. Ejecting from the cockpit, he had enough time to fly clear of his falling T-38, but too little time for his chute to open and bring him down safely.

Lovell wrote up his report, filed it with NASA and the military, and the officials accepted it without question. The next day, the investigation into the death of Ted Freeman was closed, and NASA mourned the first, freakish loss of one of its astronauts.

The Freeman inquiry was a challenge for Lovell, and solving the riddle of the astronaut's death had a distinct, if dark, satisfaction to it. This kind of investigation was essentially mortician's work, however, and when Borman was later chosen to find the cause of the Grissom, White, and Chaffee disaster, Lovell was not inclined to grumble. As it turned out, the investigation was even more grueling than anyone had imagined. As the commission huddled in their conference room, and the members of the twenty-one other panels camped out in nooks and offices around Houston and the Cape, Congress conducted its own outraged hearings, combing through the NASA organizational chart to determine whose job it had been to prevent accidents like this and how things could have been bungled so badly.

It soon became clear to all of the groups that the improvements to the command module were going to be extensive and that all of the gripes by astronauts and NASA engineers in previous years had some merit. George Low, one of the Agency's assistant administrators, set up a command-module change board, to bring control and order to the redesign and provide an open forum for the astronauts to demand changes they thought were essential. The contractors too, motivated partly by guilt, partly by a white-hot terror of another disaster, and partly — indeed, perhaps mostly — by a professional desire to deliver the space-worthy vehicle they had promised NASA they would deliver, opened their own doors to the Apollo pilots, giving them access to any aspect of any operation they wanted to explore.

Wally Schirra, Donn Eisele, and Walt Cunningham, the three men with the most immediate interest in the soundness of the next Apollo to roll off the line, took full advantage of this offer, prowling the factory floors in Downey, California, like a flying wedge, checking various components of the spacecraft as they were built.

"Any of you guys have a problem or a logjam, tell me and we'll blow it," Schirra told Cunningham and Eisele somewhat grandiosely as he dispatched them around the North American Aviation plant where the command module was manufactured.

Borman, as NASA's official, if less flamboyant, emissary to North American, began to chafe at this interference from Schirra and his juniors, and eventually phoned the Agency brass, demanding that his fellow astronauts be kept on a tighter rein. The fire, Borman argued, had been caused at least in part by chaos and conflicting engineering signals within NASA, and the last thing the men conducting the redesign needed now was dozens of voices calling for dozens of changes in a spacecraft with millions of parts. NASA agreed, Schirra backed off, and the Apollo improvements proceeded in a more orderly way.

With Borman as point man and the rest of the pilots now backing him up more quietly, the astronauts got nearly everything they had been lobbying for in a new, safer spacecraft. They had wanted a gas-operated hatch that could be opened in seven seconds, and they got it; they had wanted upgraded, fireproof wiring throughout the ship, and they got it; they had wanted nonflammable Beta cloth in the spacesuits and all fabric surfaces, and they got it. Most important, they had wanted the fire-feeding, 100 percent oxygen atmosphere that swirled through the ship when it was on the pad to be replaced by a far less dangerous 60-40 oxygen-nitrogen mix. Not surprisingly, they got that too.

When it was later pointed out to Schirra that Borman's low-key approach had evidently been the right one, that the fixes the pilots were demanding had been achieved just as easily — perhaps more easily — without a lot of crabbing and cantankerousness, Schirra was unmoved. "We all just spent a year wearing black armbands for three very good men," he liked to say. "I'll be damned if anybody's going to spend next year wearing one for me."

The modifications being made to the Apollo spacecraft were not the only changes NASA explored in the wake of the fire. Also scrutinized were the missions those ships would be sent on. Though John Kennedy had been dead since 1963, his grand promise — or damned promise, depending on how you looked at it — to have America on the moon before 1970 still loomed over the Agency. NASA officials would have considered it a profound failure not to meet that bold challenge, but they would have considered it an even greater failure to lose another crew in the effort. Accordingly, chastened Agency brass began making it clear, publicly and privately, that while America was still aiming for the moon before the end of the decade, the breathless gallop of the past few years would now be replaced by a nice, safe lope.

According to the tentative flight schedule, the first manned Apollo flight would be Schirra's Apollo 7, intended to be nothing more than a shakedown cruise of the still-suspect command module in low Earth orbit. Next would come Apollo 8, during which Jim McDivitt, Dave Scott, and Rusty Schweickart would go back into near-Earth space to test-drive both the command module and the lunar excursion module, or LEM, the ugly, buggy, leggy lander that would carry astronauts down to the surface of the moon. Next, Frank Borman, Jim Lovell, and Bill Anders would pilot Apollo 9 on a similar two-craft mission, this time taking the ships to a vertiginous altitude of 4,000 miles, in order to practice the hair-raising, high-speed reentry techniques that would be necessary for a safe return from the moon.

After that, things were wide open. The program was scheduled to continue through Apollo 20, and, in theory, any mission from Apollo 10 on could be the first to set two men down on the moon's surface. But which mission and which two men were utterly unsettled. NASA was determined not to rush things, and if it took until well into the Apollo teens before all the equipment checked out and a landing looked reasonably safe, then it would have to take that long.

In the summer of 1968, two months before Apollo 7 was scheduled for launch, circumstances in Kazakhstan — southeast of Moscow — and in Bethpage, Long Island — northeast of Levittown — conspired to scramble this cautious scenario. In August, the first lunar module arrived at Cape Kennedy from its Grumman Aerospace plant in Bethpage, and in the assessment of even the most charitable technicians, it was found to be a mess. In the early checkout runs of the fragile, foil-covered ship, it appeared that every critical component had major, seemingly insoluble problems. Elements of the spacecraft that were shipped to the Cape unassembled and were supposed to be bolted together on site did not seem to want to go together; electrical systems and plumbing did not operate as specified; seams, gaskets, and washers that were designed to remain tightly sealed were springing all manner of leaks.

Some glitches, of course, were to be expected. In ten years of building sleek, bullet-shaped spacecraft designed to fly through the atmosphere and into orbit, no one had ever attempted to build a manned ship that would operate exclusively in the vacuum of space or in the lunar world of one-sixth gravity. But the number of glitches in this gimpy ship was more than even the worst NASA pessimists could have imagined.

At the same time the LEM was causing such headaches, CIA agents working overseas picked up even more disturbing news. According to whispers coming from the Baikonur Cosmodrome, the Soviet Union was making tentative plans for a flight around the moon by a Zond spacecraft sometime before the end of the year. Nobody knew if the flight would be manned, but the Zond line was certainly capable of carrying a crew, and if a decade of getting sucker-punched by Soviet space triumphs had indicated anything, it was that when Moscow had even the possibility of pulling off a space coup, you could bet they'd give it a try.

NASA was stumped. Flying the LEM before it was ready was clearly impossible in the cautious atmosphere that now pervaded the Agency, but flying Apollo 7 and then launching nothing at all for months and months while the Russians promenaded around the moon was not an attractive option either. One afternoon in early August, 1968, Chris Kraft, deputy director of the Manned Spacecraft Center, and Deke Slayton were summoned to Bob Gilruth's office to discuss the problem. Gilruth was the overall director of the Center and, according to the scuttlebut, had been meeting all morning with George Low, the director of Flight Missions, to determine if there was some plan that would allow NASA to save face without running the risk of losing more crews. Slayton and Kraft arrived in Gilruth's office, and he and Low got straight to business.

"Chris, we've got serious problems with the upcoming flights," Low said bluntly. "We've got the Russians and we've got the LEM and neither one is cooperating."

"Especially the LEM," Kraft responded. "We're having every kind of trouble it's possible to have."

"So it couldn't be ready by December?" Low asked.

"No chance," said Kraft.

"If we wanted to fly Apollo 8 on schedule, what could we do with just the command-service module that will further the program?"

"Not much in Earth orbit," Kraft said. "Most of what we can do with that we're already planning to do on 7."

"True enough," Low said tentatively. "But suppose Apollo 8 didn't just repeat 7's mission. If we don't have an operative LEM by December, could we do something else with the command-service module alone?" Low paused for a moment. "Like orbit the moon?"

Kraft looked away and fell silent for a long minute, calculating the incalculable question Low had just asked him. He looked back at his boss and slowly shook his head no.

"George," he said, "that's a pretty difficult order. We're having a hell of a struggle getting the computer programs ready just for an Earth-orbit flight. You're asking what I think about a moon flight in four months? I don't think we can do it."

Low seemed strangely unperturbed. He turned to Slayton. "What about the crews, Deke? If we could get the systems ready for a lunar mission, would you have a crew that could make the flight?"

"The crew isn't a problem," Slayton answered. "They could get ready."

Low pressed him. "Who would you want to send? McDivitt, Scott, and Schweickart are next in line."

"I wouldn't give it to them," Slayton said. "They've been training with the LEM for a long time, and McDivitt's made it clear he wants to fly that ship. Borman's crew hasn't been at it as long, plus they're already thinking about deep-space reentry, something they'd need for a mission like this. I'd give it to Borman, Lovell, and Anders."

Low was encouraged by Slayton's response, and Kraft, infected by the enthusiasm of the other men in the room, began to soften. He asked Low for a little time to talk to his technicians and see if the computer problems could be resolved. Low agreed, and Kraft left with Slayton, promising an answer in a few days. Returning to his office, Kraft hurriedly assembled his team around him.

"I'm going to ask you a question, and I want an answer in seventy-two hours," he said. "Could we get our computer problems unraveled in time to get to the moon by December?"

Kraft's team vanished, and returned not in the requested seventy-two hours, but in twenty-four. Their answer was a unanimous one: yes, they told him, the job could be done.

Kraft got back on the phone to Low: "We think it's a good idea," he told the director of Flight Missions. "As long as nothing goes wrong on Apollo 7, we think we ought to send Apollo 8 to the moon over Christmas."

On October 11, 1968, Wally Schirra, Donn Eisele, and Walt Cunningham rode into orbit around the Earth aboard Apollo 7; eleven days later they plopped down in the Atlantic Ocean. The media applauded the mission wildly, the president phoned his congratulations to the crew, and NASA declared that the flight had achieved "101 percent" of its objectives. Inside the Agency, flight planners set about the task of sending Frank Borman, Jim Lovell, and Bill Anders to the moon just sixty days later.

The buildup to the launch of Apollo 8 was stage-managed brilliantly by NASA. Just two days before Apollo 7 was launched atop its 224-foot-tall Saturn 1-B rocket, the Agency also rolled out the Saturn 5, the 363-foot monster of a booster that would be necessary to lift the spacecraft above the atmosphere and fling it toward the moon. NASA tried to play down the event — heck, a rocket had to come out of its hangar *sometime* — but it escaped few people's notice that the rollout occurred when cameras from all over the world were already present for Apollo 7's liftoff.

The event caused the newspapers to sizzle: "U.S. Prepares Moon Shot in December," announced the *New York Times.* "Apollo 8 Ready to Circle Moon," blared the *Washington Star,* adding in smaller type that the flight "was, and still is, billed officially as a second Earth orbital flight."

NASA played the story as coyly as it could, conceding that a moon mission was a possibility for Apollo 8, but *only* a possibility; no decisions would be made until after Apollo 7 splashed down safely. Borman, Lovell, and Anders, of course, had long known the mission to the moon was a nearly done deal, and Lovell, for one, had been delighted with the development. While the high-orbit shakedown of the lunar module had had its merits, Lovell frankly saw the mission as a little duller than he might have liked. As command module pilot, he would have had the responsibility of staying behind in the Apollo spacecraft while Borman and Anders put the LEM through its paces. With the mission now changed to a LEM-less lunar orbit, the in-flight duties of the three crewmen would change dramatically; and with Lovell officially designated the navigator for this first translunar journey, his duties might be the most challenging of the trio's.

The reaction of Borman, the mission commander, was a bit more measured. Trained as a fighter pilot and known for his lightning quick reflexes and exceptional decision-making skills, Borman was one of the best pure pilots NASA had. But he also had a cautious streak.

The Air Force colonel and Gemini 7 veteran was routinely ribbed by his fellow astronauts for the careful route he took when flying his T-38 from Houston to Cape Canaveral. A strict aircraft safety rule required the pilots to fly over land while they made the trip, never straying out over the Gulf of Mexico. Yet most of the men — who made their living risking their lives in unproven planes — chafed at so hypercautious a regulation, and regularly defied it, cutting across stretches of the Gulf if they thought it would save time. Borman, however, generally played it straight, choosing a drier, if more circuitous, route along the coasts of Texas, Louisiana, Mississippi, Alabama, and finally the Florida peninsula itself. No one ever suggested that this roundabout flight path was indicative of a failure of courage, and indeed it wasn't. Rather, it was roundly accepted that the man who had aggressively sought admission to America's astronaut corps and had circled the Earth 206 times with Jim Lovell in 1965 simply felt that there was never any reason to choose a risky option when a safer one would also do.

Bill Anders, the novice member of the team, reacted to the moon assignment with the same mixed feelings as Borman, but for different reasons. As lunar module pilot, Anders had looked forward to being the resident expert on the experimental lander and to overseeing most of the test maneuvers that would help certify the ship fit to fly. Now, with the lander grounded, he would find himself with considerably less to do, concentrating mostly on supervising the health of the service module's main engine and the status of the spacecraft's communications and electrical system. It was important work, but compared with piloting a LEM at an altitude of 4,000 miles, it didn't come close. Lovell kidded Anders when the flight change was made. "Basically," he said, "we need you to sit there and look intelligent."

As was the case with all such mission assignments, as soon as even a tentative flight plan was set, the crews were permitted, in fact encouraged, to talk it over with their wives. On the August day that Frank Borman, Jim Lovell, and Bill Anders first learned they would be visiting the moon that December, Lovell's first thoughts were not of history or posterity or the grand sweep of human exploration, but of Acapulco. In recent years, a hotelier named Frank Branstetter had befriended the astronauts and had made it a point to set aside a number of rooms at Las Brisas, his Acapulco resort, for the families of crewmen just returning from missions. Lovell had been too busy to take Branstetter up on his invitation following the Gemini 12 mission, but this winter — nearly two years after the flight — the astronaut, his wife, and their four children were at last going to make the trip. Branstetter was looking forward to having the family down, and Marilyn Lovell, for one, was eager to go. It was left to her husband to inform her that their plans would have to change.

"I've been thinking about Acapulco," Lovell told his wife when he arrived home from the Manned Spacecraft Center that evening. "I'm not so sure it's a good idea."

"Why not?" Marilyn asked, more than a bit annoyed.

"I don't know. I just don't think I feel like going."

"Well don't you think it's a little late for that? You already promised the kids, we've already made the reservations —"

"I know, I know. But I thought Frank and Bill and I might go somewhere else instead."

"Like where?"

"Oh, I don't know," Lovell said with studied nonchalance. "Maybe the moon."

Marilyn stared at him, wordless. Since 1962 she had been anticipating this moment with a sort of dreamy dread. Lovell allowed her to collect herself and then, as he had in 1965 before Gemini 7 and in 1966 before Gemini 12, explained the promise of the mission, and also the perils. Before those earlier flights, both Lovells had known the risks would be considerable. Jim Lovell and Frank Borman were to spend two weeks aloft in Gemini 7, longer than any astronaut had attempted before. Once there, they would participate in a tricky rendezvous with Wally Schirra and Tom Stafford aboard Gemini 6 — a stunt no American crew had ever even considered. Gemini 12, although a mere four-day mission flown without any other accompanying manned ships, would present perils of its own: the docking with the unmanned, and unreliable, Agena spacecraft; the five and a half hours of space walking Buzz Aldrin would attempt in the middle of the mission. Both flights were high-risk ventures at best, but both at least had historical precedents. Jim Lovell would not be the first American to fly into orbit, or even the second or the third. He would be the eleventh — if anyone was counting anymore — and his wife could at least take some solace from the fact that the previous ten men had all come home to their wives none the worse for their experiences.

But Apollo 8 would be different. There were no precedents for the trip Jim Lovell would be making this time; there were no other men who had survived it before him. As he eased his wife into a chair, he described some of the details of the flight: how the ship would have to whip itself up to an unprecedented 25,000 miles per hour to escape from Earth orbit; how it would have to rely on a single engine, without a backup, to settle into lunar orbit; how it would have to rely on another burn of the engine to head home again; how it would have to reenter the Earth's atmosphere through a narrow corridor barely 2.5 degrees wide if it was going to survive its fiery plunge. Marilyn nodded and took it all in and eventually, as she had in the past, gave her sober approval.

Valerie Anders, word around the Agency had it, reacted to Bill's news with a similar guarded O.K. Susan Borman, however, was said to have responded differently. As far as Susan was concerned, the stories went, Apollo 8 was an undue risk, and she didn't especially care for the fact that her husband had been chosen to command it. While wives could do little to change flight assignments, they could make their displeasure felt within the tight NASA tribe. Susan, rumor had it, had chosen Chris Kraft as the object of her displeasure, and made it clear that even if Frank survived this harebrained flight, Kraft should not expect another civil word from her.

On the morning Apollo 8 was launched, December 21, the doubts and the acrimony were at least outwardly forgotten. Borman, Lovell, and Anders were sealed in their spacecraft just after 5:00 A.M. in preparation for a 7:51 launch. By 7:00 the networks' coverage began and much of the country was awake to witness the event live. Across Europe and in Asia, audiences numbering in the tens of millions also tuned in.

From the moment the Brobdingnagian Saturn 5 booster was lit, it was clear to TV viewers that this would be like no other launch in history. To the men in the spacecraft — one of whom had never flown in space before and two of whom had ridden only the comparatively puny, 109-foot Gemini-Titan — it was clearer still. The Titan had been designed originally as an intercontinental ballistic missile, and if you were unfortunate enough to find yourself strapped in its nose cone — where nothing but a thermonuclear warhead was supposed to be — it felt every bit the ferocious projectile it was. The lightweight rocket fairly leapt off the pad, building up velocity and g forces with staggering speed. At the burnout of the second of its two stages, the Titan pulled a crushing eight g's, causing the average 170-pound astronaut to feel as if he suddenly weighed 1,360 pounds. Just as unsettling as the rocket's speed and g's was its orientation. The Titan's guidance system preferred to do its navigating when the payload and missile were lying on their sides; as the rocket climbed, therefore, it also rolled 90 degrees to the right, causing the horizon outside the astronauts' windows to change to a vertigo-inducing vertical. Even more disturbing, the Titan had a huge range of ballistic trajectories programmed into its guidance computer, which aimed the missile below the horizon if it was headed for a military target or above the horizon if it was headed for space. As the rocket rose, the computer would continually hunt for just the right orientation, causing the missile to wiggle its nose up and down and left to right, bloodhound-fashion, sniffing for a target that might be Moscow, might be Minsk, or might be low Earth orbit, depending upon whether it was carrying warheads or spacemen on that particular mission.

The Saturn 5 was said to be a different beast. Despite the fact that the rocket produced a staggering 7.5 million pounds of thrust — nearly nineteen times more than the Titan — the designers promised that this would be a far smoother booster. Peak gravity loads were said to climb no higher than four g's, and at some points in the rocket's powered flight, its gentle acceleration and its unusual trajectory dropped the gravity load slightly *below* one g. Among the astronauts, many of whom were approaching forty, the Saturn 5 had already earned the sobriquet "the old man's rocket." The promised smoothness of the Saturn's ride, however, was until now just a promise, since no crew had as yet ridden it to space. Within the first minutes of the Apollo 8 mission, Borman, Lovell, and Anders quickly learned that the rumors about the painless rocket were all wonderfully true.

"The first stage was very smooth, and this one is smoother!" Borman exulted midway through the ascent, when the rocket's giant F-1 engines had burned out and its smaller J-2 engines had taken over.

"Roger, smooth and smoother," Capcom answered.

Less than ten minutes later, the gentle expendable booster completed its useful life, dropping its first two stages in the ocean and placing the astronauts in a stable orbit 102 miles above the Earth.

According to the mission rules for a lunar flight, a ship bound for the moon must spend its first three hours in space circling the Earth in an aptly named "parking orbit." The crew uses this time to stow equipment, calibrate instruments, take navigational readings, and generally make sure their little ship is fit to leave home. Only when everything checks out are they permitted to relight the Saturn 5's third-stage engine and break the gravitational hold of Earth.

For Frank Borman, Jim Lovell, and Bill Anders, it would be a busy three hours, and as soon as the ship was safely in orbit, they knew they'd have to get straight to work. Lovell was the first of the trio to unbuckle his seat restraints, and no sooner had he removed the belts and drifted forward than he was struck by a profound feeling of nausea. The astronauts who flew in the early days of the space program had long been warned about the possibility of space sickness in zero g, but in the tiny Mercury and Gemini capsules, where there was barely room to float up from your seat before bonking your head on the hatch, motion-related queasiness was not a problem. In Apollo there was more space to move around, and Lovell discovered that this elbow room came at a gastric price.

"Whoa," Lovell said, as much to himself as in warning to his crewmates. "You don't want to move too fast."

He eased his way gently forward, discovering — as centuries of remorseful drinkers with late-night bed spins had learned — that if he kept his eyes focused on one spot and moved very, very slowly, he could keep his churning innards under control. Easing his way about in this tentative way, Lovell began to negotiate the space directly around his seat, failing to notice that a small metal toggle protruding from the front of his spacesuit had snagged one of the metal struts of the couch. As he moved forward the toggle caught, and a loud pop and hiss echoed through the spacecraft. The astronaut looked down and noticed that his bright yellow life vest, worn as a precaution during liftoffs over water, was ballooning up to full size across his chest.

"Aw, hell," Lovell muttered, dropping his head into his hand and pushing himself back into his seat.

"What happened?" a startled Anders asked, looking over from the right-hand couch.

"What does it look like," Lovell said, more annoyed with himself than his junior pilot. "I think I snagged my vest on something."

"Well, unsnag it," Borman said. "We've got to get that thing deflated and stowed."

"I know," Lovell said, "but how?"

Borman realized Lovell had a point. The emergency life vests were inflated from little canisters of pressurized carbon dioxide that emptied their contents into the bladder of the vest. Since the canisters could not be refilled, deflating the vest required opening its exhaust valve and dumping the CO_2 into the surrounding air. Out in the ocean this was not a problem, of course, but in a cramped Apollo command module it could be a bit dicey. The cockpit was equipped with cartridges of granular lithium hydroxide that filtered CO_2 out of the air, but the cartridges had a saturation point after which they could absorb no more. While there were replacement cartridges on board, it was hardly a good idea to challenge the first cartridge on the first day with a hot belch of carbon dioxide let loose in the small cabin. Borman and Anders looked at Lovell and the three men shrugged helplessly.

"Apollo 8, Houston. Do you read?" the Capcom called, evidently concerned that he hadn't heard from the crew for a long minute.

"Roger," Borman answered. "We had a little incident here. Jim inadvertently popped one life vest, so we've got one full Mae West with us."

"Roger," the Capcom replied, seemingly without an answer to offer. "Understand."

With their 180 minutes of Earth orbit ticking away and no time to waste on the trivial matter of a life vest, Lovell and Borman suddenly hit on an answer: the urine dump. In a storage area near the foot of the couches was a long hose connected to a tiny valve leading to the outside of the spacecraft. At the loose end of the hose was a cylindrical assembly. The entire apparatus was known in flying circles as a relief tube. An astronaut in need of the relief the system provided could position the cylinder just so, open the valve to the vacuum outside, and from the comfort of a multi-million-dollar spacecraft speeding along at up to 25,000 miles per hour, urinate into the celestial void.

Lovell had availed himself of the relief tube countless times before, but only for its intended purpose. Now he would have to improvise. Struggling out of his life vest, he wrestled it down to the urine port, and with some finessing managed to wedge its nozzle into the tube. It was a forced fit, but a workable one. Lovell gave the high sign to Borman, Borman nodded back, and while the commander and the LEM pilot went through their pre-lunar checklist, Lovell coaxed his life vest back to its deflated state, patiently correcting the first blunder he had committed in nearly 430 hours in space.

The rocket burn that carried the Apollo 8 spacecraft out of Earth orbit three hours later was as uneventful as the launch itself. When the booster was lit, the ship slowly accelerated from 17,500 miles per hour to 25,000 and gradually straightened its trajectory from an earthly circle to a long, moon-bound line. From here, the astronauts knew, things would proceed serenely. As the ship moved farther and farther from Earth, the gravity of the planet would tug it insistently backward. For two days, the spacecraft would steadily lose speed, dropping to 20,000 miles per hour, then 10,000, and finally, at the five-sixths mark between the Earth and the moon, to a snail's-pace 2,000. At this point, the gravity of the jumbo planet would give way to the gravity of the rocky satellite, and the ship would begin to accelerate again. Until then, things could get very quiet in a moonbound craft, and it would be up to the astronauts and the ground crew to keep one another sharp. The morning after Apollo 8 was launched, Houston called up to the ship for a little pass-the-time chatter.

"Let me know when it gets to be breakfast time," the Capcom said just after 9 A.M. on the first full day of the flight. "I've got a newspaper to read up to you."

"Good idea," Borman said. "We never did get the news."

"You *are* the news," the Capcom said with a laugh.

"Come off it, come off it," Borman said.

"No kidding," Houston insisted. "The flight to the moon is occupying prime space on both newspaper and television. It's *the* news story. The headlines of the *Post* say, 'Moon, Here They Come.' In other news, eleven GIs that have been detained five months in Cambodia were released yesterday and will make it home for Christmas; a suspect in the Miami kidnapping has been captured; and David Eisenhower and Julie Nixon were married yesterday in New York. He was described as 'nervous.'"

"Right," said Anders.

"The Browns took Dallas apart yesterday, 31 to 20," Houston went on. "And we're sort of curious: Who do you like today, Baltimore or Minnesota?"

"Baltimore," Lovell said.

"Also, some really big news: the State Department announced only a few minutes ago that the Pueblo crew will be released at 9 P.M. tonight."

"Sounds good," Lovell said. Then, glancing at his instruments, he offered a bit of breaking news that, for the men involved in this conversation, had even greater significance. "Onboard calculations indicate that Apollo 8 at 25 hours is 104,000 miles from home."

"Yes," Houston said, "our plot board shows a similar number."

"Mighty nice view from out here," Borman said.

For most of its trip to the moon, the view the Apollo 8 astronauts had was of the distant lunar target growing ever larger in front of them. After leaving Earth orbit, the astronauts took a few rapturous sightings of the receding planet, then turned their spacecraft around to fly in a proper, nose-forward attitude. Strictly speaking, a nose-forward attitude was not necessary in outer space, where Newtonian rules kept a vehicle moving in the same direction no matter where its prow was pointed. But style and custom and a pilot's taste for tidiness generally dictated a forward-facing ship, so that was how the astronauts flew. After their second full day, however, as the spacecraft approached the lunar neighborhood, the crew would have to turn around once again.

Flying along at a speed that was climbing toward 5,000 miles per hour, Apollo 8 would be moving too fast to be captured by the relatively weak lunar gravity. Left to itself, the ship would approach the moon, arc around its far side, and then whip back toward Earth like a pebble slung from a slingshot. The phenomenon was known as the free-return trajectory, and while the automatic round trip could help provide the astronauts with a quick ride home in the event that their engine failed, it was a real nuisance to a crew that wanted not just to barnstorm the back of the moon, but to settle into orbit around it. In order to overcome the free-return slingshot, the spacecraft would have to rotate 180 degrees and then, facing rump-forward, fire the service propulsion engine, with its 22,500 pounds of thrust, until the craft slowed just enough to allow the moon's gravitational field to take hold.

The maneuver, known as lunar orbit insertion, or LOI, was a straightforward one, but it was fraught with risks. If the engine burned for too short a time, the ship would go into an unpredictable — perhaps uncontrollable — elliptical orbit that would take it high up above the moon when it was over one hemisphere and plunge it down again when it was over the other. If the engine burned too long, the ship would slow too much and drop not just down into lunar orbit but down onto the moon's surface. Complicating matters, the engine burn would have to take place when the spacecraft was behind the moon, making communication between the ship and the ground impossible. Houston would have to come up with the best burn coordinates it could, feed the data up to the crew, and trust them to carry out the maneuver on their own. The ground controllers knew exactly when the spacecraft should appear from behind the massive lunar shadow if the burn went according to plan, and only if they reacquired Apollo 8's signal at that time would they know that the LOI had worked as planned.

It was at the 2-day, 20-hour, and 4-minute mark in the flight —
when the spacecraft was just a few thousand miles from the moon
and more than 200,000 miles from home — that Capcom Jerry Carr
radioed the news to the crew that they were cleared to roll the dice
and attempt their LOI. On the East Coast it was just before four in
the morning on Christmas Eve, in Houston it was nearly three, and
in most homes in the Western Hemisphere even the fiercest lunar-
philes were fast asleep.

"Apollo 8, this is Houston," Carr said. "At 68:04 you are go for
LOI."

"O.K.," Borman answered evenly. "Apollo 8 is go."

"You are riding the best one we can find around," Carr said, trying
to sound encouraging.

"Say again?" Borman said, confused.

"You are riding the best bird we can find," Carr repeated.

"Roger," Borman said. "It's a good one."

Carr read the engine burn data up to the spacecraft and Lovell, as
navigator, tap-tapped the information into the onboard computer.
About half an hour remained before the spacecraft would slip into
radio blackout behind the moon, and, as always at times like these,
NASA chose to let the minutes pass largely in unmomentous silence.
The astronauts, well drilled in the procedures that preceded any engine
burn, wordlessly slid into their couches and buckled themselves in
place. Of course, if anything went wrong in a lunar orbit insertion,
the disaster would go well beyond the poor protection a canvas seat
belt could provide. Nevertheless, the mission protocol called for the
crew to wear restraints, and restraints were what they would wear.

"Apollo 8, Houston," Carr signaled up after a long pause. "We
have got our lunar map up and ready to go."

"Roger," Borman answered.

"Apollo 8," Carr said a bit later, "your fuel is holding steady."

"Roger," Lovell said.

"Apollo 8, we have you at 9 minutes and 30 seconds till loss of
signal."

"Roger."

Carr next called up five minutes until loss of signal, then two
minutes, then one minute, then, at last, ten seconds. At precisely the
instant the flight planners had calculated months before, the spacecraft
began to arc behind the moon, and the voices of Capcom and crew
began to fracture into crackles in one another's ears.

"Safe journey, guys," Carr shouted up, fighting to be heard through the disintegrating communications.

"Thanks a lot, troops," Anders called back.

"We'll see you on the other side," Lovell said.

"You're go all the way," Carr said.

And the line went dead.

In the surreal silence, the crew looked at one another. Lovell knew that he should be feeling something, well, profound — but there seemed to be little to feel profound about. Sure, the computers, the Capcom, the hush in his headset all told him that he was moving behind the back of the moon, but to most of his senses, there was nothing to indicate that this monumental event was taking place. He had been weightless moments ago and he was still weightless now; there had been blackness outside his window moments ago and there was blackness now. So the moon was down there somewhere? Well, he'd take it as an article of faith.

Borman turned to his right to consult his crew. "So? Are we go for this thing?"

Lovell and Anders gave their instruments one more practiced perusal.

"We're go as far as I'm concerned," Lovell said to Borman.

"Go on this side," Anders agreed.

From his middle couch, Lovell typed the last instructions into the computer. About five seconds before the scheduled firing time a display screen flashed a small, blinking "99:40." This cryptic number was one of the spacecraft's final hedges against pilot error. It was the computer's "are you sure?" code, its "last chance" code, its "make-certain-you-know-what-you're-doing-because-you're-about-to-go-for-a-hell-of-a-ride" code. Beneath the flashing numbers was a small button marked "Proceed." Lovell stared at the 99:40, then at the Proceed button, then back at the 99:40, then back at the Proceed. Then, just before the five seconds had melted away, he covered the button with his index finger and pressed.

For an instant the astronauts noticed nothing; then all at once they felt and heard a rumble at their backs. A few feet behind them, in giant tanks tucked into the rear of the spacecraft, valves opened and fluid began flowing, and from three nozzles three different chemicals swirled together in a combustion chamber. The chemicals — hydrazine, dimethylhydrazine, and nitrogen tetroxide — were known as hypergolics, and what made hypergolics special was their tendency to detonate in each other's presence. Unlike gasoline or diesel fuel or liquid hydrogen, all of which need a spark to release the energy stored in their molecular bonds, hypergolics get their kick from the catalytically contentious relationship they have with one another. Stir two hypergolics together and they will begin tangling chemically, like gamecocks in a cage; keep them together long enough, and confine their interaction well enough, and they will start releasing prodigious amounts of energy.

At Lovell's, Anders's, and Borman's backs, such an explosive interplay was now taking place. As the chemicals flashed to life inside the combustion chamber, a searing exhaust flew from the engine bell at the rear of the ship, and ever so subtly the spacecraft began to slow. Borman, Lovell, and Anders felt themselves being pressed backward in their couches. The zero g that had become so comfortable was now a fraction of one g, and the astronauts' body weight rose from nothing to a handful of pounds. Lovell looked at Borman and flashed a thumbs up; Borman smiled tightly. For four and a half minutes the engine burned, then the fire in its innards shut down.

Lovell glanced at his instrument panel. His eyes sought the readout that was labeled "Delta V." The "V" stood for velocity, "Delta" meant change, and together they would reveal how much the speed of the ship had slowed as a result of the chemical brake the hypergolics had applied. Lovell found the number and wanted to pump a fist in the air — 2,800! Perfect! 2,800 feet per second was something less than a screeching halt when you were zipping along at 7,500, but it was exactly the amount you'd need to subtract if you wanted to quit your circumlunar trajectory and surrender yourself to the gravity of the moon.

Next to the Delta V was another readout, one that only moments before had been blank. Now it displayed two numbers, 60.5 and 169.1. These were pericynthion and apocynthion readings — or closest and farthest approaches to the moon. Any old body whizzing past the moon could get a pericynthion number, but the only way you could get pericynthion *and* apocynthion was when you weren't just flying by but actually circling the lunar globe. Frank Borman, Jim Lovell, and Bill Anders, the numbers indicated, were now lunar satellites, orbiting the moon in an egg-shaped trajectory that took them 169.1 miles high by 60.5 miles low.

"We did it!" Lovell was exultant.

"Right down the pike," Anders said.

"Orbit attained," Borman agreed. "Now let's hope it fires tomorrow to take us home again."

Achieving orbit around the moon, like disappearing behind it minutes earlier, was a bit of an academic experience for the astronauts. Once the engine had quit firing and the crew had become weightless again, there was nothing beyond the data on their dashboard to confirm what they had achieved. The moon was just five dozen miles below them, but the spacecraft's upward-facing windows had not permitted the astronauts a look. Borman, Lovell, and Anders were three men who had backed into a picture gallery and had not yet turned around to see what was inside. Now, however, they had the luxury — and, with reacquisition of ground contact still twenty-five minutes away, the undisturbed privacy — to conduct their first survey of the body whose gravity was holding them fast.

Borman grabbed the attitude control handle to the right of his seat and vented a breath of propellant from the thrusters arrayed around the outside of the spacecraft. The ship glided into motion, rolling slowly counterclockwise. The first 90 degrees of the roll tipped the weightless astronauts onto their sides, with Borman at the bottom, Lovell in the middle, and Anders at the top of the stack; the next 90 moved them upside down, so the moon that had been below them was all at once above. It was into Borman's left-hand window that the pale gray, plastery surface of the land below first rolled, and so it was Borman whose eyes widened first. Lovell's center window was filled next, and finally Anders's. The two crewmen responded with the same gape the commander had.

"Magnificent." someone whispered. It might have been Borman; it might have been Lovell; it might have been Anders.

"Stupendous," someone answered.

Gliding beneath them was a ravaged, fractured, tortured panorama that had been previously glimpsed by robot probes but never before by the human eye. Ranging out in all directions was an endless, lovely-ugly expanse of hundreds — no, thousands; no, tens of thousands — of craters, pits, and gouges that dated back hundreds — no, thousands; no, millions — of millennia. There were craters next to craters, craters overlapping craters, craters obliterating craters. There were craters the size of football fields, craters the size of large islands, craters the size of small nations.

Many of the ancient pits had been catalogued and named by astronomers who first analyzed the pictures sent back from probes, and after months of study these had become as familiar to the astronauts as earthly landmarks. There were the craters Daedalus and Icarus, Korolev and Gagarin, Pasteur and Einstein and Tsiolkovsky. Scattered about the terrain were also dozens of other craters that had never been seen by human or robot. The spellbound astronauts did what they could to take this all in, pressing their faces against their five tiny windows and, for the moment at least, forgetting altogether the flight plan or the mission or the hundreds of people in Houston waiting to hear their voices.

From over the advancing horizon, something wispy started to appear. It was subtly white and subtly blue and subtly brown, and it seemed to be climbing straight up from the drab terrain. The three astronauts knew at once what they were seeing, but Borman identified it anyway.

"Earthrise," the commander said quietly.

"Get the cameras," Lovell said quickly to Anders.

"Are you sure?" asked Anders, the mission's photographer and cartographer. "Shouldn't we wait for scheduled photography times?"

Lovell gazed at the shimmery planet floating up over the scarred, pocked moon, then looked at his junior crewmate. "Get the cameras," he repeated.

On Christmas Eve day, Americans woke up to the news that three of their countrymen were in orbit around the moon. At the Borman, Lovell, and Anders homes in Houston, reporters choked the sidewalks and trampled the lawns in a way they hadn't since the Mercury days. Little information was released about the wives' and children's holiday plans, though all intended to attend Christmas services.

The small bit of excitement the families provided did not occur until early the next morning, Christmas day, when a Rolls-Royce from the Neiman Marcus department store pulled up at the front drive of the Lovell household. A NASA Public Affairs officer met the car, exchanged a few words with the driver, and then, to the surprise and indignation of the reporters being kept away from the house, showed him to the door, where the driver handed a large box to Marilyn Lovell. The box was gift-wrapped in royal blue foil and decorated with two Styrofoam balls, one painted sea blue, the other a mottled, vaguely lunar color. Circling the moon ball was a tiny white plastic spacecraft. Marilyn tore open the foil and pushed aside the star-patterned tissue paper inside the box. Beneath it was a mink jacket and a gift card that read, "Merry Christmas and love, from the Man in the Moon."

For the remainder of that morning, Marilyn Lovell went about her household chores dressed in lounging pajamas and mink jacket. Later that day, when she and her children went out for Christmas services, she switched to an appropriate churchgoing dress, but the jacket stayed on. It was only when she left the house and stepped into the Houston heat that the reporters standing outside saw what the man in the Rolls-Royce had brought.

But on Christmas Eve, the press's attention was focused nearly a quarter of a million miles away, where the astronaut who weeks earlier had bought the coat and arranged for its delivery was spending the day circling the moon in a tidy orbit that had been trimmed to a perfectly round sixty miles. The crew's work assignments for their ten planned revolutions included taking extensive photographs of the Earth and the moon, making measurements of the moon's gravitational field, and mapping potential landing sites and the topographic features that surrounded them.

Among the surface details the crew was supposed to survey were so-called initial points, lunar landmarks that members of future missions could use in beginning their final approaches. Scanning the Sea of Tranquillity, an ancient, bone-dry lava plain targeted for the first manned landing, Borman, Lovell, and Anders took note of a winding mountain range just southwest of the crater Secchi. The overall formation of the range had long since been mapped by astronomers on Earth, but the individual peaks had been far too small to be seen by telescope. These kinds of fine surface details were just the sort of features that crews would need as they navigated their way down from orbit. At the edge of the jagged range, abutting the Sea of Tranquillity, Lovell discovered a small, strangely triangular mountain that he was certain was tiny enough never to have attracted attention but was distinctive enough to be easily recognizable by future crews that came this way.

"Ever see that peak before?" Lovell asked Borman, pointing to the little formation.

"Not that I can recall."

"How about you?" he asked Anders, the arbiter of all matters topographical.

"Nah," Anders said. "I'd have remembered one shaped that way."

"Then I found it," Lovell said with a smile, "and I'm going to name it. What do you guys think of 'Mount Marilyn'?"

To NASA's administrators, just as important as Apollo 8's scientific tasks were its public relations assignments. The Agency planned two live television broadcasts from lunar orbit, one for early morning on Christmas Eve and a longer, prime-time show that evening. The morning broadcast drew an impressive audience, but with the country preoccupied with last-minute Christmas plans, it did not break any records. The evening broadcast, beamed as it was to one hundred million households, was another matter entirely. All three networks preempted their programming to carry the show, meaning that most viewers would either be watching the program from the moon or nothing at all. The broadcast began at 9:30, and the nation, as well as much of the rest of the planet, stopped to watch it.

"Welcome from the moon, Houston," Jim Lovell said to the people at NASA and, by implication, to the world. The image that flickered on television sets as he began to talk was a floating white ball hanging suspended against a colorless background. Below it, a long, gentle arc curved downward before vanishing off the edge of the screen.

"What you're seeing," said Anders as he steadied the camera and braced his buoyant body against the bulkhead of the ship, "is a view of the Earth above the lunar horizon. We're going to follow along for a while and then turn around and give you a view of the long, shadowed terrain."

"We've been orbiting at sixty miles for the last sixteen hours," Borman said while Anders pointed the lens downward at the surface, "conducting experiments, taking pictures, and firing our spacecraft engine to maneuver around. And over the hours, the moon has become a different thing for each one of us. My own impression is that it is a vast, lonely, forbidding expanse of nothing that looks rather like clouds and clouds of pumice stone. It certainly would not be a very inviting place to live or work."

"Frank, my thoughts are similar," Lovell said. "The loneliness up here is awe inspiring. It makes you realize just what you have back on Earth. The Earth from here is an oasis in the vastness of space."

"The thing that impressed me the most," Anders took over, "was the lunar sunrises and sunsets. The sky is pitch black, the moon is quite light, and the contrast between the two is a vivid line."

"Actually," Lovell added, "the best way to describe the whole area is an expanse of black and white. Absolutely no color."

The flight plan called for the broadcast to last twenty-four minutes, during which time the ship would glide across the lunar equator from east to west, covering about 72 degrees of its 360-degree orbit. The astronauts were to take this time to explain and describe, point and instruct, and try to convey through words and grainy pictures what they were seeing. The effort they made was a noble one.

"This area is not very heavily cratered, so it must be new," one of them would say.

"This crater is the delta-rim variety . . ."

"There is a dark area over here which could possibly be an old lava flow . . ."

"Coming into view are some interesting old double-ring craters . . ."

"Running along the edge of that mountain is a sinuous rill with right-hand turns."

The astronauts went on and on, and at home, audiences watched the new pictures and heard the new words and took in as much as their senses and skepticism would allow them to. Finally it was time for the show to sign off. For weeks before the flight, the three astronauts had debated the best way to conclude a broadcast from one world back to another on the eve of the holiest day in the Christian calendar. Shortly before launch day, a decision was reached, and attached to the back of the onboard flight operations manual was a sheet of paper (fireproof, of course, always fireproof these days) with a short script typed on it. Anders, pointing the television camera out the window with one hand and holding the script with the other, said, "We are now approaching the lunar sunrise, and for all the people back on Earth, the crew of Apollo 8 has a message we would like to send to you.

"In the beginning," he began, "God created the Heaven and the Earth. And the Earth was without form, and void; and darkness was upon the face of the deep." Anders read slowly for four lines, then passed the paper on to Lovell.

"And God called the light Day, and the darkness He called Night. And the evening and the morning were the first day." Lovell read four lines of his own and handed the paper to Borman.

"And God said, Let the waters under the Heaven be gathered together unto one place, and let the dry land appear." Borman continued until he reached the end of the passage, concluding with, "And God saw that it was good." When the final line was done, Borman put down the paper.

"And from the crew of Apollo 8," his voice crackled down through 239,000 miles of space, "we close with good night, good luck, a merry Christmas, and God bless all of you, all of you on the good Earth."

On television sets, the image of the lunar surface abruptly vanished, replaced first by color bands, then by static, then by newsmen recapping rhapsodically what they and the rest of the world had just seen. In the spacecraft, however, things were a good deal less lyrical. As soon as the program concluded, Frank Borman and his crew were all business, back on the line to the controllers in Houston.

"Are we off the air?" Borman asked Capcom Ken Mattingly.

"That's affirmative, Apollo 8, you are," Mattingly answered.

"Did you read everything that we had to say there?"

"Loud and clear. Thank you for a really good show."

"O.K.," Borman said. "Now Ken, we'd like to get all squared away for trans-Earth injection. Can you give us some good words like you promised?"

"Yes, sir. I have your maneuver and then we will run through a systems brief."

Just as Jerry Carr had done for the LOI burn, Mattingly read up the data and coordinates for the trans-Earth injection, or TEI, burn. Once again, Lovell typed the figures into his computer, the astronauts strapped themselves into their couches, and Houston fidgeted in silence as the minutes ticked away to loss of signal. Unlike the LOI burn, the TEI burn would require the ship to be pointed forward, adding feet per second to its speed rather than subtracting them. Also unlike the LOI burn, during TEI there would be no free-return slingshot to send the ship home in the event that the engine failed to light. If the hydrazine, dimethylhydrazine, and nitrogen tetroxide did not mix and burn and discharge just so, Frank Borman, Jim Lovell, and Bill Anders would become permanent satellites of Earth's lunar satellite, expiring from suffocation in about a week and then continuing to circle the moon, once every two hours, for hundreds — no, thousands; no, millions — of years.

The crew slipped into radio silence, and the controllers sat quietly and waited. Somewhere behind the lunar mass, the giant service propulsion engine either was or wasn't firing, and Houston wouldn't know one way or the other for forty minutes. Mission Control sat in silence for this two thirds of an hour, and as the last seconds ticked away, Ken Mattingly began trying to raise the ship. "Apollo 8, Houston," he said. There was no response.

Eight seconds later: "Apollo 8, Houston." No response.

Twenty-eight seconds later: "Apollo 8, Houston."

Forty-eight seconds later: "Apollo 8, Houston."

For one hundred more seconds the controllers sat in silence, and then, all at once: "Houston, Apollo 8," they heard Lovell call exultantly into their headsets, his tone alone confirming that the engine had burned as intended. "Please be informed, there is a Santa Claus."

"That's affirmative," Mattingly called back, audibly relieved. "You are the best ones to know."

The spacecraft splashed down in the Pacific at 10:51 A.M. Houston time on December 27. It was before dawn in the prime recovery zone, about one thousand miles southwest of Hawaii, and the crew had to wait ninety minutes in the hot, bobbing craft before the sun rose and the rescue team could pick them up. The command module hit the water and then rotated upside down, into what NASA called the stable 2 position (stable 1 was right side up). Borman pressed a button inflating balloons at the top of the spacecraft cone, and the ship slowly righted itself. From the time the crew climbed out and stepped before the television cameras, it was clear that the national ovation that would greet them would surprise even publicity-savvy NASA. Borman, Lovell, and Anders became overnight heroes, receiving award after award at one testimonial dinner after another. They became *Time* magazine's Men of the Year, addressed a joint session of Congress, rode in a New York City ticker-tape parade, met outgoing President Lyndon Johnson, met incoming President Richard Nixon.

The honors were deserved, but in a surprisingly fleeting couple of weeks, they ended. When the crew of Apollo 8 returned, the nation had satisfied itself that it could get to the moon; the passion now was to get *on* the moon. In the wake of the mission's triumph, the Agency decided that it would need just two more warm-up flights to prove the soundness of its equipment and its flight plan. Then, sometime in July, Apollo 11 — the lucky Apollo 11 — would be sent out to make the descent into the ancient lunar dust. Neil Armstrong, Michael Collins, and Buzz Aldrin would make the trip, and at the moment it looked like it would be Armstrong who would take the historic first step.

Following Apollo 11, there would be nine more lunar landings, and

Lovell, now one of the most experienced men on the astronaut roster, figured he had a pretty good shot at commanding one of them. Sure enough, when the pilot manifests were later handed down, Lovell, along with two rookies, Ken Mattingly and Fred Haise, found themselves named as backup crew for Apollo 11 and prime crew for Apollo 14, scheduled to land on the moon in October 1970. In less than two years, Lovell would travel back to the rocky planetoid he had just left and at last get the lunar walk he had joined the program for in the first place. After that, he would retire.

As it turned out, there was a problem with these plans. The flight before Lovell's, Apollo 13, was to be flown by Alan Shepard, Stuart Roosa, and Edgar Mitchell. Shepard, the first American in space, had become a national icon on May 5, 1961, when he went aloft in his tiny Mercury capsule for a fifteen-minute suborbital mission. Since then, he had been grounded because of a stubborn inner-ear problem affecting his balance. Eager to be returned to active flight status, Shepard had recently undergone a new surgical procedure to correct the disorder and, after intense lobbying within the Agency, got himself assigned to a lunar mission. With a nine-year gap between flights, however, Shepard soon realized that he would need more time to get up to speed. Before the crew manifests were set, Deke Slayton approached Jim Lovell and asked if he would mind terribly changing his plans. How would Lovell feel about giving Apollo 14 to Shepard and taking 13 instead? It would mean a lot to Al, Deke said, and it would go a long way toward ensuring the success of both missions.

Lovell shrugged. Sure, he said. Why not? Frankly, he confided to Slayton, he was looking forward to getting back to the moon, and going six months sooner than he'd expected was fine with him. One landing was essentially as good as another, and what could possibly be the difference between Apollo 13 and Apollo 14 except, perhaps, the numeral?

3

Spring 1945

IT WAS the brass and glass doors of the reception room that told the seventeen-year-old he was in the wrong place. Oh, he'd had other clues: no mom-and-pop chemistry shop would be located in a sky-scraper in the heart of the Michigan Avenue business district for example. No modest storekeeper would display the word "Incorporated" so prominently in the name of his business. No, this didn't look at all like the hobby shop for backyard inventors the boy had expected to find here, yet the listing in the phone book did say "Chemicals," and chemicals were what he needed today. Having taken the train to Chicago from his aunt's house in Oak Park just for this, he would've felt silly turning around now.

Pushing open the doors and stepping into the deep-pile carpet of the office, he found himself standing at one end of a large room facing, at a great distance, an intimidating mahogany desk. The woman behind the desk, looking as if she had never seen a jar of chemicals in her life, noticed the boy standing hesitantly just inside the door.

"Can I help you, young man?" she asked.

"Uh, I wanted to buy some chemicals," he said.

"Can you tell me where you're from?"

"Milwaukee," he answered, cautiously crossing the room. "I'm just visiting family outside Chicago."

"No," she said with the tiniest trace of a smile. "I mean, are you representing anyone?"

"Absolutely." He brightened. "Jim Siddens and Joe Sinclair."

"And they're your employers?"

"My friends."

The proto-smile again. "May I ask your name?"

"James Lovell."

"James Lovell," she said, and wrote the name down with seeming gravity. "Just a moment, James — uh, Mr. Lovell. I'll check to see if one of our sales people is available." She started to rise. "If I'm able to catch one of them, can I tell them what it is you're interested in buying?"

"Nothing much. Just some potassium nitrate, sulphur, and charcoal. A couple pounds at the most."

The woman vanished behind a paneled wooden door that shut behind her with a *whoosh,* and after a minute or so returned. "Most of our sales folks are busy," she said. "But Mr. Sawyer can see you."

Lovell was escorted behind the door to an inner office, where the promised Mr. Sawyer sat behind a decidedly smaller desk. "Son," Mr. Sawyer said when the teenager was seated in his office, "I don't know how you got our name, but you know, we don't sell chemicals by the pound here, we sell them by the railroad car."

"Uh, yes sir, I was afraid of that. But you must have just a little on hand, don't you?"

"I'm afraid not. All of our chemicals are shipped out of our warehouses. And even if we did have some here . . . Well, do you know what potassium nitrate, sulphur, and charcoal make when they're mixed together in the right proportions?"

"Rocket fuel?"

"Gunpowder."

This didn't make sense. Lovell was sure he had written down the ingredients properly. When he and Siddens and Sinclair approached their chemistry teacher, they had been very explicit about the fact that they wanted to build a working rocket. At first they wanted to build a liquid-fuel model, just like Robert Goddard and Hermann Oberth and Wernher von Braun had built. But when they started sawing iron pipes to make their combustion chamber and cannibalizing model airplanes for their spark plugs and eyeing number 10 tin cans as possible fuel tanks, they realized they might be in above their heads. Instead, their chemistry teacher had recommended a solid-fuel rocket made of little more than a cardboard mailing tube, a wooden nose cone and fins, and some powdered fuel stuffed in the bottom. He had given them the recipe for the fuel, but he had never said a thing about it really being gunpowder. Mr. Sawyer, however, assured Lovell that gunpowder was what it was, and escorted the teenager out of the chemical company office, chemicals not in hand.

Back in Milwaukee a few days later, Lovell confronted his science teacher. "Of course I know it's gunpowder," the teacher said. "The stuff's been around for two thousand years. I would hope I'd have gotten word of it by now. But if you mix it and pack it just right, it'll burn but won't explode."

With the teacher's guidance, Lovell, Siddens, and Sinclair built their rocket — a three-foot featherweight affair — packed the bottom with what they hoped were the right proportions of powder, and fitted it with a fuse. The next Saturday, they carried the missile out to a large, empty field, propped it against a rock, and pointed it skyward. Lovell, wearing a welder's protective helmet, was the self-appointed launch director, and while Siddens and Sinclair stood at a presumably safe distance, he lit the fuse — a drinking straw filled with gunpowder — and then, like centuries of "launch directors" before him, ran like hell.

Lovell performed his job flawlessly, if nervously. Crouching with his friends, he watched agape as the rocket he had just ignited smoldered for an instant, hissed promisingly, and, to the astonishment of the three boys, leapt from the ground. Trailing smoke, it zigzagged into the air, climbing about eighty feet before it wobbled ominously, took a sharp and surprising turn, and with a loud crack exploded in a splendid suicide.

Smoking bits of missile came fluttering back to the ground, leaving debris a dozen or so feet around. The boys ran up to the launch site to catch some of the drifting remains, as if looking at these burned bits would reveal what went wrong. Nothing was immediately apparent, but it seemed evident that even with the chemistry teacher's guidance, the packing of the powder was still not right, causing the chemicals to behave like the gunpowder they really wanted to be. If there was any consolation to the failed rocketeers, it was in the knowledge that with only a slightly different proportion of powders or a slightly shabbier packing, the detonation would have taken place not eighty feet away from them during the rocket's flight, but just inches away during its ignition — something that generations of luckless, and late, launch directors had also learned before them.

For Siddens and Sinclair, high school juniors who had the good sense to be looking forward to careers in the booming postwar fields of construction and manufacturing, the launch and death of the rocket that day was little more than a lark. For Lovell it was something else entirely. For several years now, he had been immersing himself in the lore of rockets, ever since stumbling across a pair of basic rocketry books that traced the evolution of the science around the world, with special emphasis on the United States (where Goddard provided one great face for rocketry's Mount Rushmore), the Soviet Union (where Konstantin Tsiolkovsky provided another), and Germany (where Oberth and von Braun rounded out the group).

Before he even reached his teens, Lovell had decided he wanted to make rocket science his life. As he moved through high school, he realized that this would not be so easy. There was little you could learn in Milwaukee's secondary schools that would qualify you for a career as preposterous as rocketry, and the only place you *could* learn about it, in college, was out of reach. Lovell's father had died five years earlier in an automobile accident, and his mother had spent the past half decade working as hard as she could just to keep the family in clothes and food. Any education beyond what was available in the public schools was out of the financial question.

Entering his final year in high school, Lovell began to consider one final option, the military. His uncle had graduated from Annapolis in 1913, had flown as one of the first naval aviators in the antisubmarine units in World War I, and had always enthralled his nephew with stories of biplanes and dogfights and taking to the sky on wood-and-canvas wings. Although a career flying military planes was not the same as a career building rockets, it did still involve flying. What's more, to the degree that any organized rocket research existed in the United States, it was the military that was doing the work. Early in his senior year, Lovell applied to the Naval Academy and a few months later received a response informing him that he had been chosen as a third alternate. The selection was flattering but little more: Lovell would be offered a position at Annapolis only in the absurdly unlikely event that disaster simultaneously befell the guys who finished first, second, and third.

Face to face with what was looking increasingly like his non-future, Lovell was rescued by the very organization that had just rejected him: the Navy. Only weeks before high school graduation season, a Navy recruiter made the rounds of the Milwaukee schools, talking up a program called the Holloway Plan. Hungry for fresh aviators after World War II, the service was launching a program in which high school graduates would be given two free years of an undergraduate engineering education, followed by flight training and six months of active sea duty with the humble rank of midshipmen. They would then be commissioned as ensigns in the regular Navy, but before beginning their service, they would be allowed to finish their two remaining years of college and earn their degrees. Promptly after graduation, they would commence their military careers as naval aviators.

The plan sounded great to Lovell and he leapt at the chance to sign up. A few months later he enrolled as a freshman at the University of Wisconsin, his tuition bills now being paid by the Department of the Navy.

From March 1946 through March 1948, Lovell studied engineering at Wisconsin. During this period he also reapplied for admission to the Naval Academy, this time at the insistence of a far more compelling agency — his mother. The head of the Lovell household was pleased that her son was going to college, but this whole business of interrupting his education to begin Navy training just didn't sit right with her. Suppose there was some kind of national emergency before he got his degree? Couldn't he, like so many other soldiers and sailors in the world wars, wind up trapped on a ship or stuck in a foxhole for the duration of the conflict, getting older and older and delaying his education longer and longer while the war or crisis dragged on? The whole thing seemed too risky.

Lovell appeasingly submitted a second application to Annapolis, but he had little real hope; admission to the academy was, he assumed, just as long a long shot as it had been two years earlier. While waiting for the anticipated rejection, he reported to the naval air station in Pensacola, Florida, to begin his flight training. Before he was even done with his pre-flight work, the long shot came in. On his way to class one day, Lovell was intercepted by a yeoman who handed him a dispatch. He was directed to report promptly to the Naval Academy, to be sworn in as an Annapolis midshipman. Strictly speaking, the "orders" were not really orders: Lovell was entitled to decline and continue his Holloway flight training. But he had to decide right away. The flight instructors at the Florida school, all young Marines back from the war, had no doubt about what choice he should make.

"Look, Lovell," one flier said to him, "what do you want to do this for? You're already a midshipman, you're already halfway through with your education, and most important, you're about to start flying. You wanna throw all that away, start your entire education over as a freshman, and not get into the cockpit of a plane for at least four more years?"

"But suppose a war or something happens," Lovell said. "Suppose we get detoured here and can't get back to school for years."

"You're not going to get detoured. All that's going to happen is you're going to go to Annapolis and wind up two years behind the other guys here."

The argument made sense, and Lovell decided that he would, to his own great surprise, tell the Naval Academy no thanks. Before he could send out his dispatch, however, word came down that he was wanted by Captain Jeter, the commander of the pre-flight school. Jeter was an old Navy salt who had been training pilots since, it seemed, the seventeenth century or so, and made it his business to know everything that was going on in his school.

"So you've heard from the Naval Academy, Midshipman Lovell," Jeter said when Lovell presented himself.

"Yes, sir."

"And they want a decision from you right away?"

"Yes, sir."

"What's your thinking at the moment?"

"Well, sir," Lovell began, pleased that he could tell the commander that he wasn't leaving flight school, that his head hadn't been turned by the glitter of Annapolis, "the way I see it, I'm already a midshipman in flight training and I already have two years of college. I can't see where the Naval Academy is going to get me anywhere closer to my goals than I am right now."

Jeter seemed to agree with that, but he also seemed to chew on it for a long moment. Then he said, "Lovell, are you happy with the Navy so far?"

"Yes, sir."

"You're pretty sure you want to make the Navy your career?"

"Yes, sir."

"Then get up to the Naval Academy, son," the commander said sternly, "and get yourself the best education you can when someone offers it to you."

Within days Lovell was packed and gone, honorably discharged as a Holloway Plan midshipman and sworn back in as an Annapolis midshipman — busted voluntarily from novice aviator back to plebe. Later that year the nation of Korea, torn by civil war, split into the Democratic People's Republic of Korea in the north and the Republic of Korea in the south. The escalating tension required the United States to increase its complement of active military forces, including the apprentice fliers who had signed up for the recently created Holloway Plan. Many of the new aviators were sent straight to duty overseas, and most fought valiantly in the eventual war. Though the Navy decorated the fliers generously, the majority of them were unable to resume their education for at least seven years.

Lovell thrived at Annapolis, immersing himself in as much science and engineering as he could, all the while keeping an eye on developments in the field of rocketry. By this time Wernher von Braun, inventor of the V-2 rocket, had been safely transplanted from Germany's Peenemünde to America's New Mexico and was successfully launching a two-stage vehicle, code-named Operation Bumper, that reached a record height of 250 miles and returned pictures clearly showing the curvature of the Earth. For rocket enthusiasts across the country, this was heady stuff. Two hundred and fifty miles wasn't just the edge of space, it was well into space. At some point (and who was to say this wasn't it?) you weren't so much going *up* anymore as you were going *out*. Rocketeers were dizzy with the promise of it all.

Fledgling midshipman Jim Lovell could only follow these developments peripherally. He had a nearly impossible four years ahead of him, during which he wouldn't have time for a lot of woolly-headed fantasizing about space travel. You could bilge at the academy at any point in your career there, but it was the first year that had the highest attrition rate. Make it through that with your sanity intact and you had a pretty good chance of going all the way.

Happily for Lovell, he would not have to slog through that first twelve-month stretch — or the remaining thirty-six months, for that matter — alone. Like most other midshipmen, when he went to Annapolis, he left a girl back home. Marriage was a prohibited condition for Annapolis students, the thinking being that apprentice sailors who were supposed to be living and breathing the ways of the military would not have time for frivolities like family. Yet going the entire four years without any romantic distraction was not desirable either. Subject an average nineteen-year-old to the workload of the average Naval Academy student, take away the diversion of a girl to write to, whose picture he could hold on to when the grind got unbearable, and you've got a nineteen-year-old headed less for a naval commission than a crackup. Having a sweetheart back home, not on site, was just the way the elders of the academy liked it.

Girlfriends of midshipmen were known then and forever as "drags," a term meant to imply less an unwelcome encumbrance than an elegant accoutrement. Drags visited Annapolis only during scheduled academy events like supervised tea dances and hops, and stayed together in glamorous, gossipy gaggles at places like Ma Chestnut's boarding house just off campus. The midshipmen preened and planned for their dates, but were permitted to be alone with them off the academy grounds only at the end of the evening when they escorted them back to the boarding houses. Only forty-five minutes was allowed for the trip, enough time for a slow walk, a moony goodbye, and nothing more. The midshipmen took advantage of every bit of the allotted three quarters of an hour, lingering at Ma Chestnut's or the other houses for as long as prudence and the threat of demerits would permit and then sprinting back through the academy gates in a breathless pack — the Flying Squad, the faculty indulgently dubbed them — just as the forty-fourth minute gave way to the forty-fifth.

Lovell's drag during his academy years was Marilyn Gerlach, a Milwaukee State Teachers College student he had met three years earlier, when he was a high school junior and she was a freshman. The two had gotten to know each other through shy eye contact in the school cafeteria line, where Lovell worked behind the counter to earn a free lunch, and Marilyn appeared every day at noon, talking and laughing with the other freshman girls. Lovell had only a passing interest in the giggly thirteen-year-old — she *was* a freshman, after all — until junior prom rolled around and he found himself dateless. Leaning over the succotash and meat loaf and calling above the din of students in the cafeteria line, Lovell asked the much younger girl if she might like to come to the older kids' cotillion.

"I don't even know how to dance," she shouted in response — telling the truth but hoping it would come out sounding coy and hard to get.

"That's O.K.," he said, "I'll teach you," having no idea how he planned to manage that.

The date worked out well, the relationship flourished, and the two stayed together when Lovell went off to the nearby University of Wisconsin and later to distant Annapolis. A year after arriving at the Naval Academy, Lovell wrote Marilyn a letter explaining that a lot of the other midshipmen were engaged to be married after graduation, but — funny thing — they all seemed to be engaged to East Coast girls. Something about the geographic proximity, he broadly hinted, just made the relationships work better. He was bringing this up to her for no particular reason, of course; he just thought she'd be interested.

As it turned out, Marilyn Gerlach was very interested, and within two months she had packed her bags, moved to Washington, D.C., transferred her college credits to George Washington University, and gotten herself a part-time job at Garfinckel's department store. Three years later, she sat in Dahlgren Hall on the Annapolis campus as Midshipman Lovell and the rest of the class of 1952 whooped, embraced, threw their caps in the air, and became graduates of the U.S. Naval Academy. Three and a half hours after that, the newly commissioned officer and his hometown girl stood in St. Anne's Episcopal Cathedral in the historic center of Annapolis and became Ensign and Mrs. James A. Lovell Jr.

Of the 783 students in the graduating class of 1952, only 50 were selected immediately for naval aviation. With an eye toward this moment of truth, Lovell had made his love for all things airborne as obvious as he could throughout his four years, even writing his senior thesis on the unheard-of topic of liquid-fuel rocketry — a thesis that Marilyn dutifully typed, all the while fretting that her husband-to-be might have served himself and his grade point average better by picking a more conventional topic, like military history. Nevertheless, the thesis earned Lovell both the high grade and the high profile he sought, and when the lucky half-hundred were selected to go on to flight school, he was among them.

Flight training took fourteen months, and after it was over, the Navy asked the graduates where they'd like to be assigned. Hoping to settle on the East Coast, Lovell volunteered for the naval air station at Quonset Point, near Newport, Rhode Island. Not yet familiar with how the military operated, he had assumed that this choice might actually have some bearing on where he would eventually be sent. The Navy worked otherwise, however, and after processing his request and acknowledging his preference, promptly dispatched him to Moffett Field, near San Francisco.

When the rookie ensign arrived on the West Coast with wife and newly acquired wings, he was assigned to Composite Squadron Three, an aircraft carrier group specializing in the white-knuckle, high-wire business of carrier night flying. Piloting a jet off the deck of a rolling aircraft carrier and then bringing it in for a landing when the ship was a couple thousand feet below and had shrunk to the size of a Scrabble tile was one of the hardest jobs in naval aviation. Trying to pull off the same maneuvers at night, often in questionable weather and when shipboard lights were doused to simulate wartime conditions, was just asking for trouble. In the 1950s night flying off carriers was still in its infancy, and on any given day it was only the least lucky pilots who got tapped for after-dark duty and had to endure the catapult shots into the blackness while their buddies gathered belowdecks to watch a movie.

Jim Lovell learned his night-flying skills off the friendly California coast, but it wasn't until six months later, on a frigid February evening off the coast of still-occupied Japan, that he made his first night flight into alien skies over an alien sea. The pilot was more than a little rusty, and the conditions for flying were less than ideal. There was no moon, a light cloud cover erased all stars, and with the loss of both, the horizon vanished too.

Happily, the maneuver the skipper had scheduled for his pilots that night was a relatively uncomplicated one. The flight plan called for four F2H Banshees to take off on a combat air patrol above the carrier, the USS Shangri-La. Night combat exercises typically involved a rendezvous in the air at 1,500 feet after taking off from the deck, then flying over the task force at 30,000 feet for about ninety minutes. Afterward, the pilots would descend and come in for a landing. Though there would be no carrier beacons to light the fliers' way back in, the ship would be beaming a radio signal to the Banshees at 518 kilocycles. The signal would attract the needle of their automatic direction finders like a divining rod, and all the men would have to do is follow where the ADF pointed and the carrier would be dead ahead. It was a simple, straightforward exercise in piloting, and with any luck the aviators would be back on board before the evening's movie reached the second reel. Almost from the start, however, things began to go sour.

Lovell was the first of the four fliers into the air, followed by his

66

teammates Bill Knutson and Daren Hillery. As was the custom in these exercises, the team leader, Dan Klinger, would be the fourth and last to leave the deck. But no sooner had Klinger begun to gun his engines than the cloud cover, which had been threatening to grow worse, made good on that threat, dropping and thickening into near opacity. Klinger was told to cut his engines and stay aboard, and in the sky Lovell, Knutson, and Hillery, who had just begun their rendezvous, were hailed by radio.

"November Papas," the ship announced, using the flight crew's call sign, "weather is lousy and the exercise has been canceled. Rendezvous and then orbit the ship for 30 minutes at 1,500 feet. We'll bring you aboard when you burn down some fuel."

In his cockpit, Lovell smiled despite himself. It would be something of both a rite of passage and a relief to put his first night flight successfully behind him. But as with all things dreaded, there would also be an entirely different kind of relief in avoiding, if only for one evening, the whole nasty business. Lovell knew that he would be ordered to come back up here soon enough and start the exercise all over again, but for now he could put it out of his mind and fly on home.

As procedure dictated, Lovell flew out ahead of the ship for two or three minutes, then turned 180 degrees and doubled back so that his teammates could join him on his wing. But when he reached the spot where the planes and the ship ought to be, neither were to be seen. He glanced at his altimeter: 1,500 feet. He glanced at his ADF: carrier dead ahead. Yet Lovell saw only a bowl of blackness about him.

"November Papa One, this is Two," Knutson suddenly called in Lovell's headset. "We don't see you. Can you tell us where you are?"

"I haven't reached Home Plate yet," Lovell responded.

"Well, Three has joined up with me," Knutson said. "We're orbiting Home Plate at 1,500 feet and waiting for you."

Lovell was confused. He looked at his altimeter and ADF again and everything seemed in order. He glanced at the control knob for the ADF: it was indeed set to receive at 518 kilocycles. He tapped the glass cover of the instrument. The needle continued to point true. What Lovell didn't know — couldn't know — was that a tracking station on the Japanese coast was also broadcasting a homing signal at 518 kilocycles. His wingmen had been fortunate enough to lock on to the ship's beacon before the coast locked on to them. Through sheer electronic happenstance, Lovell's direction finder was picking up the shore signal instead, which was steadily, insistently leading him away from the ship and into a night that was becoming more and more unfriendly.

"Home Plate," he called to the carrier, hoping that if nothing else the ship's radar might have him on its scope, "are you painting me?"

"Negative," the Shangri-La replied.

Lovell was wearing a rubberized flight suit, called a "poopy suit," which was designed to protect him if he had to ditch in the icy waters of the Sea of Japan. All at once he didn't feel so calm. Sweat began to drip down his chest in the hot, poreless suit and run down his sides and legs.

"Home Plate," he said, "I seem to have lost my wingmen somehow. I'm going to reverse course and see if I can pick them up again."

"Roger, November Papa One. Take your time and find them."

Lovell turned the plane 180 degrees and the ADF needle responded, pointing to the tail of the plane and indicating that the invisible carrier and the two invisible pilots were now behind him. Lovell cursed; the ADF was never wrong. But maybe, he thought, just maybe, the homing signal frequency had been changed and he never got the word. Strapped to his left leg was a knee board with a list of the latest communications frequencies the pilots had been issued just before manning their aircraft. All of the pilots wore knee boards when they went up, but Lovell's was somewhat different from most. The novice pilot had always been troubled by how hard it was to make out the tiny numbers on the flight plan sheets in the darkness beneath the instrument panel and, in idle moments during the long trip to the Far East, had collected spare parts from the supply office and invented an ingenious little light that he attached to his knee board. Plug the wire from the light into the plane's electrical receptacle, throw a switch, and the knee board would light right up.

Lovell had been proud of his invention, and this was his first chance to try it out. Grabbing the plug, he snapped it into the receptacle and threw the switch. As soon as he did, however, there was a brilliant flicker of light — the unmistakable sign of an overloaded circuit shorting itself out — and instantly, every bulb on the instrument panel and in the cockpit went dead.

Lovell's heart went timpanic. His mouth went dry. He looked around himself and could see absolutely nothing; the blackness outside the plane had suddenly come inside. Tearing off his oxygen mask, he gulped a breath or two of cabin air and thrust a penlight in his mouth to shine on the instruments. The silver-dollar-sized beam the tiny flashlight produced danced across the dashboard, dimly illuminating one needle or dial at a time. Lovell checked the readings as best he could and then fell against his seat to consider what he should do next.

A pilot in Lovell's kind of distress had a couple of possible options, neither of which was even remotely attractive. He could declare an emergency and ask that the ship's lights be turned on. The skipper would probably comply, but the embarrassment would be incalculable. Suppose this was a real night maneuver in a real war? Excuse us, Mr. Enemy Vessels, would you turn around while we switch our lights on? Seems one of our fliers misplaced the carrier. Uh-uh, can't go that route. Alternatively, he could declare the same emergency but turn the other way and try to find an airfield in Japan. At least he'd be over land instead of an icy, inky sea. But with his ADF in question and the lights out in his cockpit, he'd probably never find a runway and instead be forced to ditch the plane and make a parachute landing.

Lovell took the penlight out of his mouth, switched it off, and scanned the darkness. Down below him at about two o'clock, he thought he noticed a faint greenish glow forming a shimmery trail in the black water. The eerie radiance was barely visible and would have been lost to Lovell altogether had the blackness in the cockpit not acclimated his eyes to the darkness. But the sight of it made his heart leap. He was certain he knew what the strange radiance was: a cloud of phosphorescent algae churned into luminosity by the screws of a cruising carrier. Pilots knew that a spinning propeller could light up organisms in the water, and this could help them locate a missing ship. It was one of the least reliable and most desperate methods of bringing a lost plane home safely, but when all else failed, it could sometimes do the trick. Lovell told himself that all else had indeed failed, and with a fatalistic shrug he peeled off in pursuit of the dim green streak.

When he arrived just ahead of the spot in the sea and leveled off at 1,500 feet, he found to his delight that his two wingmen were there waiting for him. He was ecstatic to see the circling planes, but he knew it would not pay to let on just how ecstatic.

"We thought we'd lost you for good," Hillery called to Lovell over the airwaves. "Glad to see you decided to join us."

"Had a couple of instrumentation problems," the invisible pilot radioed back from his lightless cockpit. "No big deal."

Although the flight team had at last rendezvoused, Lovell's problems were nowhere near over: he still had to plant his darkened plane back on the carrier deck. A continuous scan of his altimeter and air-speed indicators would be essential for a safe landing, but Lovell's dim penlight did not illuminate both at the same time.

As the last plane to reach Home Plate, Lovell's was now third in a three-plane formation, meaning he would be the last to leave the sky and head for the deck. The trio of planes flew up the starboard side of the ship, and Lovell watched as first one, then the other of his teammates peeled off for their downwind leg. He heard Snapper Control, the man helping the landing signal officer, calling to the other two fliers when they were abeam of the ship's fantail, telling them to start their approaches. Dropping down to 150 feet, they swung behind the ship and descended steadily until they closed in on the deck and touched down without incident. Lovell, now on his own downwind leg and once again alone in the darkness, envied them their completed landings and their lighted cockpits, and with penlight firmly in teeth, he heard Snapper Control calling him to make his approach. Keeping one eye open for the carrier's approaching fantail while keeping the other on his instruments was no small feat, but Lovell felt he was managing. Suddenly, when he was closing fast on the ship, holding at an altitude that his last glance at the altimeter had told him was 250 feet, he noticed a red light out the left side of his canopy, floating just beneath his left wing.

What it could be he had no idea. There was certainly no plane flying anywhere between him and the water, nor was there some smaller boat or lighted buoy floating in the wake of the carrier. With a start, Lovell realized what he was looking at. The light was a reflection of his own wingtip running light, twinkling in the rolling water, which he now discovered was not a safe 250 feet below him but a bare 20 feet. His altimeter confirmed this shocking realization. Lovell was cruising along practically at wavetop level, his wheels nearly surfing the sea, heading straight for a watery wipeout or an explosive crackup against the flat stern of the mammoth carrier.

"Pull up, November Papa One, pull up!" Snapper Control screamed in his ears. "You're way too low!"

Lovell yanked back on his stick, pushed the throttles forward, and the Banshee screamed back up to 500 feet. He banked over the carrier again, came around one more time, and headed into the approach groove for another try. This time he was 500 feet up.

"You're way high, November Papa One, way high," the landing signal officer yelled at him. "You can't approach from that altitude!"

Lovell knew, however, that this altitude might be the best he was going to get. With his penlight beam dancing around his instruments and the memory of the carrier's giant stern looming in front of him like a black wall, he figured he'd rather risk dropping down on the ship from too high up than running into its rear from too low down. As the deck drew closer and closer, Lovell began falling like a rock from 500 feet to just 150. From there he went into a virtual free fall until, with a spine-compressing thud, he slammed down on the deck of the flattop, blowing two tires and skidding forward. Finally his tailhook caught the last cross-deck cable and he came to a violent stop.

Lovell shut down his engines and dropped his head into his hands. The plane handler came running up to the jet, and the ashen pilot slowly unbuckled himself, climbed out of the cockpit, and eased his body down to the deck on shaky legs.

"Glad to see you decided to come back aboard," the handler said.

"Yeah," came the hoarse response. "Glad to be back."

Going belowdecks, Lovell prepared for a debriefing with his team leader, but he was waylaid by the flight surgeon, carrying a small bottle of brandy. "You don't look too good," the ship's doctor said. "Have a little medicine. On me." Lovell took the bottle the doctor proffered and swallowed its contents in a single gulp.

When Lieutenant Junior Grade Lovell met with Lieutenant Commander Klinger, he did his best to describe the problems with his ADF, the misjudged altitudes during his approach and, reluctantly, the little invention that had blacked out his cockpit. The skipper listened with seeming sympathy, nodded with seeming understanding and, when Lovell was all done, pulled out the night-flight assignments for the following night. With a smile and a flourish, he wrote "Lovell" at the top of the list.

"You fell off the horse," the skipper said. "And you're getting back on."

As he was instructed, Lovell flew up into the black the very next evening. This time his ADF found the ship easily; this time he made his approach flawlessly; and this time he pulled off the landing without incident. But this time his marvelous lighted knee board stayed behind.

Jim Lovell eventually grew comfortable with the carrier pilot's high-wire life, ultimately amassing 107 landings and going on to become an instructor in a flock of new airplanes, including the FJ-4 Fury, the F8U Crusader, and the F3H Demon. By 1957, however, the job of patrolling the peacetime Pacific, training fliers for air wars that did not seem likely ever to take place, began to lose its luster. Near the end of that year, when the opportunity to apply for transfer arose, the pilot — who was nearing thirty and by now was the father of a three-year-old girl and a two-year-old boy — submitted a request for one of the service's most hazardous assignments: the U.S. Navy Aircraft Test Center in Patuxent River, Maryland.

Lovell was excited by the prospective change of assignments. While it took considerable skills to pilot military jets that had already been certified fit for use, it took even more to be the one who would do the certifying in the first place. Flying unproven, experimental aircraft in the skies above southern Maryland was, Lovell figured, about as close to the cutting edge of aeronautics as he could hope to get, and when his transfer request was approved, he quickly packed up his family and prepared to head east. But even before he left California, the cutting edge of his new career seemed to dull a bit.

On October 4, 1957, the Soviet Union blindsided Washington and the rest of the West with the news that it had successfully placed a twenty-three-inch robotic ball called Sputnik into a 560-mile-high orbit above the Earth. The sphere weighed only 184 pounds, which was about all Moscow's old R-7 booster could lift. Just a month later, however, Soviet engineers followed with a more powerful booster and a much bigger Sputnik, this one weighing 1,119 pounds.

Red-faced, the United States had to do something quick. The following month, American engineers rushed a little chopstick-like Vanguard rocket out to a launch pad, topped it with a six-inch cherry of a satellite, lit the fuse, and hoped for the best. The Vanguard smoldered promisingly on the pad for a few seconds, rose a few inches, and then blasted itself to hot smithereens. The cherry dropped to the ground, rolled away, and came to rest at the edge of the pad's concrete apron, where it beeped its silly radio signal to the humiliated launch directors in the blockhouse. The world laughed itself sick at the West's debacle, and American newspapers led the charge, ho-ho-ing for days about Yankee ingenuity and its remarkable new "Stayputnik" satellite.

Lovell followed the developments and did not find the jokes especially funny. Wasn't it the United States that had all those great Germans out in White Sands? Wasn't it the United States that had flown Operation Bumper more than a decade earlier? Then what were we doing looking so ridiculous now? The problem was a disturbing one, but not one that a naval aviator like Lovell could spend a lot of time fretting about. He was going off to test airplanes — something that America at least appeared to be able to build reasonably well. He had no business filling his head with a lot of rubbish about rockets. Besides, the ones he cared about most always seemed to blow up.

4

SY LIEBERGOT was accustomed to ratty data. He didn't like it —
nobody did. But he was accustomed to it.

Liebergot, like every other controller, lived and died by the data on
his screen. To the untrained eye, the glowing glyphs that were the stuff
of Liebergot's day would make not a shred of sense. But to a controller,
the numbers on the monitor meant that the little canister of people he
had helped fling a quarter of a million miles away from home was
either doing fine and buttoned up tight, which was very good, or *not*
doing fine and *not* buttoned up tight, which was very bad. If things
were not doing fine, it meant that the people in the canister might
never return from the celestial ether they had intended simply to visit,
and the folks on the ground would want to know if it was your
glowing glyphs that had started acting funny and if maybe you should
have spotted them earlier. So when the data on the screens started
getting ratty, Liebergot and everyone else got uneasy.

It's not as if nobody knew what the occasional rattiness was about.
In fact, they could even predict it. It would happen when an Apollo
spacecraft orbiting the moon disappeared behind the far side. It would
happen when a Gemini capsule orbiting the Earth passed between the
friendly footprints of two tracking stations. It would happen when a
Mercury capsule broke out of orbit and went shrieking into the atmosphere at 17,000 miles per hour, trailing a cloud of hot, angry,
signal-garbling ions.

In all these cases, transmissions streaming from the ship would be turned pretty much to hash, but before they disappeared altogether, they'd get, well, kind of ratty. Maybe your screen glyphs would tell you that the cabin pressure had suddenly dropped to zero; or maybe it would tell you that a hydrogen tank had just blown a seal, eating itself in a pressure explosion and taking part of the ship with it; or that a couple of fuel cells had just gone south on you; or that the heat shield had slipped; or that the thrusters were busted. Most likely they weren't, most likely it was just ratty data — but if they *were* right, that could be it for the canister. The problem was, you'd never know for sure what the story was until the Gemini was in touch with the next station, or the Mercury had cleared its ion storm, or the Apollo had come around to the sunny side of the street.

Liebergot was as good as anyone at interpreting this stuff — and he should have been. He first came to NASA in 1964, and by 1968 was working his own console at Mission Control in Houston. During the sixties there was no greater place for a scientist to work, no facility that more completely represented the heart, the soul, the very forebrain of the scientific world than this big, forbidding, thrilling room.

Liebergot was in charge of what was known as the electrical and environmental command console, or EECOM. EECOM controllers were responsible for the power and life-support systems of the command-service module, keeping them up and humming from the moment of launch to the moment of recovery. It was NASA that came up with the title EECOM, but Liebergot and his peers liked to think of themselves as cook-and-perk men. They were the ones who monitored the internal organs of the ship, who kept its juices and gases bubbling and flowing, and who, in the end, were responsible for keeping the mechanical organism alive in a place that it really had no business being.

Over the first year and a half of the manned Apollo program, the people who worked the Mission Control consoles accomplished remarkable things, learning to ply the translunar highway like a familiar old trading route. Four times they had sent crews to the moon — twice, on Apollos 11 and 12, they had set them down on the moon — and four times they had brought them safely home. Liebergot, like most of the other people in the room, had worked all of those flights, and had begun to feel that there was little he and his colleagues couldn't anticipate, from liftoff to moonwalk to splashdown, and little they couldn't handle. In the winter and spring of 1970, when the

Agency was making plans for Jim Lovell, Ken Mattingly, and Fred Haise's Apollo 13 mission, the controllers knew that they'd need every bit of this skill.

As the NASA brass envisioned it, the new mission was going to be one long bruiser of a flight. Apollos 11 and 12, the first two lunar landings, had been targeted to touch down in two of the moon's friendliest spots, the Sea of Tranquillity and the Ocean of Storms. While such desert-flat plains make hospitable runways, to a geologist they're a snooze — miles and miles of rocks and dust, all made of roughly the same stuff and all roughly the same age.

If you want the really good booty, you've got to head for the lunar hills. So different was the geological makeup of the moon's highlands from that of its lowlands that the higher ground even reflected sunlight more brightly, offering a shimmering, come-hither beacon to explorers staring up from Earth. On Apollo 13 NASA planned to answer the call. Targeted for touchdown on the third lunar landing was a place known as the Fra Mauro range, a stretch of rugged, Appalachian-type mounds 110 miles east of the Apollo 12 landing site. Not only would Fra Mauro yield interesting samples, but the job of reconnoitering the site and finding a safe touchdown spot would provide a valuable test of both the skills of the astronauts and the maneuverability of the lunar module.

More harrowing than Apollo 13's destination was the route the spacecraft would take to get there. On all of NASA's previous lunar missions, the crews had flown to the moon on the free-return trajectory that would guarantee them an automatic trip home in the event that their service module engine failed. On Apollo 13 this would not be possible. While the terrain of Fra Mauro made the landing site a perilous one, the lunar lighting at the time of day the crew was scheduled to arrive would make it more perilous still.

The current flight plan called for the ship to get to the moon when the sun was at such an angle that the telltale shadows Fra Mauro's boulders and hills would ordinarily cast would be erased. Without shadows, these topographic obstacles would be far harder for the pilots to see. Changing the ship's trajectory so the crew would arrive when the shadows were longer would be a simple matter, requiring only a quick engine burn on the outward coast to the moon. But once the engine was fired, the fragile free-return trajectory would be destroyed. If Apollo 13 failed to go into orbit around the moon, the new trajectory would still whip it around in the direction of Earth, but cause it to miss the home planet by some forty thousand miles.

To train for such a risky mission, both the crew and the Mission Control team that would support them put in almost unheard-of hours. The preferred way to bring the men working the Mission Control consoles up to speed was to run flight simulations. During a typical simulation, the control room would be activated exactly as it would if a real flight was taking place — consoles fully manned, screens full of data, headsets full of chatter, tracking screens at the front of the room lit and flashing. The only difference would be that all of the signals would come not from space but from a double row of consoles behind a glass wall on the right side of the main room. This was where the simulation supervisors, or Simsups, sat. Their job was to run ersatz flights and throw controllers simulated problems to see how fast they could come up with solutions. A controller's performance in these artificial situations could have a very real bearing on his future at the Agency.

One afternoon a few weeks before the launch of Apollo 13, Liebergot and the rest of the controllers were at their consoles, monitoring routine data from a routine phase of a so-far routine sim. The simulation being run was known as a fully integrated one. This meant that while the mission was a sham and the spaceship was a sham, the astronauts involved were the genuine articles. Nearby on the Johnson Space Center compound was the crew training building, equipped with working mockups of both the command and lunar modules. In residence today were Lovell, the commander of the mission, Mattingly, the command module pilot, and Haise, the LEM pilot. As on all simulations — as well as on the flight itself — the controllers could hear all the banter between the astronauts and the Capcom, but they could not break into the loop to say anything themselves. They could communicate on a separate loop with both the flight director, who sat at a console in the third row of Mission Control, and with one of several three- or four-man backroom support teams. These teams had consoles of their own with which they would track the flight and help their particular controller solve problems.

The portion of the flight plan the controllers and crew were running was that period about one hundred hours after launch when Lovell and Haise would be down on the lunar surface inside the spindly, spartan LEM, and Mattingly would be station-keeping sixty miles above, in the relative rumpus room of the command module. It was at times like this in any landing mission that the EECOM's workload was the lightest, because the mother ship just didn't have much to do, and because of the loss of signal that occurred every time it slipped behind the back of the moon. As long as your craft was functioning smoothly when it disappeared, the forty-minute blackout every two hours gave you a chance to stretch a little, take your eyes off the screen, and plan for any upcoming maneuvers.

As one of today's simulated blackouts began, Liebergot was checking his screen when he noticed something funny: a barely perceptible drop-off in the cabin pressure reading. This flutter — no more than a bump in the pounds per square inch data — was visible for barely a second before the ship vanished behind the moon, annihilating the readings entirely. Liebergot and his backroom team were on the air to each other instantly.

"Did you see that cabin pressure?" the backroom asked.

"Saw it," said Liebergot.

"How much did it go down?"

"About a tenth of a p.s.i., no more."

"Not much," the backroom said. "What do you think?"

"It's probably nothing," Liebergot answered.

"Ratty data?"

"I'm sure. Right before loss of signal. What else could it be?"

Liebergot and his backroom relaxed, confident that ratty data was the right explanation. In an authentic flight, ratty data *would* have been the right explanation. But in this flight, the Simsups had decided that ratty data would be the wrong explanation. For forty minutes of blackout, Liebergot and his support team did nothing about the oxygen anomaly, convinced that what they had seen was merely a harmless illusion. Then the ship came out of blackout, and Ken Mattingly's voice called out across the simulated void.

"We had a sudden depressurization here, Houston," he said. "Cabin pressure is down to zero and at the moment I'm on suit pressure. I'm guessing there's a leak in the bulkhead, but I don't know."

Liebergot went cold. The blip in pressure had been real. This was a test aimed squarely at the EECOM, and he had failed it. The Simsups — the damn Simsups — had screwed him but good. Lovell, Mattingly, and Haise had not been in on the game. Mattingly had been suddenly thrown the problem — not in the form of a real pressure loss in the simulator, of course, but in the form of a cabin pressure needle plummeting to zero — and he did the only thing he could do: put on his suit, button it up, and wait for reacquisition of signal. Only Liebergot and his backroom had been given any warning, and they had done . . . precisely nothing.

Liebergot waited for a response on the communications loop from the flight director. If the director had still been Chris Kraft, the man who oversaw Mission Control through Mercury and Gemini, Liebergot figured he would be finished, bilged. Kraft didn't crap around. You lose a ship, even a phony ship, and you just might lose your skin. In this case, Liebergot hadn't actually lost the ship, but he had lost something almost as valuable: forty minutes, in which he and his backroom could have come up with solutions to the catastrophe that the signal had warned them about.

Kraft had left the flight director's job some time ago, moving upstairs into NASA management. In his place was Gene Kranz, the crewcut, square-faced, ex–Korean War aviator who came to NASA before Mercury and rose slowly and steadily until, at the outset of Apollo, he became lead flight director.

To the men in the room, Kranz was still a bit of an enigma. Running Mission Control from his consecrated console, he seemed every bit the military man he had once been. His instructions were terse and always clear; his tone rarely brooked any nonsense. The one nonregulation indulgence he allowed himself concerned his clothes. During lunar flights that could run on for days and even weeks, four rotating teams of console crews would work in Mission Control, each one headed by a different flight director. The teams were designated by color, and Kranz's had been dubbed the White Team. The lead flight director took a competitive pride in the talents of his crew, and during flights had lately made it a point to wear a white vest over his regulation white shirt and black tie, as a sort of unabashed team emblem. The vest made Kranz seem more approachable, if not loveable, and the controllers who worked for him enjoyed their boss's one mild eccentricity. Today was just a simulation day, however, and Kranz's vest was nowhere in sight. Even if it had been, Liebergot suspected it would hold no protective magic. The entire control room heard Mattingly radio down his problem; the entire control room heard the Capcom radio back a "roger." And the entire room waited to hear how Kranz would respond.

"All right," the flight director said after a seemingly endless pause, "let's work the problem."

Liebergot let out a breath. This, he knew, was Kranz-speak for "I'm sparing your ass," and he fell to work at his console with a gusto that was equal parts relief and gratitude. But salvaging the simulated mission was not an easy matter. Liebergot and the other controllers decided to try out a little-practiced survival plan in which the LEM would blast off for an immediate docking with the mother ship and then remain attached to it, serving as a kind of lifeboat into which the astronauts could crowd themselves until approaching Earth, when they would crawl back into the command module, jettison the LEM, and reenter the atmosphere. The lifeboat idea had been kicking around since the early days of the Apollo program in 1964, and a few such maneuvers had even been practiced in early 1969, when the Apollo 9 astronauts flew the first LEM in Earth orbit. However, nobody seriously believed it would ever have to be used.

Kranz let the lifeboat exercise run for a few hours, until he was convinced that the controllers and astronauts had learned the survival protocols and — not incidentally — that Liebergot had learned a lesson. Finally though, they aborted the sim and went on to another, less fanciful one. This, of course, made sense. Only a few weeks remained before the launch, and there were plenty of scenarios to rehearse that were a lot more likely to occur than a dead command module and a lifeboat LEM.

For all its promise, the mission of Apollo 13 was never one that seized the imagination of the country. For pure drama, there were plenty of other things you could pay attention to in the spring of 1970 than the adventures of — what was it by now? — the fifth and sixth men who would walk on the moon. On April 9, two days before the scheduled launch, the *New York Times* made no mention of the mission, devoting front-page coverage to the U.S. Senate's surprise rejection of President Nixon's latest Supreme Court nominee, Judge G. Harrold Carswell.

Elsewhere in the news that week was the announcement of an eleven-month high in Southeast Asian casualty figures; a decision by the Massachusetts Supreme Judicial Court to delay the release of the Mary Jo Kopechne inquest results; the introduction of an ingeniously packaged women's hosiery product, L'eggs; a revelation by Paul McCartney that he was experiencing "personal difficulties" with the other three Beatles and had decided to leave the band; and the opening of the baseball season, one of the last that could include the headline "Tigers Set Back Senators." The first significant mention of Apollo 13 in the *Times* that week was on April 10, the day before the flight — on page 78, the weather page.

To the degree the mission stirred any public interest at all, much of it involved an almost morbid fascination with the numeral of this particular Apollo. All of the Mercury flights had used the number 7 in their names — Faith 7, Friendship 7, Sigma 7 — in honor of the seven astronauts who made up the team. Manned Gemini capsules had started counting at Gemini 3, but stopped after ten flights at Gemini 12. Manned Apollo missions had started at Apollo 7, and with fourteen manned flights planned, NASA knew it would have to confront an Apollo 13 eventually.

Bringing one of humanity's greatest scientific endeavors eyeball to eyeball with one of its most enduring superstitions had an irresistible appeal, and most people applauded the hubris, the c'mon-I-dare-you arrogance, of flying the mission anyway, and even embroidering a big, loud "XIII" on the patches of the suits the astronauts would be wearing throughout the flight. During the weeks before the launch, the public went on a sort of 13 scavenger hunt, looking for numerological omens portending disaster for the mission. (The flight was scheduled to begin on April 11, 1970, or 4/11/70 — add a four, two ones, a seven and a zero, and you get 13. Liftoff was planned for 1:13 Houston time, which, if that wasn't bad enough, is 13:13 military time. If the launch took place on schedule, the ship would pass into the moon's gravitational field on April 13.)

NASA found all this voodoo laughable in the extreme, and so did Lovell. As far as the commander of the mission was concerned, his trip to Fra Mauro was a scientific expedition, no more, no less. There was no room for a lot of superstitious claptrap, and the motto he chose for the official mission patch reflected that belief. Harking back to his Annapolis days, Lovell borrowed the Navy's motto, *Ex tridens scientia* ("From the sea, knowledge") and changed it slightly, to *Ex luna scientia*. To Lovell, the acquisition of knowledge seemed like a pretty good reason to make a lunar trip.

The preparations for Apollo 13 went flawlessly — so much for the question of bad luck, Jim Lovell liked to point out — until seven days before the mission, when Charlie Duke got sick. Duke was the LEM pilot in the backup crew, which also included commander John Young and command module pilot Jack Swigert. Duke had caught a case of German measles from one of his children and unknowingly exposed Young, Swigert, Lovell, Mattingly, and Haise to it. Blood tests proved that the rest of the backup crew, as well as Lovell and Haise, had been exposed to the disease before and thus carried protective antibodies. Mattingly, however, was not so immunized and was therefore in real danger of coming down with the sickness.

In cases like this, NASA's rules were simple: a potentially ill crew member could never be trusted at the helm of a spacecraft, so Mattingly would have to be bumped from the flight. Lovell, who had been training with this crew for the better part of a year, went ballistic. Now? You want to change the crew *now*, a week before liftoff, because of some possible bug? At the crew meeting in Houston where the decision was announced, Lovell went to bat for his command module pilot.

"How long is the incubation period for this thing?" the commander asked the flight surgeon.

"About ten days to two weeks," the doctor answered.

"So he'd be healthy at liftoff?" Lovell said.

"Yes."

"And healthy when we get to the moon?"

"Yes."

"Then what's the problem?" Lovell said. "If he starts running a fever when Fred and I are down on the surface, he can have that whole time to get over it. If he's not better by then, he can just sweat it out on the flight home. I can't think of a better place to have the measles than in a nice cozy spaceship."

The flight surgeon stared incredulously at Lovell, waited for him to finish his pitch, then thumbed Mattingly out of the lineup.

Though Lovell was fiercely loyal to his command module pilot, his new crew member was no slouch. At thirty-eight, Jack Swigert had previously been known mostly for being the only unmarried astronaut ever accepted into the NASA corps. In the early 1960s — when image was all, and aptitude sometimes seemed to come second — this was unthinkable. But as the nation's attitudes loosened up in the late 1960s, so did NASA's. The tall, crewcut Swigert had the reputation — good-naturedly tolerated by the Agency — of a rambunctious bachelor with an active social life. Whether this was true or not was unknown, but Swigert did what he could to perpetuate the image. His Houston apartment included a fur-covered recliner, a beer spigot in the kitchen, wine-making equipment, and a state-of-the-art stereo system.

NASA was willing to indulge all these less-than-upright distractions because Swigert was also a highly competent, fiercely confident pilot. He had trained devotedly for his understudy role on 13 and, after being shifted up to the prime crew, was put through a meat grinder of more-rigorous drills. During the course of the previous year, the original crew members had become so accustomed to working with one another that Lovell and Haise had even learned to interpret the nuances and inflections in Mattingly's voice — a valuable skill at those moments in the flight when the two LEM pilots would have to rely on the command module pilot's shouted commands alone to steer their lander to a safe rendezvous. After Mattingly was bumped, it took several days of simulator drills before NASA and the astronauts themselves were convinced that the new prime crew could work together as efficiently as the old one had.

Just forty-eight hours before liftoff, Swigert was certified fit to fly. The only remaining problem the flight planners faced now was the need to manufacture a new commemorative plaque, to be attached to the outside of the LEM. Already bolted in place on the lander's forward leg was a decorative panel bearing the engraved names of the three prime crewmen. In its place, a new, snap-on plaque would have to be milled, reflecting the last-minute change in personnel. The only problem Swigert himself now faced — as the newspapers delighted in reporting — was that with all the sudden hubbub, he had forgotten to file his federal income tax form. The return was due, of course, on April 15, which was four days after launch or about the time this particular taxpayer would be in orbit around the moon. Swigert decided simply to put the problem out of his head, figuring he would work something out when he got home. Mattingly, of course, was now free to file his return at his leisure.

The third member of the Apollo 13 crew was lunar module pilot and former Marine flier Fred Haise. At thirty-six, Haise was the youngest of the trio, and his black hair and angular features made him seem younger still. Though married, with three children and a fourth on the way, Haise was still known to his friends by the excruciatingly youthful nickname "Pecky," a moniker he picked up playing a woodpecker in a first-grade play. Unlike Lovell and Swigert, Haise found flying an acquired taste. What he really liked about space travel was the exploration, the science, the research. One NASA scientist referred to him as a "drilling fool," a reference to the almost preternatural pleasure Haise got out of the geological equipment he and Lovell would use to extract core samples from the lunar surface. The description was not necessarily what you'd look for in an astronaut during the early, daredevil Mercury days. But it was exactly what you'd look for in someone wearing a pressure suit with *Ex luna scientia* embroidered on its front.

Apollo 13 was launched on schedule, at 13:13 Houston time on April 11, and three hours later blasted out of Earth orbit toward the moon. For Swigert and Haise, neither of whom had ever been in space before, the experiences of launch and orbit and translunar burn were inexpressably novel. For Lovell, making his fourth trip atop a rocket (and his second atop the immense Saturn 5), it was little more than a return to business. On the first full day of the mission, the lunar veteran, now occupying the exalted left-hand seat that Frank Borman had claimed a year and a half earlier, called down to Earth for some of the idle banter that he and Borman and Bill Anders had come to look forward to during their week in space in 1968.

"Hello there, Houston, 13," Lovell said.

"13, Houston, go ahead," answered the Capcom. As on all flights, the Capcoms assigned to work this one were other astronauts, the thinking being that three men in a can zipping along at 25,000 miles an hour would just as soon do their talking to a fellow traveler and not some technician who had never graduated beyond the coach seat of a commercial jet. The Capcom today was Joe Kerwin, one of NASA's greener novices. Kerwin had never yet flown in space, but the flight manifests all said that one day he would, and that's what counted.

"We'd almost forgotten," Lovell said to Kerwin. "We'd like to hear what the news is."

"O.K., there's not a whole lot to it," Kerwin said. "The Astros survived, 8 to 7. The Braves got five runs in the ninth inning, but they just made it. They had earthquakes in Manila and other areas of the island of Luzon. West German Chancellor Willy Brandt, who witnessed your launch from the Cape yesterday, and President Nixon will complete a round of talks today. The air traffic controllers are still out, but you'll be happy to know the controllers in Mission Control are still on the job."

"Thank goodness for that." Lovell laughed.

"Also," Kerwin went on, "some truck lines are being struck in the Midwest, some schoolteachers have walked off the job in Minneapolis. And of course today's favorite pastime across the country . . ." Kerwin paused for dramatic effect. "Uh-oh, have you guys completed your income tax?"

Swigert, sitting in the center couch, cut into the loop. "How do I apply for an extension?" he asked, his voice businesslike, unamused. Kerwin, who knew he had hit his mark, laughed. "Joe, it ain't too funny," Swigert protested. "Things happened kinda fast down there and I do need an extension." A few other controllers could now be heard laughing on the line. "I'm really serious," Swigert said. "I didn't get mine filed."

"You're breaking up the room down here," Kerwin said.

"Well," grumbled Swigert, "I may be spending time in another quarantine when we get back, besides the medical one they're planning for us."

"We'll see what we can do, Jack," said Kerwin. "In the meantime, the uniform of the day for you guys will be in-flight coverall garments with swords and medals, and tonight's movie, shown in the lower equipment bay, will be John Wayne, Lou Costello, and Shirley Temple in 'The Flight of Apollo 13.' Over."

That the crew and the ground could spend so much of their time engaging in such ship-to-shore small talk still occasionally amazed Lovell. There would be no movie on the flight, of course, and there would be no swords or medals or uniform of the day. But playful references to the slow-paced life aboard a roomy, cruising Navy ship was not lost on the ex–Annapolis man. The joke in the old Mercury days was that astronauts didn't climb into their capsules, they put them on. The spacecraft were preposterously small and uncomfortable, and the missions lasted an average of just eight and a half hours. In the Gemini capsule where Lovell had cut his orbital teeth, there was about twice the interior space, but also twice the number of occupants.

As Lovell had discovered in Apollo 8, and as Haise and Swigert were now learning, NASA's lunar ships were an entirely different engineering animal. The Apollo command module was an eleven-foot-tall cone-shaped structure, nearly thirteen feet wide at the base. The walls of the crew compartment were made of a thin sandwich of aluminum sheets and an insulating honeycomb filler. Surrounding that was an outer shell of a layer of steel, more honeycomb, and another layer of steel. These double bulkheads — no more than a few inches thick — were all that separated the astronauts inside the cockpit from the near-absolute vacuum of an outside environment where temperatures ranged from a gristle-frying 280 degrees Fahrenheit in sunlight to a paralyzing minus 280 degrees in shadow. Inside the ship, it was a balmy 72.

The astronauts' couches lay three abreast, and were actually not couches at all. Since the crew would spend nearly the entire flight in a state of weightless float, they needed no padding beneath them to support their bodies comfortably; instead, each so-called couch was made of nothing more than a metal frame and a cloth sling — easy to build and, most important, light. Each couch was mounted on collapsible aluminum struts, designed to absorb shock during splashdown if the capsule parachuted into the sea — or, in the case of a mistargeted touchdown, onto land — with too much of a jolt. At the foot of the three cots was a storage area that served as a sort of second room (Unheard of! Unimaginable in the Gemini and Mercury eras!) called the lower equipment bay. It was here that supplies and hardware were stored and the navigation station was located.

Directly in front of the astronauts was a big, battleship-gray, 180-degree instrument panel. The five hundred or so controls were designed to be operated by hands made fat, slow, and clumsy by pressurized gloves, and consisted principally of toggle switches, thumb wheels, push buttons, and rotary switches with click stops. Critical switches, such as engine-firing and module-jettisoning controls, were protected by locks or guards, so that they could not be thrown accidentally by an errant knee or elbow. The instrument panel readouts were made up primarily of meters, lights, and tiny rectangular windows containing either "gray flags" or "barber poles." A gray flag was a patch of gray metal that filled the window when a switch was in its ordinary position. A striped flag like a barber pole would take its place when, for whatever reason, that setting had to be changed.

At the astronauts' backs, behind the heat shield that protected the bottom of the conical command module during reentry, was the twenty-five-foot, cylindrical service module. Protruding from the back of the service module was the exhaust bell for the ship's engine. The service module was inaccessible to the astronauts, in much the same way the trailer of a truck is inaccessible to the driver sitting in the cab. (Since the windows of the command module faced forward, the service module was invisible to the astronauts as well.) The interior of the service module cylinder was divided into six separate bays, which contained the entrails of the ship — the fuel cells, hydrogen tanks, power relay stations, life-support equipment, engine fuel, and the guts of the engine itself. It also contained — side by side, on a shelf in bay number four — two oxygen tanks.

At the other end of the command module–service module stack, connected to the top of the command module's cone by an airtight tunnel, was the LEM. The four-legged, twenty-three-foot tall craft had an altogether awkward shape that made it look like nothing so much as a gigantic spider. Indeed, during Apollo 9, the lunar module's maiden flight, the ship was nicknamed "Spider," and the command module was called by an equally descriptive "Gumdrop." For Apollo 13, Lovell had opted for names with a little more dignity, selecting "Odyssey" for his command module and "Aquarius" for his LEM. (The press had erroneously reported that Aquarius was chosen as a tribute to *Hair,* — a musical Lovell had not seen and had no intention of seeing. The truth was, he took the name from the Aquarius of Egyptian mythology, the water carrier who brought fertility and knowledge to the Nile valley. Odyssey he chose because he just plain liked the ring of the word, and because the dictionary defined it as "a long voyage marked by many changes of fortune" — though he preferred to leave off the last part.) While the crew compartment of Odyssey was a comparatively spacious affair, the lunar module's crew compartment was an oppressively cramped, seven-foot eight-inch sideways cylinder that featured not the five portholes and panoramic dashboard of the command module but just two triangular windows and a pair of tiny instrument panels. The LEM was designed to support two men, and only two men, for up to two days. And only two days.

NASA was extremely proud of this pair of spacecraft and liked to show them off. Since the triumphant success of the Apollo 8 broadcasts two Christmases ago, crews had continued to go aloft with television cameras stowed in their equipment bays and with time for live broadcasts written into their flight plans. The practice reached its peak of popularity during the Apollo 11 moon landing in the summer of 1969, when stations around the globe carried Neil Armstrong and Buzz Aldrin's first tentative moonwalk and most of the world stood still to watch it. But by the time Apollo 13 rolled around, the world had lost interest. A little after the two-day mark in the mission, the crew was scheduled for its first TV show, but none of the networks intended to carry it. The broadcast was set to begin at 8:24 P.M. on Monday, April 13, in the time slot belonging to NBC's *Rowan & Martin's Laugh-In* and CBS's *Here's Lucy*. ABC had programmed the 1966 movie *Where Bullets Fly*, followed by *The Dick Cavett Show*.

Viewers across the nation had shown little interest in having any of these programs preempted by the show from space, and even in Mission Control, NASA technicians themselves were only half interested. The broadcast would begin an hour and a half before the afternoon-evening shift punched out, and most of the men at the consoles were already looking forward to finishing their workday and stopping for a drink at the Singin' Wheel, a red-brick, antiques-stuffed saloon just off the Space Center compound.

NASA and the Apollo crew nevertheless decided to go ahead with the show and to make the feed available to any stations that might want to tape bits of it for their eleven o'clock news shows. A little coverage, they figured, was better than no coverage at all. Besides, the wives of the astronauts had come to look forward to these periodic broadcasts, and nobody in NASA wanted to tell them that the custom would be discontinued. Already tonight, the controllers in Houston could see that Marilyn Lovell and two of her four children, sixteen-year-old Barbara and eleven-year-old Susan, had settled into the cushioned seats in the glassed-in VIP gallery at the rear of Mission Control. Also with them was Mary Haise, the wife of the first-time astronaut, preparing to watch as her husband's image was beamed down from space.

The program that nobody but Marilyn, Barbara, Susan, Mary, and the controllers saw began with a choppy, murky image of Fred Haise drifting up toward the tunnel connecting the command module and the LEM. Lovell was reclining on Swigert's couch in the middle of the command module, operating the camera. Swigert had shifted left to Lovell's couch.

"What we plan to do for you today," Lovell said to nobody but Houston, "is start out in spaceship Odyssey and then take you on through the tunnel to Aquarius. Your TV operator is now resting on the center couch looking at Fred, and Fred will now transport himself into the tunnel, and we'll show you a little bit of the landing vehicle."

Haise obliged for the camera, floating through the cone of the command module and emerging into the LEM, descending head-first through the ceiling like a cross-dimensional traveler entering another world through whatever time-space portal happened to be available. Lovell floated slowly after him.

"One thing I noticed, Jack," said the upside-down Haise to his Capcom, "was that starting upright in the command module and heading down into Aquarius, there's an orientation change. Even though I'd practiced it in the water tank, it's still pretty unusual. I find myself now standing with my head on the floor when I get down into the LEM."

"That's a great picture, Jim," Jack Lousma, the Capcom, encouraged the commander. "You got the light just right."

Lovell entered the LEM, flipped his position, and descended feet-first onto a large bulge in the floor of the module. "For the sake of all the people back home," said Haise, "housed inside this can under Jim's feet is the LEM ascent engine, the engine that we use to get off the moon. Immediately adjacent to the engine cover here, I have my hand on a white box. This happens to be Jim's backpack, which will supply oxygen and water for cooling while on the lunar surface."

"Roger, Fred, we see it," said Lousma. "The picture's coming through real good and your description is good. We see Jim's got the camera oriented the way we like to look at it, so keep talking."

Lovell and Haise cheerfully complied, sending their good picture and good descriptions back to Earth. While the show proceeded in its folksy way, much of Mission Control was busy with other things. On the closed communications loop intended for the people at the consoles only, most of the controllers were planning maneuvers the crew would perform as soon as they signed off the air. Kranz, the flight director, led the discussions, refereeing requests, setting priorities, and determining which exercises were essential and which could wait. The chatter on this loop would have made decidedly less sense to earthbound observers than the TV show intended for their consumption.

"Flight, EECOM," Liebergot called into the loop.

"Go ahead, EECOM," Kranz said.

"At 55 plus 50, we would sure like to have a cryo stir. All four tanks."

"Let's wait until they get settled down some more."

"Roger."

"Flight, GNC," signaled Buck Willoughby, the guidance, navigation, and control officer.

"Go ahead, GNC."

"We would like to reenable the other two quads for the maneuver."

"You want them to enable C and D, right?"

"Right."

"You want them to disable A and B?"

"No."

"OK, all four quads."

"Flight, INCO," said the instrumentation and communications officer.

"Go, INCO."

"I would like to confirm the configuration of their high gain now. We would like to know what track mode they're in."

"O.K., let's just stand by one there."

The maneuvers Houston was planning for the crew, for all their technochatter sound, were fairly routine ones. The INCO's reference to the "high gain" concerned the service module's main antenna, which had to be transmitting on a particular frequency and set at a particular angle depending on the position and trajectory of the spacecraft. Charged with the round-the-clock job of monitoring the ship's communications systems, the INCO needed to check periodically to make sure everything was oriented as it should be. The business with the quads concerned the four clusters of attitude-control thrusters arrayed around the service module which moved the ship from one position to another. The crew was going to make some navigational shifts after the TV show, and the GNC wanted all four sets of thrusters up and running.

The other exercise, the "cryo stir" Liebergot requested, was perhaps the most routine of all. The service module was equipped with not only two oxygen tanks but also two hydrogen tanks, all of which maintained their gases in a hypercold, or cryogenic, state. The temperature, which in the case of the oxygen tanks could drop as low as minus 340 degrees Fahrenheit, kept the gases at what is known as supercritical density — a chemically queer condition in which a material is not quite a solid, not quite a liquid, and not quite a gas, but something slushily in between. So well were the tanks insulated that if they were filled with ordinary ice and placed in a 70-degree room, it would take eight and a half years for the ice to melt down to water just above the freezing point, and another four years for the water to rise to room temperature. That's what the designers liked to claim, in any case, and since nobody would ever actually run this test, NASA took them at their word.

The real magic of the cryogenic tanks, however, was not what happened to the oxygen and hydrogen while they were still inside the vessels, but what happened when they were channeled out. The tanks were connected to three fuel cells equipped with catalyzing electrodes. Flowing into the cells and reacting with the electrodes, the two gases would combine and, in a happy coincidence of chemistry and technology, produce a trio of byproducts: electricity, water, and heat. From just two gases, the cells would produce three consumables no life-sustaining spacecraft could do without.

Although the oxygen and hydrogen tanks were equally important in keeping the ship alive and thrumming, the oxygen tanks were especially precious because they also contained virtually all of the crew's supply of breathable air. Each of the two tanks was a sphere twenty-six inches in diameter, holding 320 pounds of oxygen at a pressure of up to 935 pounds per square inch. Immersed in the tank, like fingers testing the temperature of a tubful of bath water, were two electrical probes. One, running the length of the tank from top to bottom, was a combination quantity gauge and thermostat; the other, adjacent to it, was a combination heater and fan. The heater was used to warm and expand the oxygen in case the pressure in the tank dropped too low. The fans were used to stir the stuff up — something an EECOM would request at least once a day, since supercritical gases tend to stratify, confounding the tanks' quantity probes.

While Liebergot waited for his stir and the other controllers planned further procedures, the crew continued its television tour. On the large monitor at the front of Mission Control, a milky image of the moon appeared, evoking memories of the Apollo 8 broadcasts, when the whole world had been watching.

"Out the right window now," said Lovell, the narrator, "you can see the objective, and I'll zoom in on it and see if this brings it in better."

"It's beginning to look a little bigger to us now," said Haise. "I can see quite distinctly some of the features with the naked eye. So far, though, it's still looking pretty gray, with some white spots."

Lovell then swung the camera back inside the LEM. On the screen, Haise appeared to be making adjustments to a large cloth sling of some kind. "Now we can see Fred engaged in his favorite pastime," Lovell explained.

"He's not in the food locker, is he?" asked Lousma.

"That's his second-favorite pastime," said Lovell. "Now he's rigging his hammock for sleep on the lunar surface."

"Roger. Sleeping and then eating."

Lovell pushed away from Haise and began to drift back toward the tunnel. "O.K., Houston," he said, "for the benefit of the television viewers, we've just about completed our inspection of Aquarius, and now we're proceeding back into Odyssey."

"O.K., Jim. We think you ought to conclude it from here now, but what do you think?"

"Anytime you want to terminate, we're all set to go," Lovell agreed. Having played to a nearly empty room for some twenty-seven minutes now, he allowed more than a little relief to creep into his voice. "We've just got to put the cabin repress valve in."

"Roger," Lousma said.

The repress valve was a lunar module control used to help maintain equal pressure between the two spacecraft. Hearing the exchange, Haise helpfully turned the valve, causing a sudden hiss and thump to rock both ships. Holding the camera, Lovell visibly flinched. Earlier in the mission, the commander had begun to suspect that his over-exuberant crewmate sometimes used the repress valve more than was strictly necessary, deriving mischievous pleasure from the startle effect it had on his two crewmates. Here in their third full day, the joke had grown the tiniest bit frayed.

"Every time he does that," Lovell said candidly, "our hearts jump in our mouths. Jack, anytime you want to terminate TV, we're all set to go."

"O.K., Jim," Lousma concluded. "It's been a great show."

"Roger," said Lovell. "Sounds good. This is the crew of Apollo 13, wishing everyone there a nice evening. We're just about ready to close out Aquarius and get back for a pleasant evening in Odyssey. Good night."

And the projection screen went blank.

In Houston, Marilyn Lovell smiled. Her husband looked fine, if a little scraggly with a three-day growth of stubble, and his voice sounded level and serene. Though he certainly wouldn't have let on during a TV show if anything about the mission had been causing him concern, he would not have been able to keep at least a small edge out of his voice either. But Marilyn heard no edge tonight. Her husband was evidently happy with this flight so far and, she assumed, was looking forward to its lunar climax. She herself was glad it was almost halfway over, and was looking forward to the Pacific splashdown. Marilyn glanced at her watch, said a quick goodbye to the NASA Public Affairs officer who had been watching the broadcast with her, and she and Mary Haise left for their homes to make sure their children got to bed on time.

Down on the floor of Mission Control, Lousma looked over the list of maneuvers the crew would have to carry out before getting their own chance to turn in for the night. As Capcom, he had at least some control over when the astronauts would be told to perform each task, and he figured he'd give them a few minutes to get their camera stowed and return to their couches before radioing up the instructions for the cryo stir, the thruster maneuvers, and the antenna readings.

Before Lovell could get out of the tunnel or Haise could get out of the LEM, however, the controllers and crew had to return immediately to business. On the command module pilot's console a yellow warning light flashed on, indicating that there might — *might* — be a problem with the pressure in the cryogenic system. At the same time, a corresponding signal appeared on Liebergot's console. Scanning the data on his screen, Liebergot saw that the alarm was caused by a low-pressure reading in one of the hydrogen tanks, a tank that had been presenting intermittent problems for the past two days. If the cryo tanks or their quantity sensors were getting even a little balky, it was as good an indication as any that all four needed a good stir. As Lovell floated back to his left-side couch and Swigert shimmied over to his rightful position in the center, Houston radioed up its instructions.

"We'd like you to roll right to 060 and null your rates."

"O.K., we'll do it," Lovell responded.

"And we'd like you to check your C4 thrusters."

"O.K., Jack."

"And we've got one more item for you when you get a chance. We'd like you to stir up your cryo tanks."

"O.K.," Lovell said. "Stand by."

As Lovell prepared for the thruster adjustments and Haise finished closing down the LEM and drifted through the tunnel back toward

Odyssey, Swigert threw the switch to stir all four cryogenic tanks. Back on the ground, Liebergot and his backroom monitored their screens, waiting for the stabilization in hydrogen pressure that would follow the stir.

Of all the possible disaster scenarios that astronauts and controllers consider in planning a mission, few are more ghastly — or more capricious, or more sudden, or more total, or more feared — than a surprise hit by a rogue meteor. At speeds encountered in Earth orbit, a cosmic sand grain no more than a tenth of an inch across would strike a spacecraft with an energetic wallop equivalent to a bowling ball traveling at 60 miles per hour. The punch that was landed would be an invisible one, but it could be enough to rip a yawning hole in the spacecraft's skin, releasing in a single sigh the tiny pressure pocket needed to sustain life. Outside Earth orbit, where speeds could be faster, the danger was even greater. When Apollo astronauts first began traveling to the moon, one thing they dreaded most but spoke of least was the sudden jolt, the sudden tremor, the sudden boot in the bulkhead that indicated their highest of high-tech projectiles and some meandering low-tech projectile had, in a statistically absurd convergence, found each other like the pairs of fused bullets that once littered the battlefields of Gettysburg and Antietam, and had, like the bullets, done each other some serious damage.

In the sixteen seconds following the beginning of the cryo stir, the astronauts of Apollo 13 were executing their next maneuvers and awaiting additional commands when a bang-whump-shudder shook the ship. Swigert, strapped in his seat, felt the spacecraft quake beneath him; Lovell, moving about the command module, felt a thunderclap rumble through him; Haise, still in the tunnel, actually saw its walls shift around him. It was nothing that Haise and Swigert had ever experienced before; nor was it anything that Lovell, with his three prior flights and weeks spent in the cosmic deep, had come across either.

Lovell's first impulse was to be pissed off. Haise! This had to be Haise and his bloody repress valve! Once, maybe, the joke was funny. But twice? Three times? Even allowing for a rookie's misplaced exuberance, this was pushing things too far. The commander turned toward the tunnel, to find the eyes of his crewman and scald them with an angry glare. But when the two men's glances locked, it was Lovell who was brought up short. Haise's eyes were huge, unexpectedly huge, saucer-wide and white on all sides. These weren't the crinkly, merry eyes of someone who had just gotten off another good one at the expense of the boss and was awaiting a smiley rebuke. Rather, they were the eyes of someone who was frightened — truly, wholly, profoundly frightened.

"It wasn't me," Haise croaked out in answer to the commander's unasked question.

Lovell turned to his left to look at Swigert, but he got nothing. He saw the same confusion here, the same answer here, the same *eyes* here. Over Swigert's head, high up in the center section of the command module's console, an amber warning light flashed on. Simultaneously, an alarm sounded in Haise's headphone and another warning light, on the right-hand side of the instrument panel where the electrical systems were monitored, began to glow too. Swigert checked the panels and saw that there appeared to be an abrupt and inexplicable loss of power in what the crew called main bus B — one of two main power distribution panels that together provided juice to all of the hardware in the command module. If one bus lost power, it meant that half the systems in the spacecraft could suddenly go dead.

"Hey," Swigert shouted down to Houston, "we've got a problem here."

"This is Houston, say again please," Lousma responded.

"Houston, we've had a problem," Lovell repeated for Swigert. "We've had a main B bus undervolt."

"Roger, main B undervolt. O.K., stand by, 13, we're looking at it."

Sy Liebergot heard this exchange and, like all of the other controllers in the room, immediately started to scan his console. Even before he could, however, a voice screamed into his headset.

"What's the matter with the data, EECOM?" It was Larry Sheaks, one of three men in the EECOM backroom who monitored environmental readings and helped Liebergot manage any anomalies. Following Sheaks, the voice of George Bliss, another EECOM engineer, piped up: "We got more than a problem."

Liebergot looked over his monitor and his breath caught. Everywhere, it seemed, his readouts had gone into the tank. These aren't the numbers you get on a real flight, he thought. These are the implausibly bad numbers some smart little Simsup sent you during training when he wanted to see if you were paying attention.

But this wasn't training. The first and worst reading Liebergot noticed — right next to the hydrogen numbers he had been watching so closely an instant before — was the data for the spacecraft's two main oxygen tanks. On his readout, tank number two, which held half the oxygen for the entire ship, had suddenly ceased to exist. The data had simply fallen to zero, vanished, or, as controllers liked to put it, had just gone away.

"We lost O_2 tank two pressure," Bliss confirmed.

Liebergot scanned his screen and discovered more bad news. "O.K. you guys, we've lost fuel cell one and two pressure."

For an instant Liebergot felt secretly sick. According to what he was hearing in his headset and seeing on his screen, most of Odyssey's power system, not to mention half its atmospheric system, had gone on the fritz. The diagnosis was horrible, but it was by no means a conclusive one. It was possible that nothing had gone wrong with the equipment at all, that what was really broken were the sensors. Maybe they were spitting out faulty data that made it seem as if there were a problem. This did happen from time to time, and before jumping to conclusions, any good EECOM would exhaust the easy possibilities first.

"We may have had an instrumentation problem, Flight," Liebergot said to Kranz. "Let me add them up."

"Rog," Kranz said.

Up in their still-rocking, still-shuddering Odyssey, Lovell, Swigert, and Haise couldn't hear this exchange, but their instrument panel indicated it might be true. Haise pushed his way out of the tunnel and returned to his couch to check his electrical data, and saw that main bus B appeared to have rallied.

He sighed. "O.K. Right now, Houston," he said, "the voltage is looking good." Then he added, a little edgily, "We had a pretty large bang associated with the caution and warning there."

"Roger, Fred," Lousma said, unruffled — as if "large bangs" were a virtual commonplace during lunar missions.

"In the interim here," added Lovell, "we're starting to go ahead and button up the tunnel again."

The equanimity in Lovell's voice belied the urgency with which this "buttoning up" was taking place. Swigert unbuckled himself from his couch and dashed through the lower equipment bay and into the tunnel. All three astronauts were thinking the same thing: this was probably a meteor. Since the command module seemed to be in reasonably good shape, it was likely that the LEM had been hit. If that was so, they'd want to slam the hatch and seal the tunnel as quickly as possible, to prevent what could be a rapidly depressurizing lander from sucking the oxygen out of the command module, through the tunnel, and releasing it into space.

Swigert wrestled the hatch into place but couldn't get it locked down. He tried again and failed. He tried a third time and still couldn't make it work. Lovell floated into the tunnel, nudged Swigert out of the way, and tried it himself. True enough, the hatch didn't seem able to lock. After a couple of attempts, he threw up his hands and put the problem aside. If the integrity of the LEM had been compromised, the two craft would surely have depressurized by now. If there had been a meteor, it had evidently not damaged the crew compartments of either the LEM or the command module.

"Forget the hatch," Lovell said to Swigert. "Let's just take it out and tie it down out of the way."

Swigert nodded, and Lovell swam out of the tunnel, through the equipment bay, and back toward his couch to see if he could learn anything else from his instrument panel. Right away, he had more good news for Mission Control: while the readings for oxygen tank two may have gone through the floor in Houston, they had gone through the roof in the spacecraft. On Lovell's instrument panel, the quantity needle for the tank was so high it was off the top of the scale. Though this was probably not a terribly precise reading, it was a hell of a lot closer to what the O_2 level should have been than the "empty" signal that was showing up on the EECOM's screens. Lovell reported this happy data to Lousma, who responded with a noncommittal "roger."

At the moment, "roger" was about as specific as Lousma could afford to get. Assuming this wasn't an "instrumentation problem," as Liebergot had suggested hopefully, nothing going on in the spacecraft made much sense. Technically, a problem in an oxygen tank, a fuel cell, and a bus could all happen simultaneously, since the tanks fed O_2 to the fuel cells and the fuel cells in turn fed power to the bus. As a practical and statistical matter, however, it was extremely unlikely. The oxygen tanks were built with the fewest number of parts possible, making the likelihood of a breakdown as slim as possible. Even if one tank did fail, the other tank would still be more than adequate to power all three cells. And as long as all three fuel cells were operating, both buses would continue operating too. The probability of any one of these components failing was down in the multi-multi-decimal places. The probability of one tank, two fuel cells, and one bus failing at the same time was off the numerical charts.

Making matters worse, throughout the main room of Mission Control, other controllers were calling in other irregularities. An instant after the jolt that shook Odyssey, Bill Fenner, the guidance officer, or GUIDO, one of the men responsible for planning the spacecraft's trajectory, came on the line to announce that he had detected a "hardware restart" aboard the ship. This referred to the process by which an onboard computer would sense an undefined glitch brewing somewhere in the depths of the spacecraft, take a sort of deep computer breath, and set out on a data hunt to determine what had gone wrong. In a ship with as many perplexing problems as Odyssey currently had, a hardware restart would not be unexpected. However, the computer seemed to believe that the source of the bang the crew reported lay somewhere *within* the ship rather than outside it. This appeared to rule out a meteor hit; but if it wasn't a space rock that shook the ship, what was it?

Seconds after the bang, the instrumentation and communications officer had signed on the loop with a problem of his own.

"Flight, INCO," he said.

"Go, INCO," Kranz responded.

"We switched to wide beam width about the time he had that problem."

"O.K. You say you went to wide beam there?"

"Yes."

"See if you can correlate the times," Kranz said. Then, for clarity and certainty, he repeated: "Get the time you went to wide beam, INCO."

This was worth repeating, because the INCO had reported that when the mysterious jolt shook Odyssey, the ship's radio had taken it upon itself to quit transmitting through its high-gain antenna and switch to four smaller, omnidirectional antennas mounted about the service module. A spacecraft's radio should no more arbitrarily change its own antennas than a television set should change its own channels.

To some people in the room, the antenna problem at least was actually a cause for relief. This *had* to be an instrumentation problem. To have an oxygen tank, a fuel cell, and a bus all crap out was unlikely enough, but to suggest that at the same time an antenna had begun switching stations was just too much. It was as if an auto mechanic had given your brand-new car a once-over and then called you to say that your battery, your generator, and your starter were all shot and, oh yeah, your tires had suddenly gone flat, your radiator had burst, and your doors had popped their hinges. You just might suspect that the problem lay less with the car than with the mechanic.

Kranz, more than most, suspected this might be the case, and signed on to the loop to probe Liebergot.

"Sy, what do you want to do?" he asked. "Have you got a sick sensor–type problem there or what?"

Lousma was wondering the same thing, and severed his ground-to-air link long enough to ask Kranz, "Is there any kind of lead we can give them? Are we looking at instrumentation, or have we got real problems?"

On the EECOM loop, there were doubts too.

"Larry, you don't believe that O_2 tank pressure, do you?" Liebergot asked Sheaks.

"No, no," Sheaks answered. "Manifold's good, environmental control system is good."

Much of what fueled the controllers' skepticism was that the readouts up in Odyssey did not parallel those on the ground. After all, Lovell, Swigert, and Haise had already made it clear that according to *their* data, the bus and the O_2 tank were now fine. If the numbers aren't lining up, why believe the bad ones?

Up in the ship, however, the rosy readings that drove these hopes now began to change. Haise, who hadn't stopped scanning his instruments since the trouble started, caught a glimpse of his bus readouts, and his temporarily high spirits fell. According to Odyssey's sensors, main bus B, which had appeared to have rallied, had crashed again. Worse, bus A's readings had begun to fail too. The sick bus, it seemed, was dragging the healthy one down with it. At the same time, Lovell looked over his oxygen tank and fuel cell readings and got even worse news: oxygen tank two, which a moment before had read full to bursting, was reading dry as a bone. Most disturbing, the fuel cell readouts on Odyssey's instrument panel were now as poor as they were on Liebergot's screens, with two of the three cells putting out no juice at all.

At the sight of this last reading, Lovell could have spit. If the fuel cell data were accurate, he could kiss his trip to Fra Mauro goodbye. NASA had a lot of unbreakable rules when it came to lunar landings, and one of the most unbreakable ones was: If you don't have three in-the-pink fuel cells, you don't go anywhere. Technically, one cell would probably be enough to do the job safely, but when it came to something as fundamental as power, the Agency liked to have a big fat fluffy cushion, and for NASA even two cells weren't cushion enough. Lovell caught Swigert's and Haise's attention and pointed to the fuel cell readings.

"If these are real," Lovell said, "the landing's off."

Swigert started radioing the bad news down to the ground. "We've got a main bus A undervolt showing," he said to Houston. "It's about twenty-five and a half. Main bus B is reading zip right now."

"Roger," Lousma said.

"Fuel cell one and three are both showing gray flags," Lovell said, "but both are showing zip on the flows."

"We copy," replied Lousma.

"And Jack," Lovell added, "O_2 cryo number two tank is reading zero. Did you get that?"

"O_2 quantity is zero," Lousma repeated.

Bad as these developments were, Lovell had yet another problem to contend with. Better than ten minutes after the initial bang, his spacecraft was still swaying and wobbling. Each time the command-service module and the attached LEM moved, the thrusters would fire automatically to counteract the motion and try to stabilize the ships. But each time they appeared to have succeeded, the ships would start lurching again and the thrusters would resume their firing.

Lovell now took hold of the manual attitude controller built into the console, to the right of his seat. If the automatic systems couldn't bring the ships to heel, perhaps a pilot could. Lovell was concerned about keeping the spacecraft under control for more than aesthetic reasons. Apollo ships on the way to the moon did not simply fly straight and true, with the command module's nose pointed properly forward and the LEM attached to it like a big, ungainly hood ornament. Rather, the ships rotated slowly like a 1 r.p.m. top. This was known as the passive thermal control, or PTC, position and was intended to keep the ships evenly barbecued all around, preventing one side from cooking in the glare of the unfiltered sun and the other side from frosting over in the deep freeze of shadowed space. The thruster convulsions of Apollo 13 had shot the graceful PTC choreography all to hell, and unless Lovell could regain control, he faced the real danger of ultra-high and ultra-low temperatures seeping through the ship's skin and damaging sensitive equipment. But no matter how Lovell worked his manual thrusters, he could not seem to settle his spacecraft down. No sooner had he stabilized Odyssey than it would drift off line again.

For a pilot who had taken to space three times before, with little more than nuisance problems from his equipment, this was getting to be intolerable. The electrical system in Lovell's smoothly functioning craft had gone on the fritz, the safe harbor of home was shrinking in his rear-view mirror at better than 2,000 miles every hour, and now he faced even greater danger because *something* — who knew what — kept shoving his ship this way and that.

The commander let go of the attitude controller, punched open his seat restraint buckle, and floated up to the left-hand window to see if he could determine what was going on out there. It was the oldest pilot's instinct in the world. Even when he was nearly 200,000 miles from home, in a sealed spacecraft surrounded by the killing vacuum of space, what Lovell really needed was a simple walk-around, a chance to make one slow, 360-degree circuit of his ship, to eyeball the exterior, kick the tires, look for damage, sniff for leaks, and then tell the folks on the ground if anything was really wrong and just what had to be done to fix it.

However, he had to settle for a look out the side window, in the hope that whatever problem Odyssey might have would somehow make itself clear. The odds of diagnosing the ship's illness this way were long, but as it turned out, they paid off instantly. As soon as Lovell pressed his nose to the glass, his eye caught a thin, white, gassy cloud surrounding his craft, crystallizing on contact with space, and forming an irridescent halo that extended tenuously for miles in all directions. Lovell drew a breath and began to suspect he might be in deep, deep trouble.

If there's one thing a spacecraft commander doesn't want to see when he looks out his window, it's something venting from his ship. In the same way that airline pilots fear smoke on a wing, space pilots fear venting. Venting can never be dismissed as instrumentation, venting can never be brushed off as ratty data. Venting means that something has breached the integrity of your craft and is slowly, perhaps fatally, bleeding its essence out into space.

Lovell gazed at the growing gas cloud. If the fuel cells hadn't killed his lunar touchdown, this certainly did. In a way, he felt strangely philosophical — risks of the trade, rules of the game, and all that. He knew that his landing on the moon was never a sure thing until the footpads of the LEM had settled into the lunar dust, and now it looked as if they never would. At some point, Lovell understood, he'd mourn this fact, but that time was not now. Now he had to tell Houston — where they were still checking their instrumentation and analyzing their readouts — that the answer did not lie in the data but in a glowing cloud surrounding the ailing ship.

"It looks to me," Lovell told the ground uninflectedly, "that we are venting something." Then, for impact, and perhaps to persuade himself, he repeated: "We are venting something into space."

"Roger," Lousma responded in the mandatory matter-of-factness of the Capcom, "we copy your venting."

"It's a gas of some sort," Lovell said.

"Can you tell us anything about it? Where is it coming from?"

"It's coming out of window one right now, Jack," Lovell answered, offering only as much detail as his limited vantage point provided.

The understated report from the spacecraft tore through the control room like a bullet.

"Crew thinks they're venting something," Lousma said to the loop at large.

"I heard that," Kranz said.

"Copy that, Flight?" Lousma asked, just to be sure.

"Rog," Kranz assured him. "O.K. everybody, let's think of the kinds of things we'd be venting. GNC, you got anything that looks abnormal on your system?"

"Negative, Flight."

"How about you, EECOM? You see anything with the instrumentation you've got that could be venting?"

"That's affirmed, Flight," Liebergot said, thinking, of course, of oxygen tank two. If a tank of gas is suddenly reading empty and a cloud of gas is surrounding the spacecraft, it's a good bet the two are connected, especially if the whole mess had been preceded by a suspicious, ship-shaking bang. "Let me look at the system as far as venting is concerned," Liebergot said to Flight.

"O.K., let's start scanning," Kranz agreed. "I assume you've called in your backup EECOM to see if we can get some more brain power on this thing."

"We got one here."

"Rog."

The change on the loop and in the room was palpable. No one said anything out loud, no one declared anything officially, but the controllers began to recognize that Apollo 13, which had been launched in triumph just over two days earlier, might have just metamorphosed from a brilliant mission of exploration to one of simple survival. As this realization broke across the room, Kranz came on the loop.

"O.K.," he began. "Let's everybody keep cool. Let's make sure we don't do anything that's going to blow our electrical power or cause us to lose fuel cell number two. Let's solve the problem, but let's not make it any worse by guessing."

Lovell, Swigert, and Haise could not hear Kranz's speech, but at the moment they didn't need to be told to keep cool. The moon landing was definitely off, but beyond that, they were probably in no imminent danger. As Kranz had pointed out, fuel cell two was fine. As the crew and controllers knew, oxygen tank one was healthy as well. Not for nothing did NASA design its ships with backup system after backup system. A spacecraft with one cell and one tank of air might not be fit to take you to Fra Mauro, but it was surely fit to take you back to Earth.

Lovell drifted over to the center of the command module to get a read on his remaining oxygen tank and see how much of a margin of error it would provide them. If the engineers had planned it right, the crew would arrive back home with a substantial load of O_2 to spare. The commander glanced at the meter and froze: the quantity needle for tank one was well below full and visibly falling. As Lovell watched, almost entranced, he could see it easing downward in an eerie, slow-motion slide. Lovell was put in mind of the needle on a car's gas gauge. Funny how you can never actually see the thing budge; funny how it always seems frozen in place, but nevertheless makes its way down to empty. This needle, though, was decidedly on the move.

This discovery, horrifying as it was, explained a lot. Whatever it was that had happened to tank two, that event was over. The tank had gone off line or blown its top or cracked a seam or something, but beyond the very fact of its absence, it had ceased to be a factor in the functioning of the ship. Tank one, however, was still in a slow leak. Its contents were obviously streaming into space, and the force of the leak was no doubt what was responsible for the out-of-control motion of the ship. It was nice to know that when the needle finally reached zero, Odyssey's oscillations would at last disappear. The downside, of course, was that so would its ability to sustain the life of the crew.

Lovell knew Houston would have to be alerted. The change in pressure was subtle enough that perhaps the controllers hadn't noticed it yet. The best way — the pilot's instinctive way — was to play it down; keep it casual. Hey you guys, notice anything about that other tank? Lovell nudged Swigert, pointed to the tank one meter, then pointed to his microphone. Swigert nodded.

"Jack," the command module pilot asked quietly, "are you copying O_2 tank one cryo pressure?"

There was a pause. Maybe Lousma looked at Liebergot's monitor, maybe Liebergot told him off the loop. Maybe he even knew already. "That's affirmative," the Capcom said.

As near as Lovell could tell, it would be a while before the ship's endgame would play out. He had no way of calculating the leak rate in the tank, but if the moving needle was any indication, he had a couple hours at least before the 320 pounds of oxygen were gone. When the tank gasped its last, the only air and electricity left on board would come from a trio of compact batteries and a single, small oxygen tank. These were intended to be used at the very end of the flight, when the command module would be separated from the service module and would still need a few bursts of power and a few puffs of air to see it through reentry. The little tank and the batteries could run for just a couple of hours. Combining this with what was left in the hissing oxygen tank, Odyssey alone could keep the crew alive until sometime between midnight and 3 A.M. Houston time. It was now a little after 10 P.M.

But Odyssey wasn't alone. Attached to its nose was the hale and hearty, fat and fueled Aquarius, an Aquarius with no leaks, no gas clouds. An Aquarius that could hold and sustain two men comfortably, and in a pinch, three men with some jostling. No matter what happened to Odyssey, Aquarius could protect the crew. For a little while, anyway. From this point in space, Lovell knew, a return to Earth would take about one hundred hours. The LEM had enough air and power only for the forty-five or so hours it would have taken to descend to the surface of the moon, stay there for a day and a half, and fly back up for a rendezvous with Odyssey. And that air and power would last forty-five hours only if there were two men aboard;

put another passenger inside and you cut that time down considerably. Water on the lander was similarly limited.

But Lovell realized that for the moment Aquarius might offer the only option. He looked across the cabin at Fred Haise, his lunar module pilot. Of the three of them, it was Haise who knew the LEM best, who had trained in it the longest, who would be able to coax the most out of its limited resources.

"If we're going to get home," Lovell said to his crewman, "we're going to have to use Aquarius."

Back on the ground, Liebergot had discovered the falling pressure in tank one at about the same time Lovell did. Unlike the commander of the mission, the EECOM, sitting at the safe remove of a control room in Houston, was not yet prepared to give up on his spacecraft, but he did not hold out great hopes for it either. Liebergot turned to his right, where Bob Heselmeyer, the environmental control officer for the LEM, sat. At this moment, the EECOM and his lunar module counterpart could not have been in more different worlds. They were both working the same mission, both struggling with the same crisis, yet Liebergot was looking out from the abyss of a console full of blinking lights and sickly data, while Heselmeyer was monitoring a slumbering Aquarius beaming home not a single worrisome reading.

Liebergot glanced almost enviously at Heselmeyer's perfect little screen with all its perfect little numbers and then looked grimly back at his own console. On either side of the monitor were handles that maintenance technicians used to pull the screen out for repairs and adjustments. Liebergot all at once discovered that for several minutes he had been clutching the handles in a near death grip. He released the handles and shook his arms to restore their circulation — but not before noticing that the backs of both his hands had turned a cold, bloodless white.

5

Monday, April 13, 10:40 P.M. eastern time

WALLY SCHIRRA had been looking forward to a Cutty and water all evening. For the past four hours, he had been grinning and glad-handing, nursing a flat soda while the people around him got pleasantly plastered. Now it was his chance to tie one on — at least a small one — too.

Schirra didn't especially mind being the only sober soul at a black-tie bash. Or if he did mind, he had ceased to notice. This was a work night for Wally, another in a series of a million or so evenings in the barrel, and as he and the other astronauts had long since learned, drinking in the barrel was just like drinking on any other job. You just didn't do it — too big a risk of a gaffe that would find its way into a newspaper or onto the airwaves or up to the office of the NASA administrator. When the evening was over he could do what he wanted, but as long as he was here, he was on duty.

Schirra was working an American Petroleum Club function in New York. He was not only a featured party guest but also the featured speaker. Ordinarily, the ex-astronaut would not dash off to New York for just any function; but he rather liked this group and enjoyed attending their affairs. Besides, he had to be in the city anyway. Since retiring from the Agency in early 1969, Schirra had signed on with CBS to help Walter Cronkite cover all of the Apollo moon landings. His first assignment had been Apollo 11, in July 1969, then Apollo 12 in November. Just two days ago, he and Cronkite had gone on the air to cover the launch of Apollo 13. Tomorrow, Jim Lovell, Jack Swigert, and Fred Haise would be preparing for their lunar landing, and Schirra would be on hand to help broadcast that too.

But that was tomorrow. Right now, Schirra was wrapping up his duties at the Petroleum Club and making his way across town to Toots Shor's on West 52nd Street. Wally knew Toots well, and though it was late, he knew that the convivial tavern owner would probably have a pretty full house. Schirra arrived at the restaurant, made his way to the bar, and ordered his Cutty and water. As expected, the place was full. And as expected, just as the drink showed up, Toots did too, working his way across the room in what seemed like a hurry. Wally smiled a greeting and Toots, curiously, did not smile back.

"Wally, don't touch that drink," Shor said when he reached him.

"What's wrong, Toots?"

"We just got a call — all hell broke loose down in Houston."

"What happened?"

"I really don't know, but they're having some kind of problem. A *big* problem, Wally. There's a car from CBS out front for you. Cronkite's going on the air, and you're supposed to go on with him."

Schirra rushed out to the door and saw the car waiting for him. He jumped in the back seat, announced his name, and with barely a nod the driver took off across town. When the car reached CBS, Schirra raced to the studio and found Cronkite about to go on the air.

The anchorman did not look good. He called Schirra over and thrust a sheaf of wire-service copy at him. Schirra scanned the text hurriedly, and with each sentence his heart sank. This was bad. This was worse than bad. This was . . . unheard of. He had a thousand questions, but there wasn't time to ask.

"We're going on in a minute," Cronkite told him, "but you can't go on like that."

Schirra looked down and realized that he was still wearing the black tie, night-in-the-barrel outfit he had begun the evening in. Cronkite sent a runner off to his dressing room, who returned moments later with a tweedy journalist's jacket, complete with elbow patches, and a scruffy tie. Schirra stood still for a few quick dabs of makeup and then put Cronkite's clothes on over his starched, frilly tuxedo shirt. Through the shirt, the tweed itched something awful, but there was nothing Schirra could do about it now.

The stage hand waved Cronkite and Schirra over to the set, and the journalist and the astronaut took their seats. Seconds later, the red light on the camera flashed on, and television screens across the country were filled with the image of a steady Walter Cronkite and a slightly dazed Wally Schirra. Cronkite began reading his copy, and it was only then, as America learned the full scope of the crisis unfolding aboard Apollo 13, that Schirra learned it too. Within seconds, he had forgotten all about the maddening itch from the borrowed jacket.

Across town, the ice in Wally's abandoned Cutty had not even melted yet.

The ride from the Manned Spacecraft Center to the Houston suburb of Timber Cove took about fifteen minutes, but on a good night with no traffic, Marilyn Lovell could make it in eleven or twelve. Tonight was such a night, and Marilyn knew she'd be home in time to tuck her youngest child, four-year-old Jeffrey, into bed and get Susan and Barbara into the house and off to sleep at a respectable hour. Marilyn, like most NASA wives, had driven this route a thousand or so times before, but this evening she would have preferred not to make the trip.

Things were a lot easier the first three times her husband went into space, when NASA still had the television networks in a hammerlock and routinely got all the TV time it wanted. Marilyn couldn't help but feel cheated by how much had changed since then. At least when Apollo 12 had gone up five months ago, Jane Conrad had gotten to watch *some* of Pete's broadcasts between the Earth and the moon without having to run all the way over to the Space Center to do it. For that flight, the NASA bosses had still harbored hopes of retaining the huge TV audiences they had enjoyed during Apollo 11, and even tried to sweeten the public relations pot by scrapping the crude black-and-white camera Neil and Buzz had used on the surface of the moon and replacing it with a more sophisticated color job. The idea seemed like a good one, but only until the moment Al Bean and Pete actually set foot on the lunar surface and accidentally pointed their wonderful new camera toward the sun, frying its single eye like an egg and forcing the cancellation of all broadcasts for the rest of the trip. From then on, it was all downhill as far as NASA and the networks were concerned, and though Agency technicians had equipped the cameras on Apollo 13 with a stronger filter, assuring uninterrupted broadcasts back to Earth, the TV stations had essentially shrugged at the offer. Thanks to NASA, Marilyn would be able to see as much as she wanted of her husband on this trip, but thanks to the networks, she would not be able to do so from her own family room.

Marilyn pulled her car into the driveway on Lazywood Lane, cut the ignition, and glanced at her watch. It was too late to call her fourth child, fifteen-year old Jay, at St. John's Military Academy in Wisconsin, to tell him that the broadcast had gone well and that his father looked fine. Jay knew that had anything not gone well, he'd be alerted right away, but Marilyn still liked to tell him herself. Now it would have to keep until tomorrow.

Shooing Susan and Barbara toward the house, Marilyn hurried up the walk. Elsa Johnson, a friend from the Cape Canaveral area, was staying with the earthbound Lovells during the week of the lunar mission and had volunteered to sit with Jeffrey tonight; Marilyn was anxious to relieve her. Astronaut wives were profoundly grateful for friendship and company while their husbands were off on their outlandish business trips, and Marilyn did not want to impose on Elsa's generosity.

"How was Jim?" Elsa asked as soon as Marilyn came through the door, Susan and Barbara racing in ahead of her.

"Terrific," Marilyn said. "Happy and relaxed. They look like they're having fun up there. How is Jeffrey?"

"Asleep already. He nodded right off."

Marilyn hung her sweater in the closet, walked into the family room, and jumped slightly when she noticed a man sitting on her couch reading a magazine. Then she laughed at herself and waved hello. The man was Bob McMurrey, a NASA protocol officer. The wife and children of each crew member were routinely allocated at least one protocol man, whose job it was to spend the time from liftoff to splashdown living with the family, protecting them from the press and onlookers crowding their sidewalk, and explaining any unexpected developments in the flight.

Ordinarily, the job could be demanding, and McMurrey, who had been assigned to the Lovells during Apollo 8, was used to putting in long hours. For Apollo 13, however, there were no onlookers or reporters outside and, so far, no unexpected developments. McMurrey had spent the past few days much as he was spending this evening — sitting on the sofa, sipping coffee, and reading yet another magazine from the large stack next to him. On the floor nearby, the Lovells' blue merle collie, Christi, completed the domestic scene: he lay dozing at McMurrey's feet, as if accepting this stand-in paterfamilias while the real thing was away.

Marilyn had hoped for a little additional company tonight, and earlier in the day had invited her next door neighbor, Betty Benware, over for a drink; but Betty had begged off. Her husband, Bob, was the head of the Philco-Ford group that maintained the consoles and other equipment in Mission Control, and the couple had just spent two days entertaining his bosses, who had come down to see how the operation ran during an actual flight.

Apart from the protocol man, the only other direct connection Marilyn had to the Space Center during the long days of the mission was a squawk box NASA had hooked up in her bedroom three days earlier. The box served as a listen-only intercom that allowed an astronaut's wife to monitor the communications between her husband and the Capcom around the clock. Better than 90 percent of what families could ever hope to hear on this party line was incomprehensible — a lot of numbers and vectors that even the flight controllers themselves occasionally found tedious. But Marilyn and the other wives were listening less for the words than they were for the Tone — the Trouble Tone — and for this the box could be indispensable. At this time of night, with the crew already on their sleep shift, the box would be carrying only static. And with McMurrey settled comfortably in the family room with nothing to report himself, Marilyn figured it was safe to put any thoughts about the mission out of her mind and head toward the kitchen to make coffee with Elsa. Before she could get there, the front door opened and Pete and Jane Conrad walked in.

"Did you see him?" Jane asked Marilyn.

"Saw them all," Marilyn said. "They look great. Everything seems to be going exactly according to schedule."

"Jim runs a tight ship," Conrad said.

"I just wish they'd've put the broadcast on TV," Marilyn said. "Let people see what a good job they're doing."

"They'll put a minute of it on the late news," said Jane, "if only to remind everybody that they're up there."

Marilyn was about to show Pete and Jane into the kitchen for some coffee when the phone rang. McMurrey started to lift himself from the sofa to answer it, but Marilyn, who was closer, waved him off with a smile and picked up the receiver herself.

"Marilyn?" the voice on the line said tentatively. "It's Jerry Ham-

110

mack. I'm calling from over at the Center." Jerry Hammack and his wife, Adeline, lived across the street and were close friends of the Lovells. Hammack himself was head of the NASA recovery team that was responsible for plucking the Apollo command modules out of the ocean when they splashed down at the end of a mission.

"Jerry," Marilyn said, surprised, "what are you doing working so late?"

"I just wanted to let you know that you don't have anything to worry about. The Russians, the Japanese, and a lot of other countries have already offered to help in the recovery. We can bring them down in just about any ocean and have them on a carrier in no time."

"Jerry, *what* are you talking about? Have you been out drinking?"

"Hasn't anyone told you?"

"Told me what?"

"About the problem . . ."

In any small company town, news of a problem at the factory or the plant travels fast. In the suburbs of Houston, where the business was space, the factory was Mission Control, and the likelihood of a problem occurring was always unsettlingly high, it travels even faster. Nearby, in the Borman home, the phone rang at about the same time Marilyn Lovell's did. The former Apollo 8 commander listened to the news from the Space Center, hung up the receiver, and turned to Susan.

"Lovell's in trouble," Borman said. "It does *not* look good. I'm going over to NASA. You go over to the house."

Susan picked up the receiver Borman had just dropped and phoned the nearby McCullough home, where Marilyn's friend Carmie lived.

"Frank says there's a problem with the moon flight," she said. "Meet me at Marilyn's in five minutes."

At the house next to the Lovells', the Benwares got their own call from the Space Center.

"You'd better go next door," Bob said to his wife, Betty, after listening to the news. "I'd better go to work."

In the Lovell home itself, Marilyn, fresh off her breezy twelve-minute ride from the Space Center, was aware of none of this.

"What problem?" she now said to Hammack, her voice noticeably rising. "Jerry, I just saw Jim on TV. Everything was fine!" In the kitchen, Elsa and Jane turned.

"Uh, well, everything isn't fine. A few things have gone wrong."

"What things?!"

"Well . . . mostly it's a power problem," Hammack began to hedge.

"A fuel cell problem, actually. They're running out of electricity and, well, it looks like they're not going to be able to go through with the landing." In the background, Marilyn heard the second phone line ringing in the study and saw McMurrey run to answer it.

"Oh, Jerry, that's terrible," she said. "Jim's worked so hard for this. He's going to be so disappointed." She caught Jane's eye and Jane mouthed, "What happened?" Marilyn held up her hand in a wait-a-second gesture.

"Yeah, I'm sure he will be," Hammack said. "But in any case, I didn't want you to worry. We're doing everything we can over here."

Marilyn hung up and turned to Jane. "This is terrible," she said. "Something went wrong with a fuel cell and they're canceling the landing. That was the only reason Jim went back out there, and now he's just going to have to turn around and come home."

"Marilyn, I'm so sorry," Jane said. The two friends exchanged a sisterly hug, and over Jane's shoulder, Marilyn saw Conrad and McMurrey standing by the study, deep in whispered conversation. Conrad looked pale and distracted; his eyes were wide.

"Marilyn," Conrad said hoarsely, "where's the squawk box?"

"Why do you need the squawk box?" Marilyn asked.

"No one's talked to you yet?"

"Yes, I just talked to Jerry Hammack. He told me about the fuel cell problem."

"Marilyn," Conrad said quietly, "this is more than a fuel cell problem."

Conrad steered Marilyn to a chair, sat her down, and explained everything the protocol man had just told him: the disappearance of the oxygen in tank two, the problem with tank one, the venting, the gyrating, the plummeting power, the thinning air, and, worst, the mysterious bang that had started it all. Marilyn listened and felt suddenly sick. This wasn't what was supposed to happen. Before Jim went back out there, this was precisely what he had promised would *never* happen.

Marilyn pulled away from Conrad, ran to the television, and flipped it on. Instinctively, she switched it not to CBS, where family friend Wally Schirra would be working, but to ABC, where Jules Bergman, the giant of science correspondents, could be found. Almost immediately, she was sorry she had. Bergman, she discovered, was talking about the same oxygen tanks Conrad had just mentioned, the same spacecraft gyrations, the same mysterious bang. But unlike Conrad,

Bergman was talking about one other thing too: the odds. As Marilyn listened, Bergman told his audience that while nobody could predict these things precisely, there appeared to be no better than a 10 percent chance that the crew of Apollo 13 would make it home alive.

Marilyn turned from the screen and covered her face. The number the newscaster cited was bad enough, but even if he had been talking about better odds and happier outcomes, his report would still have been chilling. Though no one else in the room recognized it, Marilyn instantly noticed that Bergman, like Conrad and Hammack before him, was using the Tone.

Throughout Houston, other people who were neither inside Mission Control nor family members of the imperiled pilots were getting the news in other ways. On the roof of Building 16A at the Manned Spacecraft Center, engineer Andy Saulietes was camped out with three of his colleagues, fussing with an array of costly stargazing equipment. Tonight, like the past three nights, Saulietes and the others had been pointing a powerful fourteen-inch telescope approximately moonward, looking at the images they picked up on a nearby black-and-white television monitor. For the most part, what they were picking up was a twinkling, rapidly shrinking object that, their instruments told them, was now roughly 200,000 miles from the Earth. To the lay eye, the object would look completely unremarkable, but Saulietes and the others were acutely interested in tracking its progress.

What they were seeing was the cold, spent, tumbling third stage of Apollo 13's Saturn 5 booster, speeding away from Earth at approximately 2,000 miles per hour. The single-engine system that had made up the top third of the rocket had pushed the Odyssey and Aquarius spaceships out of Earth orbit two days earlier and was now heading toward a collision with the moon. Somewhere in a nearby trajectory, the command and lunar modules were also speeding along, but the tiny twin ships had long since passed beyond the limits of Saulietes's telescope. Indeed, as Saulietes and his colleagues squinted out into space, they could see that the third stage had itself almost vanished from the screen.

The men on the roof had an air-to-ground communications monitor with them so they could track the progress of the flight and listen for key mission events that could affect their observations. The event they were waiting for principally was a water or urine dump from the spacecraft Odyssey. When the spray of waste fluid vented from the side of the ship, it would crystallize on contact with space, forming an icy cloud of starry flecks that Wally Schirra, in one of his singular linguistic strokes, had dubbed the constellation Urion. If the cloud tonight was large enough and caught the sunlight just right, Saulietes believed he might be able to spot the spacecraft.

At about 9:35 Saulietes, focusing closely on the image coming in through his telescope and only half listening to the air-to-ground chatter, thought he heard Jack Swigert say something about a problem; moments later, he thought he heard Jim Lovell repeat the call. Saulietes didn't pay these transmissions much mind. He'd tracked Apollos 8, 10, 11, and 12 on their way to the moon, and lunar ships were forever reporting one little malfunction or another that required Houston's assistance. What did seize his attention a few minutes later, however, was the image on his TV monitor.

In the middle of the screen, a sudden and unexpected pinpoint of light appeared and steadily grew. It was right where the spacecraft ought to be, but it was way too big to be a water or urine dump, and nothing else Saulietes had seen on the previous four lunar flights could account for it. It was almost as if some huge, gassy halo had surrounded the ship, spreading out slowly for twenty-five or thirty miles. That would have to be a *lot* of urine. Saulietes reached over to his monitor and pressed the Record switch. The system would copy three or four frames of the current image, allowing him to call them up and scrutinize them later. It was unlikely that the pictures would tell Saulietes much of anything; probably there was some glitch in his telescope or his monitor causing the curious halo. If so, he wanted to get to the bottom of it quickly, before continuing to track the rest of this otherwise nominal flight.

A few miles away, in a suburban settlement not far from Timber Cove, Chris Kraft, the deputy director of the Spacecraft Center, had no greater reason than Saulietes to worry about the progress of the moon mission. Since surrendering the flight director's chair at the beginning of the Apollo program, Kraft had been able to take a less frantic approach to his work, and he didn't mind the change a bit. Having paid his dues in the high-stress trenches of Mission Control through six Mercury flights and ten Gemini flights, Kraft had been more than content, after Jim Lovell and Buzz Aldrin's Gemini 12, to turn things over to Gene Kranz and the rest of the team of flight directors who worked under him.

At the moment, Kraft was taking a shower. It was a little before 10 P.M., and the last he'd heard, things were nominal at the nearby Space Center and in the Apollo spacecraft. The crew would be turning in for the night right about now, and Kraft intended to do the same. No need to keep graveyard-shift hours when there was Gene Kranz or someone else sitting at the flight director's console. Through the bathroom door, Kraft thought he heard the phone ring once, then twice, then stop as his wife picked it up.

"Betty Ann?" the voice on the end of the line said. "It's Gene Kranz. I need to talk to Chris." The flight director's console, Betty Ann Kraft knew, had an outside phone line as well as an in-house line, and while it was not routine for the man in charge of a mission to be making calls outside the Space Center, it was not unprecedented either. Betty Ann, who had heard it all and seen it all in Kraft's years at the Agency, was unmoved to hear from Kranz.

"Gene, Chris is in the shower right now. Can I have him call you back?"

"No, I don't think you can. I need you to get him out," Kranz said. "Right away."

Betty Ann hurried to the bathroom and brought the dripping Kraft to the phone.

"Chris," Kranz said, "you'd better get over here now. We've got a hell of a problem. We've lost oxygen pressure, we've lost a bus, we're losing fuel cells. It seems there's been an explosion."

Kraft, who had known Kranz for years, did not know his protégé and successor to declare a crisis when there was no crisis or to sound urgent when there was no reason to sound urgent. More important, he sure as hell didn't know him to call for backup counsel when there was no need for backup counsel — but now he was making that call.

"Sit tight," Kraft said. "I'm on my way."

The former flight director who'd grown tired of the Mission Control hot seat threw his clothes on, raced out of his house still half wet, and jumped into his car. He made the ten-mile ride to the Space Center in less than fifteen minutes, sometimes pushing 60 on the darkened roads of the quiet suburb that had just begun drifting off to sleep.

During a crisis in any space flight, particularly a flight as complex as a lunar mission, the men in the spacecraft and the men on the ground operate in a sort of hierarchy of denial. When a ship went suddenly sour, it was the pilots who were at the center of the problem; it was they who heard the bang or witnessed the venting or saw the instrument panel readings go into the tank, and so it was they who usually had the most pessimistic view of the crisis. Though no pilot was ever anxious to abandon his ship or abort his mission, no pilot wanted to push his craft beyond where his experience and his senses told him it was able to go. Next in line were the individual console controllers in Houston. For the most part, none of these men had ever been in a spacecraft themselves, and from the beginning of their careers had only the numbers on their screens to tell them what was wrong with the ship in their charge. Unlike the men inside the spacecraft, the controllers knew that their lives, health, and immediate futures were not intimately bound up with the life, health, and immediate future of the spacecraft, and while this sometimes led them to have more faith in a sick ship than the ship truly deserved, it also provided them with a problem-solving detachment that the astronauts could not hope to have. Furthest removed from the problem, but ultimately responsible for getting it solved, was the flight director.

In addition to all of the written rules that governed a mission, the flight director operated under an unwritten rule known as down-moding. Before a mission was officially aborted, the doctrine of down-moding required the flight director to preserve as much of it as he could without endangering the lives of the astronauts. If a crew couldn't land on the moon, could they at least orbit it? If they couldn't orbit, could they at least whip around the far side to take some hurried sightings? Getting out as far as the lunar neighborhood was a complicated, expensive job, and if the primary goals of the project could not be met, it was up to the man in charge to have some secondary and tertiary goals in mind. Only when the last options for a down-moded mission were exhausted would the flight director give up the exploratory ghost and bring his crew home.

During the fifty-seventh hour of the flight of Apollo 13, as the Marilyn Lovells and Mary Haises were receiving their calls from NASA, as the Chris Krafts were racing toward the Space Center, as the Jules Bergmans were taking to the airwaves, NASA's hierarchy of denial was up and running. Gene Kranz, on his feet behind his console in Mission Control, paced and smoked as he always did at critical moments, working his communications loop like a one-switchboard operator in a town of ten thousand. At other consoles, controllers scanned their screens and analyzed their data, hoping to find some solution to the ills that were afflicting their particular part of the spacecraft. And in the spacecraft itself, the three men at the heart of it all were sweating out the crisis with a first-person investment that the men on the ground could only begin to fathom.

For Lovell, Swigert, and Haise, what was causing the most sweat, as the crisis approached the sixty-minute mark, was the continuous heaving and pitching of their spacecraft, caused by the venting from O_2 tank one. The unwanted movements were known, in the flat vernacular of the pilots, as rates, and as the controllers fought to dope out the cause of Odyssey's myriad problems and cobble together some makeshift solution, Lovell continued trying to bring the gyrations under control.

"I can't get this thing steadied," the commander growled to himself as he manipulated the thrusters, working his gun-handle controller this way and that.

"We're still getting a hell of a rate, aren't we?" Swigert said from his center seat.

"Blame that," Lovell said, indicating the glowing gas cloud outside his window with a tilt of his head.

"Watch the ball," Swigert warned, eyeing an instrument set into the dashboard. "Don't go into gimbal lock."

The instrument Swigert was monitoring so uneasily, the flight director attitude indicator — known to fliers as the 8 ball — was a small sphere crisscrossed with angle markings and nautical-like lines. The gyros that controlled it were the heart of the ship's navigation system. For a crew to find their way in space, they needed to know at all times the attitude of their spacecraft relative to any point in the celestial sky. To provide this, the spacecraft came equipped with a guidance system containing a stationary component, known as a stable element, that was inertially fixed in space relative to the stars. Arrayed around it was a series of gimbals that moved with every motion of the ship. The guidance system kept the onboard computer constantly updated on the changing attitude of the ship relative to the stable element, and hence to the stars, while the 8 ball provided the same information to the pilots.

For a vehicle that needed to refine its trajectory by fractions of a degree on the quarter-million-mile coast to the moon, the system worked exceptionally well, with one small exception. If the spacecraft inadvertently moved to a full right or full left yaw position, the gimbals had a nasty habit of lining up with one another and locking in that configuration, instantly wiping out whatever knowledge the computer had of the ship's attitude. A spacecraft with no vestibular system was no good to anyone, least of all to the pilots who were relying on it to take them home, and the 8 ball was thus designed to keep the crew alert to how close they might be coming to gimbal lock. In addition to all of the lines and angles inscribed on the ball, two nickel-sized red disks were also painted on it, 180 degrees apart. When a red disk began to float into the window, it meant the gimbals were close to lining up. When the disk moved to the center of the window, it meant the gimbals were locked, the attitude reference was lost, and so, at least in terms of navigation, was the ship.

Now, as Swigert, the spacecraft's navigator, gazed at the glass, a flash of red floated in from the right. "Red's coming into view," he warned Lovell again.

"I see it," Lovell said, his eyes darting to the instrument panel. "And I wish I didn't." He yawed the ship hard aport and the red dot disappeared.

In the control room, the guidance console's instruments picked up the same dangerous range of motion as Lovell's attitude indicator, and the GUIDO came on the line to warn Kranz.

"Flight, Guidance," he called into the loop.

"Go, Guidance," Kranz answered.

"He's getting close to gimbal lock there."

"Roger. Capcom, recommend he bring up C-3, C-4, B-3, B-4, C-1, and C-2 thrusters, and advise he's getting close to gimbal lock."

"Roger," Lousma answered, then switched to his air-to-ground loop and relayed the instructions to the spacecraft.

Lovell heard the message and nodded to Swigert, but did not acknowledge Lousma. While the commander kept an eye on the attitude indicator and glanced out the window, the command module pilot began reconfiguring the thrusters as Lousma had instructed.

"13, Houston. Do you read?" Lousma asked when he got no response.

On the right side of the cockpit, Haise, whose command module responsibilities were primarily the care and maintenance of the electrical systems, had returned to his assigned seat, where he could more closely monitor the spacecraft's worsening power problems. "Yes," the LEM pilot answered the ground, glancing across the cockpit at his crewmates. "We got it."

"Affirm," Lovell added tersely.

As Lovell and Swigert struggled with the attitude of the ship, Kranz continued to pace behind his console, juggling a hundred other problems that were competing for his attention. On his flight director's loop, the INCO called to report that he was having a nightmare of a time trying to keep the antennas aligned on the lurching, power-poor ship; the guidance and navigation officer, or GNC, called to say that he was coming dangerously close to developing a thermal imbalance, as one part of the spacecraft stayed too long in the direct glare of the sun; the EECOM reported that the welter of power and oxygen problems that had started this whole mess had not stabilized themselves, and by all signs were getting worse.

Of all of the status updates, it was these EECOM reports that claimed most of Kranz's attention. According to Sy Liebergot's forlorn bulletins, oxygen tank two, which had mysteriously vanished at 55 hours and 54 minutes into the mission, indeed appeared to be gone for good; tank one, which had begun the evening at an in-the-pink pressure of 860 pounds per square inch, was now down to close to half that and was losing pressure at better than a pound per minute; fuel cells one and three were all but gone, fuel cell two was dying fast, and as the remaining fuel cell faded, the remaining bus — main bus A — faded with it. As the spacecraft continued to operate with all its power-gobbling hardware up and running, the whole precarious system threatened to collapse under the load.

At the EECOM console and in the backroom, Liebergot and his support team of George Bliss, Dick Brown, and Larry Sheaks knew that their options were extremely limited. To prevent the electrical system from shutting itself down completely, the EECOM could always connect the spacecraft's reentry batteries to the two dead or dying buses. The batteries were prodigious energy producers and would bring the craft back to full power almost instantly. The hitch was, they would last only a couple of hours. If Liebergot put the batteries to use now, Odyssey would essentially be eating its seed corn, gobbling the little power it would need for the plunge back into Earth's atmosphere — provided it ever made it back to Earth.

If he didn't make this move, however, the problem could become even worse. When the remaining oxygen tank finally started to run dry, the ship would automatically begin making up for the loss of air and power by sipping at will from the tiny tank of O_2 in the command module that was also used for reentry. The official name for that tank was the surge tank, and its job during the hours and days of the flight leading up to reentry was to compensate for fluctuations in the main oxygen supply, taking up excess gas if the pressure in the twin tanks climbed too high or providing a puff or two of its own O_2 if the pressure dropped too low. At the end of the mission, the oxygen in the surge tank would be topped off by the surplus in the presumably healthy main tanks, providing the crew with air for reentry. But with tank two dead and tank one dying, Odyssey would eventually bleed its surge tank dry. The only answer was to connect the batteries briefly to prop up the dying bus and then begin reducing power use as fast and far as possible. This at least might ease the demands on the remaining fuel cell and delay the death of the oxygen and electrical system until a better answer could be found. At the same time the EECOM was coming to this realization, his backroom was reaching it too.

"Sy," Dick Brown said into Liebergot's headset, "I think we're going to have to throw a battery on bus B and bus A until we psych them out."

"I agree," Liebergot said. "Let's do it."

"Also," Brown added, "I think we ought to start powering down."

"Yeah," Liebergot said, then dialed up the flight director's loop. "Flight," he said a little warily.

"Go ahead," Kranz said.

"I think the best thing we can do right now is start a power-down."

"O.K.," Kranz said, "you want to power down, look at the telemetry and all that good stuff, and then come back up?"

Liebergot smiled slightly to himself. Come back up? Kranz wanted to know if this ship could come back up? No, he wanted to tell him, the way it looked now, this ship was terminal and it was never going to come back up. But Kranz's and Liebergot's jobs precluded, at least for now, a discussion like that. It was Kranz's responsibility to down-mode the mission carefully, and it was Liebergot's job to provide him with the best possible spacecraft to do it with.

"That's right," Liebergot said cooperatively.

"How much do you want to power down?"

"A total of 10 amps, Flight."

"A total of 10 amps," Kranz repeated, and whistled softly. The entire spacecraft was pulling only about 50 amps; Liebergot was suggesting pulling the plug on 20 percent of its systems. Kranz punched up his

Capcom loop. "Capcom, we recommend emergency power-down checklist one through five. We want to power down a delta of 10 amps from where we are now."

"Roger, Flight," Lousma said, and opened up the air-to-ground loop. "13, this is Houston. We'd like you to go to your checklist, the pink pages one through five. Do a power-down until we get to a delta of 10 amps."

Lovell looked at Swigert and Haise and smiled a tight smile. The commander and his crew knew that this mission, at least the way it was originally planned, was over. However, they also knew that Houston would have to reach that conclusion for themselves. It sometimes took Mission Control a while to catch up with the pilots on these things, but the power-down order was the first suggestion that the ground was coming around.

Lovell nodded to Swigert, and the command module pilot pushed off toward the lower equipment bay to retrieve the emergency checklist. Mission protocols and flight plans were typed on fireproof sheets of paper and arranged, steno-book fashion, between two cardboard covers held together by two metal rings. Books containing noncritical procedures were stowed in storage cabinets around the ship; those with more vital procedures were secured by Velcro strips to handy spots on the walls of the spacecraft. The power-down checklist was contained in one such book, and Swigert found it in the equipment bay, tore it off its fabric anchor, and carried it back up to the couches. With Haise looking over his shoulder, the command module pilot began running through the steps that would put his spacecraft partially to sleep.

"13, Houston. Did you copy our power-down request?" Lousma asked when he heard no response from Swigert or Lovell.

"Roger, Jack, we're doing it right now," Swigert said.

"It's the pink pages, emergency pages one through five," Lousma repeated, wanting to be certain the crew was certain.

"O.K.," Swigert reassured him.

"Power down until you get an amperage of 10 amps less than you have now."

"O.K.," Swigert said again; this time he said it more firmly.

As Jack Swigert began shutting down the first of the dozens of systems that the pink emergency pages instructed him to, Chris Kraft pulled into the parking lot outside Building 30, the Mission Control building, and sprinted toward the main entrance and the lobby elevator. As soon as he arrived at the third floor and entered the high-ceilinged auditorium where he had overseen so many flights over so many years, he could tell the kind of trouble this mission was in. There was a small knot of men surrounding Jack Lousma's Capcom console and larger clusters around the EECOM console — where it appeared from a distance that Seymour Liebergot was on station tonight — and Kranz's flight director's console.

Kraft approached Kranz's station with an outsider's deference that did not come easily to him. As Kranz's former mentor and now his boss, Kraft knew what his job would be here tonight — and basically it would be anything Kranz said it was. The rules for conducting a manned space flight were explicit, and as any controller knew, perhaps the most explicit, least negotiable rule of all was that the flight director was the unquestioned authority of all he surveyed. Kraft and Kranz themselves had written the rule back in 1959 when Kraft was flight director and Kranz was cutting his teeth at the Agency. The precise wording of the rule was sweeping: "The flight director can do anything he feels is necessary for the safety of the crew and the conduct of the flight regardless of the mission rules." Kraft had exercised that authority willingly and well throughout sixteen missions, and at the beginning of the Apollo program, when he passed the lead flight director's baton on to Kranz, he passed the power on with it.

Kraft made his way down the tiered, theater-like incline of the control room to Kranz's third-row console, where the flight director looked up and nodded gratefully. Kraft then moved a few feet away, plugged his headset into his own console, and dialed up the air-to-ground loop and the flight director's loop to see what he could learn. As soon as he did, he was brought up short. With the exception of the Gemini 8 abort five years ago and the Apollo 1 fire three years ago, Kraft had never heard a flight director keeping so many balls in the air at once.

"TELMU and CONTROL, from Flight," Kranz called to the environmental and navigation officers for the LEM.

"Go, Flight," Bob Heselmeyer, the TELMU, answered from a console near Liebergot's.

"Will you take a look at the pre-launch data and see if there's anything that may have started the venting?"

"Roger, Flight."

"And I want a report on that in the next fifteen minutes — quick look–type stuff."

"Roger."

"Network, from Flight," Kranz called to the technicians who looked after the computers in the Real Time Computer Complex, a ground-floor facility in the Space Center compound that housed the fastest data processors NASA owned.

"Go, Flight."

"Bring me up another computer in the RTCC, will you?"

"We've already got one machine on in the RTCC, and we've got dual CPs downstairs."

"O.K., I want another machine up in the RTCC, and I want a couple of guys capable of running delogs down there."

"Roger that."

"GNC, Flight," Kranz called.

"Go, Flight," the guidance and navigation officer said.

"Give me a gross amount of thruster propellants consumed so far."

"Roger, Flight. We're still below the limits."

"EECOM, from Flight."

"Go, Flight."

"What does the status of your buses tell you now?"

"It tells me . . . uh . . . give me about two more minutes, Flight."

"O.K. Take your time."

Listening in on the flight director's loop, Kraft was not surprised to hear Liebergot having trouble answering a routine inquiry from Kranz. Even the greenest control room staffer could see that this emergency was essentially an EECOM's emergency, and answers could not come quickly from that console tonight.

Just what was occupying Liebergot and his backroom was not immediately evident on the flight director's loop. On the EECOM channel, though, things were much clearer — and much more disturbing. The emergency power-down and the battery hookup, while relatively extreme measures to hold the disintegrating power system together, were apparently not working. On Sy Liebergot's and the backroom's screens, the readouts now revealed that the pressure in tank one was down to a mere 318 pounds per square inch, and even this remaining supply was lower than it seemed to be. The oxygen tanks needed a pressure of at least 100 p.s.i. to feed the gas through the lines and out to the remaining fuel cell. Once another 218 pounds bled off, the precious bit of gas left in the tank would be useless. Worse, the steadily falling pressure in the tank had caused the predicted cannibalization of the surge tank to begin. The ship, like an organism suffering from an autoimmune disease, had now started feeding on itself.

"Hey, Sy," Bliss said on the backroom loop, "you probably want to isolate the surge tank and just use as much cryo as you can. We've gotta save that surge."

"Is the tank going down?" Liebergot asked.

"Roger that," Bliss said emphatically.

Liebergot groaned. "Flight, EECOM," he said.

"Go, EECOM."

"Let's have 'em isolate the surge tank and save it. We'll use the cryo as much as we can."

"Uh, say that again." Kranz said skeptically.

"Let's isolate the surge tank in the command module."

"Why that?" Kranz snapped, not yet accepting the near-certain death of the ship. "I don't understand that, Sy."

"I want to use the cryos as much as possible."

"That would seem to be the opposite of what you want to do if you want to keep the fuel cells going."

"The fuel cells are fed off the tanks in the service module, Flight. The surge tank is in the command module. We want to save the surge tank, which we'll need for entry."

"O.K.," Kranz said, his voice falling. "I'm with you, I'm with you." He turned back to the loop and spoke resignedly. "Capcom, let's isolate the surge tank."

"13, Houston," Lousma called. "We'd like you to isolate your O$_2$ surge tank."

Swigert acknowledged, threw the surge tank switch on the reentry panel, and then, appreciating the momentousness of what he had done, called back to Earth for a confirmation that he had done it right.

"Is the surge tank off now, Jack?" Swigert asked.

"That's affirmative," Lousma answered.

As soon as this exchange was completed, the men on the EECOM loop, who had been listening in, settled into gloom.

"George, it looks grim," Liebergot said.

"Yes it does," said Bliss.

"It's going down. We're losing it."

"Yes we are."

On Liebergot's and Bliss's screens, the remaining oxygen tank was now below 300 pounds per square inch and falling at a rate of 1.7 pounds per minute. Working with pencil and scrap paper, Bliss performed some quick calculations. Factoring in both the current depressurization rate and the rate at which the leak was accelerating, he estimated that in one hour and fifty-four minutes the tank would fall below the critical 100 p.s.i. and from then onward be useless.

"That'll be the end of it for the fuel cells," Bliss told Liebergot somberly.

Liebergot, however, had a final alternative, but it was one he was loath to use: he could tell Flight to tell Capcom to tell the crew to shut the reactant valves on their two defective fuel cells. The reactant valves regulated the flow of oxygen from the giant cryogenic tanks into the cells themselves. If the leak that was killing tank one could not be found in the body of the tank or in the gas lines that ran from it, perhaps it was located downstream in one or both of the dead cells. Shutting off the valves either would stop the O_2 bleeding, allowing Odyssey to stabilize itself and power back up, or it would do nothing at all, allowing the controllers to give up on the ship altogether and turn to other survival plans.

The problem was, shutting the reactant valves was a decision from which there was no turning back. The valves were such delicate, precisely calibrated bits of equipment that once shut, they could not be reopened without a team of technicians to adjust them and tweak them and certify them fit to fly. Since no such technicians were available 200,000 miles from Earth, and since mission rules required three healthy fuel cells for a lunar landing, Liebergot knew that the suggestion he was planning to make would, in effect, be a formal acknowledgment that the mission was aborted. The possibility of pulling out of this crisis with enough command module function left for even a lunar orbit had long since evaporated with the venting gas, but from his modest console in his modest corner of Mission Control, Liebergot did not relish being the one to make this grim reality official. Nevertheless, as far as he could see, it was his only choice.

"Flight, EECOM," Liebergot said.

"Go ahead, EECOM."

"I want to shut off the reactant valves, starting with fuel cell three first, to see if we can stop the flows."

"You want to shut off the reactant valve to fuel cell three?" Kranz repeated for clarity.

"Roger."

If Kranz was disturbed by the enormity of this suggestion, this time he didn't show it. "Capcom," he said without emotion, "let's have them close the reactant valve to fuel cell number three. We're going to try to stop this O_2 flow."

Lousma acknowledged Kranz's order and turned to the air-to-ground loop. "O.K., 13, this is Houston. It appears to us that we're losing O_2 flow through fuel cell three, so we want you to close the reac valve on fuel cell three. You copy?"

In Odyssey, Lovell, Swigert, and Haise heard this command and paused in what they were doing. None of the three men had harbored any illusions that their mission was anything but aborted, but to hear it come up to them in the form of this simple, clipped directive, to hear it made official like this, still stopped them cold.

"Did I hear you right?" Haise, the electrical specialist, asked Lousma. "You want me to shut the reac valve on fuel cell three?"

"That's affirmative," Lousma answered.

"You want me to go through the whole smash for fuel cell shut-down?"

"That's affirmative."

Haise turned to Lovell and nodded sadly. "It's official," said the astronaut who until just an hour ago was to have been the sixth man on the moon.

"It's over," said Lovell, who was to have been the fifth.

"I'm sorry," said Swigert, who would have overseen the mother ship in lunar orbit while his colleagues walked. "We did everything we could."

At the EECOM console and in the backroom, Liebergot, Bliss, Sheaks, and Brown watched their monitors as the valve in fuel cell three was slammed shut. The numbers for oxygen tank one confirmed their worst fears: the O_2 leak continued. Liebergot asked Kranz to order that fuel cell one be shut next. Kranz complied — and the oxygen leak continued.

Liebergot looked away from his screen; the end, he knew, was at last here. Had the explosion or meteor collision or whatever else crippled the ship occurred seven hours earlier or one hour later, it would have been another EECOM on console at the time, another EECOM who would have attended this death watch. But the accident happened 55 hours, 54 minutes, and 53 seconds into the mission, during the last hour of a shift that by sheer scheduling happenstance belonged to Seymour Liebergot. Now Liebergot, through no fault of his own, was about to become the first flight controller in the history of the manned space program to lose the ship that had been placed in his charge, a calamity any controller worked his whole career to avoid. The EECOM turned to his right, toward where Bob Heselmeyer, the LEM's environmental officer, sat. As Liebergot glanced again at Heselmeyer's screen, he could not help thinking of that simulation, that terrible simulation which had nearly cost him his job a few weeks earlier.

"Remember," said Liebergot, "when we were working on those lifeboat procedures?"

Heselmeyer gave him a blank look.

"The LEM lifeboat procedures we worked on in that sim?" Liebergot repeated.

Heselmeyer still stared blankly.

"I think," said Liebergot, "it's time we dusted them off."

The EECOM steeled himself, signed back on the loop, and called to his flight director.

"Flight, EECOM."

"Go ahead, EECOM."

"The pressure in O_2 tank one is all the way down to 297," Liebergot said. "We'd better think about getting into the LEM."

"Roger, EECOM," Kranz said. "TELMU and CONTROL, from Flight," he called to the LEM's environmental and guidance officers.

"Go, Flight."

"I want you to get some guys figuring out minimum power needed in the LEM to sustain life."

"Roger."

"And I want LEM manning around the clock."

"Roger that too."

At the same time this conversation was taking place, Jack Swigert, on the center couch in Odyssey, looked at his instrument panel and discovered that while the oxygen readings might have been grim on the ground, they were downright dire in the spacecraft. Squinting through the growing darkness of his powered-down ship, where the temperature had fallen to a chilly 58 degrees, Swigert saw that his tank one pressure was down to a bare 205 pounds per square inch.

"Houston," he said, signing back on the air, "it looks like tank one O_2 pressure is just a hair over 200. Does it look to you like it's still going down?"

"It's slowly going to zero," Lousma responded. "We're starting to think about the LEM lifeboat."

Swigert, Lovell, and Haise exchanged nods. "Yes," the command module pilot said, "that's what we're thinking about too."

With an O.K. to abandon ship at last granted by the ground, the crew wasted little time in getting started. Assuming the three men were entertaining any hopes of getting home, they could not just take up residence in the LEM and let their fading mother ship sputter to a halt like a car out of gas on a country road. Rather, since Odyssey would have to be used at the end of the flight for reentry, the ship would have to be shut off one switch or system at a time so as to preserve the operation of all of its instruments and maintain the calibration of their settings. Under ideal conditions, all three men would handle the job; under current conditions, however, Swigert would have to take care of things on his own, because at the same time Odyssey was being taken off line, Aquarius would have to be brought *on* line, a two-man task that would have to be completed before the command module expired.

Lovell and Haise swam through the lower equipment bay of Odyssey and into the LEM, where they had broadcast their happy travelogue barely two hours earlier. Haise settled into his spot on the right side of the craft and surveyed the blacked-out instrument panel. Lovell floated to his station on the left.

"I didn't think I'd be back here this soon," Haise said.

"Just be happy it's here to come back to," Lovell said.

With the possibility of having a healthy ship again under his command, Lovell felt a brief surge of optimism, but Houston was about to extinguish it. At Mission Control, it was about time for the afternoon-evening shift of controllers to turn their consoles over to the night shift. According to the schedule drawn up before this flight, Glynn Lunney's Black Team would follow Gene Kranz's White Team in the four-team rotation. Lunney, in turn, would be followed after eight hours by Gerald Griffin's Gold Team, then Milt Windler's Maroon Team. Now, all over the room fresh technicians from Lunney's group were arriving at their posts, plugging their headsets into extra console jacks, and standing silently by the sides of the frazzled men who had been on duty since two o'clock that afternoon. At the flight director's console, Lunney himself prepared to relieve Gene Kranz. At the EECOM console, Clint Burton came up beside Liebergot and laid a sympathetic hand on his shoulder; Liebergot looked up, smiled weakly, pushed away from the console, and gestured toward the seat with a sorrowful shrug. Burton nodded, sat down before the screen, and as soon as he did, discovered that the situation had deteriorated even further.

"George," he said to Bliss, who was still working the backroom, "how long do we have left on that tank now?"

"Uh . . ." Bliss stalled, consulting his readouts and calculating the speed of the leak. "A little over an hour. We're getting a new rate."

"I didn't copy," Burton said incredulously, exchanging a shocked look with Liebergot at his side.

"We're getting a new rate here, Clint," Bliss repeated.

"O.K. I'd like you to calculate it as close as you can."

"Roger."

Until Bliss did his calculating, Burton did not want to transmit the new estimates up to the crew, and moments later, he was glad he hadn't. Checking the oxygen readouts, Bliss could see the leak rate accelerating from 1.7 pounds per minute toward 3 and beyond.

"EECOM," Bliss said. "We've got a little under forty minutes left in tank one." After a brief pause he came back on the line: "Leak rate's increasing all the time, EECOM. Now it looks like we've got about *eighteen* minutes left." A few moments later, Bliss's voice came into Burton's ear to tell him that the eighteen predicted minutes was now down to just seven. A minute later, the seven was down to four.

"Flight, EECOM," Burton said.

"Go ahead."

"We need to open up the surge tank. The pressure's dropping, and the rate's accelerating."

"Wouldn't you rather have them breathe off the LEM?" Lunney asked.

"We've got to get the LEM going first!" Bliss prompted Burton through his headset.

"We've got to get into the LEM first, Flight," Burton repeated.

"Capcom, get 'em going into the LEM!" Lunney ordered. "We've got to get the oxygen on in the LEM!"

"13, Houston," Lousma said to Swigert. "We'd like you to start making your way over to the LEM."

Swigert heard Lousma's command but was not inclined to act on it immediately. He knew he could live for at least a little while on the ambient air left in the command module cockpit, and he was not about to leave without completing his power-down. He responded to Lousma nonresponsively: "Fred and Jim are in the LEM already," he said.

As Swigert raced through his power-down, Lovell and Haise worked to bring the LEM to life. The first order of business was its guidance platform. Aquarius was equipped with a three-gimbal guidance system that was essentially identical to Odyssey's. Before the platform could be used, the power-up protocol called for the command module pilot, Swigert, to note the orientation and coordinates of the guidance platform in his craft and shout them through the tunnel to the commander in the LEM. The commander would then perform some quick conversion computations on each coordinate, to reflect the slightly different orientations of the LEM and the command module, and then type the converted numbers into the LEM's computer. If the calculations weren't made and the numbers weren't typed in before Odyssey lost all its power, the information in its computer would be lost forever.

Racing against the dying tank, Lovell tore a blank sheet of paper out of a flight plan and fished a pen out of the upper-arm pocket of his flight suit. Interrupting Swigert and Lousma as they volleyed power-down data to each other, Lovell requested the first few guidance coordinates and Swigert hurriedly complied. But as the commander copied the numbers onto his scrap paper and prepared to perform the necessary calculations, he was seized by a momentary and unaccustomed uncertainty. *Could* he perform the calculations properly? Would his ciphering be correct? 3 times 5 *is* 15, isn't it? 175 minus 82 *is* 93, isn't it? With the clock ticking down and so much riding on these rudimentary calculations, Lovell all at once found himself doubting his ability to add and subtract.

"Houston," Lovell said, "I've got some numbers for you, but I want you to double-check my arithmetic so far."

"O.K., Jim," Lousma said, a bit confused.

"The roll CAL angle is minus 2 degrees," Lovell said, consulting his paper. "The command module angles are 355.57, 167.78, and 351.87."

"Roger, we copy." There was silence on the line as the men at the guidance console, unbidden, checked Lovell's math and sent a thumbs-up back to Lousma. "O.K., Aquarius," he said, "your arithmetic looks good there."

Lovell signaled Haise to enter the numbers into the computer, got the remaining coordinates from Swigert, and for the next few minutes the crew worked frantically, flipping toggle switches and circuit breakers and turning every other knob and dial necessary to reconfigure the twin spacecraft. The process was a chaotic one, with the ground shouting instructions up to the astronauts, the crew shouting questions down, and both sets of transmissions often colliding on the air, conveying no useful information in either direction. Glynn Lunney, briefly lost in all the clashing babble, inadvertently ordered that the attitude-control jets be powered down in Odyssey before the corresponding jets could be powered up in Aquarius, and for one fleeting moment Aquarius was in danger of tumbling drunkenly into gimbal lock. Finally, however, the twin craft were ready — or as ready as the astronauts could get them on such inhumanly short notice — and Lovell alerted Houston.

"O.K.," he called to Lousma, "Aquarius is up and Odyssey is completely powered down according to the procedures you read to Jack."

"Roger, we copy," Lousma responded. "That's where we want to be, Jim."

In the now dark, now quiet Odyssey, Swigert took a lingering look around. *This,* if truth be told, was where he'd want to be. Among the lunar crews, there was often a bit of grumbling about which two pilots on any given flight would be assigned the job of landing on the moon and which would be given the lower-profile task of staying behind in the lunar orbiter. Some command module pilots could not help but think that the less glamorous orbital assignment was a slight, a slur on their flying abilities. After all, wouldn't NASA give its most accomplished pilots the most challenging parts of its missions?

Swigert never saw things that way. He liked his job, was proud of his job. Sure it lacked some of the drama of the commander's or the LEM pilot's, but it had its compensations. It was the command module pilot who was essentially the driver on this absurd expedition; the command module pilot who was the navigator; the command module pilot who would have to bring the two moon walkers safely to the point where they could separate their lander and begin their descent to the surface; the command module pilot who would have to be there to manage the rendezvous when they came back up. And, most dramatically, it was the command module pilot who would have to be enough of a flier to bring his spacecraft home alone if his fellow crewmen never did make it back up. Swigert had been given a wonderful ship with which to do all of these jobs, and now chance and circumstance had taken that ship away from him. Until such time as he and Lovell and Haise and NASA could figure out a way to bring the spacecraft back to life, he — like Bill Anders, the LEM-less LEM pilot on Apollo 8 — would be a command module pilot without a command module. Swigert drifted through the tunnel from the rapidly chilling Odyssey to the slowly warming Aquarius and floated down between Lovell and Haise.

"It's up to you now," he said.

Sitting at his flight director's console, Glynn Lunney allowed himself a moment of relief — but only a brief one. His crew had just transferred themselves from a ship in which they were certain not to survive the next few minutes to one in which they were likely not to survive the next few days. The improvement was real, he knew, but ultimately academic. What concerned Lunney most at the moment was not the LEM's life-support capability. The oxygen, water, and power aboard the ship either would or wouldn't be sufficient to sustain the three men for the time needed for a trip back to Earth, but it would take a while for that problem to play itself out. What concerned Lunney now was the trajectory the ship was following.

During an aborted lunar mission, there were a few ways to bring a ship in distress home to Earth. The most straightforward was the so-called direct abort, in which a crew on its way out to the moon would turn their command-service module around so that they were flying tail-forward, then fire their 22,500-pound hypergolic engine at full throttle for better than five minutes. The maneuver was designed to bring the spacecraft — which might be traveling at 25,000 miles per hour — to a complete standstill, and then get it moving just as fast in the opposite direction.

An alternative to the direct abort in deep space was the circumlunar abort. In the event a spacecraft got too close to the moon to try a deep-space maneuver, the free-return trajectory that every lunar crew since Apollo 8 had followed would whip the ship around the back of the moon and send it on a gravity-assisted, slingshot path home. The maneuver could take a lot longer than a direct abort, but it had the advantage of requiring no engine burn, no mid-course turning, in fact nothing at all for the crew to do except ride along.

On Apollo 13, the free-return option was limited. The irregular course the ship was following to reach Fra Mauro took it off the slingshot route and placed it on a flight path that would send it around the moon and whip it back *toward* home, but cause it to pass 40,000 miles over the earthly cloud tops. For situations like this, the lunar flight plan included a procedure known as a PC+2 burn. Two hours after pericynthion — or closest approach to the back side of the moon — the spacecraft would fire its engine, changing its course just enough to aim it precisely toward Earth and, not incidentally, shorten its transit time home.

NASA flight planners liked having all these options; indeed, for maneuvers as critical as return-to-Earth abort burns, they required all these options. In this case, however, it appeared they might have come up one option short. Nearly every abort protocol written into the flight plans and practiced by the crew presumed the availability of one very important piece of equipment: the command-service module's giant main engine. Getting back home required all the power the hypergolic howitzer could provide, but on Apollo 13 that howitzer was probably out of ammunition. If the bang that shook the ship hadn't busted the engine, the power-down had almost certainly made it impossible to muster enough electricity to fire it.

The LEM still had an engine, of course — actually the LEM still had *two* engines, one in the descent stage and one in the ascent stage — but the LEM was not made for flying this way. It was possible to nose the docked ships around by firing the lander's engines in tweaks and bursts, but a full-throttle, full-bore burn for something as crucial as returning to Earth? It wasn't something the engineers even wanted to consider. Unless somebody could come up with a way to bring the wounded service module's engine back to life, however, firing the LEM engine to propel both ships was going to be the only route home — and the untried maneuver would have to be dreamed up, worked out, and executed on Lunney's watch.

"O.K. everybody," Lunney said quietly to the loop at large, "we've got a lot of long-range problems to deal with."

In the Houston suburb of Timber Cove, Marilyn and Jim Lovell's house had begun to fill up with neighbors and friends of neighbors, NASA employees and their spouses, protocol officers and their assistants. First, Susan Borman would appear at the door, then Carmie

McCullough, then Betty Benware. Marilyn would see each new visitor and fleetingly wonder how all these people had heard the news that she, the wife of the man in danger, had only just learned, and then the doorbell would ring, and more people would swarm in, and Marilyn would ask herself the question again. These new arrivals joined Elsa Johnson and the Conrads and the others in deflecting the reporters, answering the constantly ringing phones, and keeping a watchful eye on the woman whose husband, if you believed Jules Bergman, was looking at a 90 percent likelihood of not living through tomorrow.

While the guests made it their business to keep a watchful eye on Marilyn, surprisingly few actually talked to her directly — a relief both to them and to her. Beyond a few pro forma reassurances, nobody had the vaguest idea what words of encouragement they could offer that would ring even remotely true, and Marilyn did not want to make them try.

The only real answers that were available came from the television, and — with the exception of a brief period an hour or so ago when Marilyn retreated to her bathroom, locked the door, knelt on the tile floor, and prayed — she had not moved from the screen. In the short time since the accident, nobody else — either inside NASA or on a network other than ABC — had offered as grim a prediction of the astronauts' likely fates as Bergman had, but Marilyn took little reassurance from that. Somehow, it had become important to her to hear every word the doomsaying newsman said, as if optimistic opinions from anyone else would carry no weight until Bergman himself retracted his dire forecast. So far, he did not seem so inclined.

"We're watching the picture here from the Manned Spacecraft Center, where a flawless flight for its first fifty-six hours has now turned into the only real emergency since the Gemini 8 flight," Bergman was saying. "This is America's twenty-third space flight, and so far it is the first that might actually endanger the lives of the astronauts. The astronauts have had to in effect bail out of the command module and move over to the lunar module. The question now is how long can the oxygen in the lunar module hold out. The LEM oxygen supply would be a maximum of forty-five hours."

Bergman cut to correspondent David Snell in Houston, who was standing in front of a wall-sized diagram of the lunar module, but Marilyn did not want to hear any more. She didn't know as much about space travel as her husband or his crewmates, but she knew enough; and she knew that forty-five hours was only about half as much time as it would take to get back to Earth. If somebody didn't figure out something soon, the one-in-ten chance Bergman was giving the crew would dwindle quickly to zero.

Suddenly Marilyn's thoughts wandered upstairs. The tumult in her family room had been going on for half an hour now, and nobody had yet gone to check on the children. During flights, the sons and daughters of astronauts were used to their homes becoming meeting grounds for the extended NASA clan, but company did not usually arrive this late at night or in these numbers, and the phone certainly did not ring with this frequency.

Marilyn, a bit dazed, summoned her neighbor Adeline Hammack and asked her if she would mind checking upstairs to see if the children had been disturbed. Adeline agreed, went up the steps, and peeked into the bedrooms. Eleven-year-old Susan was sound asleep, but her little brother Jeffrey wasn't.

"Why are so many people here?" the four-year-old asked.

Adeline sat down on the bed. "You know where your Daddy's going this week, don't you?" she asked.

"The moon," Jeffrey answered.

"And you know what he was planning to do when he got there?"

"Walk around."

"That's right. Well, it looks like something broke on the spaceship and he's going to have to turn around and come home. He's not going to get to walk around after all, but the good thing is, he'll be home even sooner than he was supposed to be. Maybe even by Friday."

"But he *said*," Jeffrey protested, raising himself from his pillow.

"Said what?" Adeline asked.

"He said he was going to bring me a moon rock."

Adeline smiled. "I know. And I know he wishes he could. But this time he probably won't be able to. Maybe when you grow up, you can go there and bring him one."

Adeline laid Jeffrey back down in his bed, quietly left his room, and tiptoed to the bedroom of sixteen-year-old Barbara. Like Susan, Barbara seemed to be deeply asleep. But unlike Susan, she didn't appear to have been that way for long. Barbara was under the covers, head on her pillow, eyes closed, but Adeline also noticed something else: in the crook of her arm, she was tightly holding on to a Bible.

6

TOM KELLY went to sleep before eleven on the evening of April 13, and he did not want to be disturbed. For the past few months, Kelly had been going to bed earlier and getting up later than he had for a long time, and he liked it just fine.

It wasn't as if Kelly had objected to the hours he'd been keeping up till then. Indeed, for nine years Kelly had been putting in ten- and twelve-hour workdays with barely a thought that there was any other way to live. It had been like that at Grumman Aerospace, in Bethpage, Long Island, since the early 1960s, when the company got the contract to build the so-called lunar excursion module, the odd, insectile ship that was supposed to put a man on the moon before 1970.

At first Grumman had not been very interested in any LEM. From the day President Kennedy announced his outrageous plan for lunar exploration, the company had set its sights on the truly big engineering prize: the Apollo command module, the mother ship that would ferry the fragile lander out to the moon and wait for it in orbit while it descended to the surface and flew back up. To the press and the taxpayers, of course, the orbiter did not have nearly as much glamour as a multilegged, crater-hopping lander. But Grumman wasn't interested in what struck the public as glamorous. Grumman was interested in what struck its stockholders as sensible, and for a company that had dividends to pay and annual reports to issue, building a workhorse orbiter that NASA could use for years — for lunar missions, Earth-orbit missions, space-station missions — made a lot more business sense than trying to design a specialized moonship that would be built for one job only, provided it could be built at all.

To be sure, Grumman was not alone in lusting after the orbiter. Also making its interest known was North American Aviation, in Downey, California. Grumman knew that North American was a formidable opponent, and when the bids for the project came in and the contracts for the work went out, it was the California colossus that got the nod to do the job. Nobody in the aerospace industry knew how many orbiters North American would get to build for the government, but with up to eight years of research and development, and potentially dozens of manned and unmanned flights ahead of it, the company, most people agreed, had hit paydirt. A year later, perhaps as a consolation prize, perhaps because North American now had its hands full with its trophy orbiter, Grumman was chosen to build the less coveted lander — receiving a contract from the government, congratulations from its competitors, and more than a few good-luck snickers from the rest of the engineering community.

In the years that followed, the snickers stopped, and ever since March of 1969, when the Apollo crew of Jim McDivitt, Dave Scott, and Rusty Schweickart took the first manned LEM into Earth orbit, separated it from the command-service module, and danced it away into its own distant orbit, the ship had been the darling of a moon-happy public. So brilliantly had the lander performed in that first flight, that NASA even decided to try a few experimental maneuvers in which the docked ships would be pushed around not by the orbiter's big blunderbuss of a service-propulsion engine but by the LEM's pistol of a descent engine. After all, you never knew when North American's trusty orbiter would need an emergency push from Grumman's little lander.

From Apollo 9 on, no American spacecraft had gone aloft without a LEM along, and the five flights in the past thirteen months had begun to take their toll on Kelly and the other Grumman workers. The company typically kept three separate crews working around the clock to monitor any flying LEM — one crew stationed in a room just off Mission Control, one in a support building nearby on the Space Center campus, and one back at Bethpage. An engineering manager like Kelly often had to be available to visit any of the three sites on any given day, and by the time Apollo 13 flew, the company knew its senior people could not be expected to maintain that pace indefinitely. As a reward for the time they put in so far, Grumman decided to send a few of its more valued men on sabbaticals to the Massachusetts Institute of Technology, where they were to spend a year catching their breath and studying industrial management. Kelly was among the first of the engineering higher-ups to be chosen for the program, and he looked forward to the change.

For the last few days, Kelly had been following the mission of Apollo 13 from his room in Cambridge, and on the evening of April 13, he knew, Jim Lovell and Fred Haise would be visiting the LEM for an initial inspection and beaming the pictures back to Earth. Kelly would have liked to watch that first proud hatch-popping, as he had on previous flights, but the networks weren't going to carry the program, and the only places he would have been able to see it were Bethpage and Houston. His Grumman colleagues, like the men at the consoles in Mission Control, would be watching the transmission, and Kelly knew they'd phone him if anything was amiss, but for someone who had been on hand when the first strut was cut on the very first LEM, this was a poor substitute. Still, in the initial months of his voluntary exile in Cambridge, Kelly figured this would have to do, and after waiting up until after the scheduled time for the LEM inspection, he went to bed.

At a little after one in the morning, Kelly's phone did ring. The engineer opened one eye, glanced at his clock, and lunged for the receiver. Groggily, he croaked a hello into the mouthpiece.

"Tom," the person on the other end of the line said, "wake up. Fast."

The voice, Kelly knew instantly, belonged to Howard Wright, another Bethpage employee also in town on the MIT sabbatical.

"Howard," Kelly said, "what's wrong?"

"We've got big problems, Tom. Real big. There's been some kind of explosion on 13. They're out of power, they're out of oxygen, and they've had to abandon ship and go over into the LEM."

"What are you talking about?" Kelly asked, all at once wide awake.

"Just what I said. Lovell, Swigert, and Haise are in deep trouble. I talked to Grumman and they want us down there right away. They're going to have a light plane waiting at Logan and we're supposed to get over there pronto."

Kelly sat up with a start and, with Wright still on the phone, turned on his bedside radio. Immediately, he could tell that his friend was right. The all-news station was carrying what appeared to be a press conference from Houston, and as Kelly twisted the radio knob, he found that all the other AM stations were carrying it too. He could hear reporters shouting questions to unidentified representatives of NASA, and from what he could make out, the answers did not sound encouraging.

". . . be able to tell us just what caused this problem?" a reporter was asking as Kelly stopped at a random station. "Can an incident like the one that happened tonight be caused if the ship was hit by a meteorite?"

"Whatever happened appears to have been quite violent," a voice answered; it sounded like Jim McDivitt, the former commander of Apollo 9 and the current director of the Apollo program office. "I'm not assuming that that's what happened, you understand, but that could have done it."

"We have not tried too much to reconstruct what happened," another voice was saying, it sounded like Chris Kraft's, "because we're more concerned for the moment with getting the situation under control."

"A question for Jim McDivitt," another reporter said (so it *was* McDivitt). "How much electrical power do we have in the LEM, and how much oxygen?"

"It depends on how we use it," said McDivitt. "We have four batteries in the descent stage of the LEM and two in the ascent stage. As for oxygen, we have forty-eight pounds in the descent tanks and one pound each in the ascent tanks."

"Compared to other emergencies, Chris," (so it *was* Kraft) a reporter called out, "for example, Scott Carpenter's overshoot, Gemini 8's stuck thruster, or John Glenn's retro pack problem, how would you classify this situation?"

There was a long pause on the air. "I would say," Kraft finally responded, "that this is as serious a situation as we've ever had in manned space flight."

Tom Kelly switched off the radio, closed his eyes, and spoke into the phone: "Howard, let's get to the airport."

Chris Kraft wasn't much in the mood to conduct a press conference tonight. He suspected he'd have to; actually, he *knew* he'd have to.

During the other emergencies the media liked to ask him about — that Carpenter flight, that Glenn flight, that Gemini 8 stuck-thruster flight — there wasn't time to dawdle with reporters. Those emergencies had happened in Earth orbit, where the astronauts were always just a half hour away from a safe splashdown, and by the time a crisis settled down enough for him to talk to the press, the capsules were already floating in the water, and the cameramen had other things to shoot besides a flight director answering questions from a dais.

Tonight events were happening much more slowly, and from the moment word got out that there was a problem aboard Apollo 13, reporters had been clamoring for statements from the men in the control room. As soon as Lovell, Swigert, and Haise moved over to Aquarius, Bob Gilruth, the Space Center director, sent forth Kraft, McDivitt, and Sig Sjoberg, the director of Flight Operations, to satisfy the media. The press conference had taken place in the Public Affairs building, a few hundred yards from Mission Control. Kraft had sprinted the quarter mile or so to the conference, and now that it was over, he ran even faster to get back.

Though the deputy director of the Space Center had been gone from Mission Control for less than an hour, as soon as he reentered he could see that the atmosphere in the room had changed dramatically. Things had quieted down considerably at the EECOM station, where the crisis that had become a death watch had now become a postmortem. The screen that had been blinking with bulletins from the dying Odyssey had now essentially gone flat-line, with zeros and blank spots appearing where oxygen and power readings had once been. Clint Burton and a handful of other technicians hovered over the console, murmuring to one another and occasionally glancing back at the screen, as if there were still some chance that the expired spacecraft would stir to life. For practical purposes, however, the activity at that console had stopped.

Elsewhere around the room, the mood was a good deal less subdued. Though Glynn Lunney's Black Team had replaced Gene Kranz's White Team, the White Team showed no signs of leaving the auditorium. At most of the consoles, the recently relieved controllers stood or crouched behind their replacements, their eyes focused on the screens they had been monitoring for the previous eight hours and their headsets plugged into auxiliary jacks reserved for visitors. At the Capcom console, astronaut Jack Lousma — who, like all Capcoms, worked a three-man rotation instead of a four-man one, in order to minimize the number of different voices on the air-to-ground loop — was for the most part being left alone to conduct his communications with the crew in peace; but at the other consoles, clusters of people stood at workstations that had been designed for one.

As it had been earlier in the evening, the biggest crush was still at the flight director's console, where Lunney was juggling traffic on the in-house loop, with Kranz pacing back and forth behind him, occasionally summoning various White Team controllers over for consultations. As Kraft approached the two flight directors and glanced at the console they were sharing, he could tell they had their hands full. Above Lunney's monitor was a series of green, amber, and red lights, each connected to one of the consoles around the room. During a launch, the controllers would use these lights to inform the flight director of the status of their systems in the brief but explosive minutes between the time the spacecraft left the pad and the time it settled into Earth orbit. A green light indicated that a controller's systems were operating normally; amber meant there was a problem, and the controller needed to talk to the flight director at once; red meant there were grounds for an abort.

After the launch phase was through, these lights became superfluous, and over time, flight directors had begun using them to help field calls that came in from around the room. A controller signing on to the loop with a question or a request during a flight would often as not be told to "go to amber" so the flight director could contemplate the problem without forgetting to call back with an answer. On Lunney's console, half of the more than two dozen lights were now amber, and as the flight director himself signed on, he was about to bring the rest of the controllers on line.

"O.K.," Lunney said to the room at large, "I'd like to get everybody up here at the moment. RETRO, GUIDANCE, CONTROL, TELMU, GNC, EECOM, CAPCOM, INCO, and FAO. I want everybody on the loop. Give me an amber, please."

The green lights on Lunney's console flashed off immediately and the amber ones flashed on, with the exception of the RETRO officer, who was involved in a discussion with his backroom. "GUIDANCE," Lunney said impatiently to the controller sitting next to the RETRO station, "get a RETRO on the loop, please."

"Go ahead," Bobby Spencer, the lead RETRO said, overhearing Lunney's request and signing on before he could be jabbed by the GUIDO.

"Look," Lunney said, "I want to go over where we stand on a

number of things. Most important, we've got an engine burn to do, which is one piece of business. We've got to get tracking and attitude set to take care of that burn. We've got to get the LEM powered down and get nonrequired stuff off line so we don't use up power unnecessarily. And we've got to get people who aren't on console working on long-range problems associated with the LEM in the lifeboat mode. TELMU, I assume you're working with all these consumables problems — O_2, water, power?"

"Roger, Flight," the TELMU said.

"Can you say anything generally now? Have we worked out a way to come home on the consumables we've got?"

"Negative, Flight."

"Are we working on it?"

"We're working on it."

"All right. I'm going to want to start hearing from you about this."

"Roger, Flight."

"CONTROL, Flight," Lunney called next.

"Go, Flight."

"We've still got to get spacecraft attitude and motion worked out before we're ready to burn the engine. Are you working on that problem?"

"Affirmative."

"Are we close to solving it?"

"Negative."

"How long do you think it's going to take?"

"I don't have an estimate right now, Flight. We're trying to get it to you as soon as possible. Grumman has given us a procedure to reconfigure the LEM's autopilot to take into account the dead command-service module. Suggest you send a crew to the simulator to see if it works."

"FIDO, Flight," Lunney said.

"Go, Flight."

"What's the closest approach to the moon we're looking at right now."

"About sixty miles, Flight."

"Recovery, Flight."

"Go, Flight."

"How do we stand on ships in splashdown sites?"

"Present efforts are to identify ships in the Atlantic and Indian oceans."

"O.K., gentlemen," Lunney said. "Those are the major subjects I'm thinking about. I want to start getting closed out on some of them. Does anybody have anything else to discuss? RETRO?"

"Negative, Flight," Bobby Spencer answered, promptly this time.

"Guidance?"

"Negative, Flight."

"GNC?"

"Negative, Flight."

"FIDO?"

"Negative, Flight."

"Capcom?"

"Negative, Flight."

"O.K., you can all go back green. But let's make sure we stay on top of these issues and just keep this thing moving right along."

Of all of the problems Lunney faced, the most complex was the burn. In the hour or so since the astronauts had moved over to Aquarius, no definite decisions had yet been made about how to propel the docked ships toward home, and with the spacecraft moving closer to the moon, at a speed climbing back up to 5,000 miles per hour, the options were quickly fading. A direct abort, if one could even be attempted, got harder and harder the farther the ships got from Earth. A PC+2 burn, if one was going to be attempted, would take a lot of planning, and the time for pericynthion was closing in fast. It would always be possible to fire the engine after the PC+2 point, but the earlier in the earthward transit a burn was attempted, the less fuel it would take to affect the trajectory; the longer the burn was delayed, the longer the engine would have to be fired.

Pacing behind Kranz, who was also pacing, Kraft knew which return route he'd choose. The service propulsion engine, he was certain, was useless. Even if there was some way of mustering enough electricity to get the engine going, Kraft was not convinced that the crippled Odyssey would be able to take the strain. No one knew the condition of the service module, but if the force of the bang had been any indication, it was possible that the sudden application of 22,500 pounds of thrust would collapse the entire back end of the spacecraft, causing both docked ships to tumble ass over tea kettle, sending the crew not back toward Earth but barrel-rolling down to the surface of the moon.

The only way home, Kraft figured, was to use the LEM's engine —

and more important, to use it right away. It would be tomorrow evening before the docked ships first passed behind the shadow of the moon, and it would be close to three hours beyond that before they reached the PC+2 milestone. Waiting the better part of a day to get the crew on its homeward trajectory seemed nonchalant at best and downright reckless at worst. What Kraft wanted to do was fire the descent engine now, get the ship back on its free-return slingshot course, and when it emerged from behind the moon and reached the PC+2 point, execute any maneuvers that might be required to refine the trajectory or increase its speed.

In the past, when Chris Kraft had an idea like this, that idea got implemented. Nowadays, though, things were different. It was Gene Kranz who dictated the direction of things, Gene Kranz who was the true *capo di tutti capi* of the control room. If Chris Kraft wanted something done, he was free to suggest it to Kranz, but he could no longer decree it. In the aisle behind the flight director's console, Kraft was about to stop Kranz's pacing and discuss his two-step burn idea when Kranz turned to him.

"Chris," he said, "I sure as hell don't trust that service module engine."

"I don't either, Gene," said Kraft.

"I'm not sure we could fire it even if we wanted to."

"I'm not either."

"No matter what else we do, I think we're going to have to go around the moon."

"Concur," Kraft said. "When do you want to burn?"

"Well, I don't want to wait till tomorrow evening," Kranz said. "How about we try a quick burn for a free return now, get that squared away, and then figure out if we want to speed them up with a PC+2 tomorrow."

Kraft nodded. "Gene," he said after a considerable pause, "I think that's a good idea."

Two rows down and one console over, Chuck Deiterich, an off-duty retrofire officer, or RETRO, standing behind his accustomed console, and Jerry Bostick, an off-duty flight dynamics officer, or FIDO, could not hear Kranz and Kraft's discussion, but they knew the options as well as their bosses. Though it was Kraft and Kranz and Lunney who would ultimately decide the ship's route home, it was Deiterich and Bostick and the other flight dynamics specialists who would have to come up with the protocols to pull the plan off. At the FIDO station, Bostick pushed his microphone out of range of his mouth, and leaned toward Deiterich.

"Chuck," he said quietly, "how do we all want to do this thing?"

"Jerry," Deiterich answered, "I don't know."

"I assume we're ruling out Odyssey's engine."

"Absolutely."

"I assume we're going around the moon."

"Absolutely."

"And I assume we want to get them on free return as quick as possible."

"Definitely."

After a moment Bostick said, "Then I suggest we get our shit together fast."

Close to a quarter of a million miles away, in the crowded cockpit of Aquarius, the men on whose behalf Bostick and Deiterich would be working had more elemental things on their minds than a return-to-Earth engine burn. Settling into his two-man spacecraft with his three-man crew, Jim Lovell had the chance to look around at the hand circumstance had dealt him. He did not like what he saw. The commander was standing at his station on the left side of the cabin, wedged between the port bulkhead on one side and a projecting shelf that held his attitude controller on the other. Haise was off to the right, squeezed uneasily between the starboard wall and his own backup attitude control. Swigert was between and behind the two pilots, perched awkwardly atop the bulge in the floor that covered the guts of the ascent-stage engine. When Lovell shifted too far to the right, he jostled Swigert, who in turn jostled Haise. When Haise shifted too far to the left, the wave rippled back in the other direction.

With three warm bodies in a space built for two, and with the electrical and environmental systems slowly stirring to life, the temperature inside the previously cold Aquarius had begun to rise — but only to a point. The power-down of Odyssey had caused the command module's thermometer to plummet almost immediately, and when Lovell had last checked the environmental readings before coming over to Aquarius, the cockpit was at 58 degrees and falling. Now, with all of the equipment in the command module shut down, its interior was growing even colder; and with the balky hatch between the ships still not in place, leaving the tunnel wide open, the temperature in the LEM was starting to fall too. Already, the gathering chill and the collective respiration of the three men were causing condensation to form on the walls and windows.

"It's not going to be easy flying this thing if we can't even see through the glass," Lovell said to no one in particular as he glanced at the fogged triangular porthole in front of him.

"We'll get them wiped off," Haise said.

"And we'll have to *keep* them wiped off. The colder it gets, the more they'll cloud up."

"Can you see anything out there anyway?" Haise asked.

Lovell wiped a bit of condensation off his window and peered through the small cleared patch. The view from Aquarius was much the same as it had been from Odyssey: a swirling cloud of oxygen ice crystals and particles of debris from whatever blast had rocked the craft. Lovell surveyed the mess for a moment.

"Just the same cloud of junk we had next door," he said.

"Well, we're not going to be able to wipe that away, are we?" Haise said dourly.

"You know," Lovell said, turning to Swigert, "if it's getting cold in here, it's going to be freezing in Odyssey. We may want to bring some food and water over before it's too late."

"You want me to get it?" Swigert asked.

"It would be a big help. Fill up as many drink bags as you can with water from the potable tank and grab some food packets along with them."

"I'm on my way," Swigert said.

Standing on the engine cover, the command module pilot bent into a partial crouch and straightened up again quickly, springing into the tunnel that led back to his ship. Entering the lower equipment bay at the foot of the couches, he stopped at the food locker, lifted the lid, and peered inside. The rations for a ten-day moon trip were nothing if not generous, and the Odyssey larder was full to bursting. There were packets of turkey and gravy, spaghetti and meat sauce, chicken soup, chicken salad, pea soup, tuna salad, scrambled eggs, corn flakes, sandwich spread, chocolate bars, peaches, pears, apricots, bacon squares, sausage patties, orange drink, cinnamon toast, brownies, and more. Each packet was secured with a strip of Velcro, color-coded to indicate each crewman's rations. The commander's Velcro was red, the command module pilot's white, and the LEM pilot's blue.

Swigert scooped up a few handfuls of packets and left them to float in the air nearby. Turning to the potable water tank, he grabbed a few drink bags and began filling them from a plastic water gun attached to a length of flexible tubing. On the first bag, however, the gun misfired and a mercury-like globule of water floated downward and splashed around Swigert's soft cloth shoes.

"Damn!" Swigert said loudly.

"What's wrong?" Haise called.

"Nothing. I just drenched my shoes."

"They'll dry," Haise said.

"They'll freeze before they dry," said Swigert.

More important to Lovell than the housekeeping chores inside his spacecraft were the conditions outside. Though he had not expected the gas and rubble released by the accident to have dissipated by now, his peek through the wet windows was still disheartening. The halo of detritus surrounding the ship was not a safety problem. Since the spacecraft and the surrounding debris were all moving at roughly the same velocity, it was unlikely that any of the particles would collide with the ship; if they did, the difference in the relative speed of the trash and the vehicle would be small enough to cause little more than a ding. Rather, it was the problem of navigation that gave Lovell the most concern.

The alignment the commander had programmed into the LEM's computer was, he hoped, good enough to give the guidance system a rough idea of the true attitude of the LEM. But in order to orient the spacecraft precisely enough for any engine burns, he would have to perform a far more exacting "fine alignment." This procedure required the commander to recognize particular stars in particular constellations outside his window and adjust his guidance platform by taking sightings on those stars with his alignment optical telescope, or AOT. With only sixty miles clearance as Odyssey and Aquarius arced around the moon, even a tiny miscalculation in orientation during the free-return burn could cause the twin ships to auger into the far side, plowing a long, permanent trench in the lunar surface.

For the better part of the last hour, Houston had been fretting about just this problem, calling up to the ship with an occasional "Aquarius, can you see any stars yet?" When Lovell looked out his window, however, he saw not only the target stars he needed to see to make his alignment but hundreds, indeed thousands, of brightly glowing false stars made up of the debris. Separating the genuine articles from the constellations of counterfeit ones would be an impossible job. Lovell decided that the only answer was to take control of the LEM's thrusters and nudge Aquarius and Odyssey around in the cloud, looking for a gap that would provide him a clear line of sight out into space.

"Hand me a towel, Freddo," Lovell said to Haise. "I want to see if I can't just maneuver out of this stuff."

Haise handed Lovell a small square of terry cloth stored in a supply kit to his right, and the commander wiped off first his own window and then his LEM pilot's. The two men took a long look out their portholes and whistled in unison.

"What a mess," Haise said.

"No worse than on this side," Lovell said.

He switched the attitude-control system to its manual setting and reached for his hand controller. As on Odyssey, there were four clusters of four thrusters spaced evenly around the outside of the craft, each of which was placed so that it could exert sufficient torque to rotate Aquarius about its center of gravity. And, as on Odyssey, the entire system was controlled with a pistol-grip device. Lovell ever so carefully pushed the handle forward, attempting a pitch-down motion. The ship lurched abruptly — and somewhat sickeningly — up to the left. If the thruster system on Odyssey had been balky, the one on Aquarius seemed out of control.

"Whoa!" Lovell said, releasing the controller. "That's a hell of a yaw."

"That's not the way that's supposed to act," Haise said.

"It's sure not the way it's ever acted before."

The problem, as Lovell and Haise realized, was the combined spacecraft's center of gravity. The LEM's attitude-control system was designed to be used only after the ship had separated from the command-service module and was drifting alone in near-lunar space. In the simulators that both Lovell and Haise had trained in, the guidance computers were programmed to mimic the mass distribution of the free-floating ship, and the pilots had learned how to twirl the spacecraft in practically any direction using only the barest puff of propellant to get the job done. The LEM that Lovell was piloting today, however, was not flying free but laboring along with the cold, dead bulk of a 63,400-pound orbiter protruding from its roof. This shifted the center of gravity dramatically upward, well into the command module or beyond, and the familiar feel of the LEM's perfectly calibrated thrusters changed completely.

In the command module, Swigert felt the twin ships suddenly sway around him and, carrying his packets of food and drink, swam back through the tunnel to see what the commander was up to.

"What's the status down here?" Swigert asked as Lovell gave the controller another gentle push and the spacecraft responded with another clumsy movement.

"Trying to get a star alignment," Haise explained.

"It's not going to be easy with that attached," Swigert said, pointing with his thumb up the tunnel toward Odyssey.

"You're telling me," Lovell said with a frustrated laugh.

As Lovell manipulated his controller, the attitude indicators on board the LEM and the angle readouts in Houston began to register the erratic movement of the ships. At the LEM consoles in Mission Control, Hal Loden, the man who oversaw the navigation systems for the lander, was alarmed when he noticed the motion of his gauges. The indicators for all three gimbals in the spacecraft jerked crazily, moving through just the range of uncontrolled motion that made it most likely they would align and lock. If the gimbals locked and the alignment Lovell had labored so hard to transfer over from Odyssey was lost, any chance of orienting the ships sufficiently for subsequent engine burns would be effectively eliminated.

"Flight, CONTROL," Loden said hurriedly.

"Go, CONTROL," Lunney answered.

"It looks like we're drifting around on the gimbal angles here. He's in mid-impulse right now and I'm assuming that's where he wants to be, but if he doesn't watch it real close, he's going to get himself into gimbal lock real quick."

"He may be looking for stars," Lunney said.

"Maybe, but it might be worth a confirmation."

"Roger," Lunney said. "Capcom, have him watch his gimbal angle."

"Roger," Lousma said, then dialed up the air-to-ground loop. "Aquarius, Houston. You're watching your gimbals there, aren't you?"

Lovell, trying to figure out a whole new way to fly his spacecraft, turned to Haise and rolled his eyes. Yes, he was watching his gimbals. *And* his thrusters. *And* his attitude indicator. *And* the cloud of junk outside his windows. Lousma had been on watch at the Capcom console since early this afternoon, and Lovell was grateful for his fellow astronaut's help. But asking a lunar pilot whether he was remembering to watch his gimbals was like asking an airplane pilot whether he was remembering to use his flaps. In both cases, you could be entirely certain, the answer was yes.

Lovell turned slowly to Haise. "Tell them," he said with subdued anger, "that we are."

Lousma, who had himself spent plenty of hours in Apollo simulators, picked up the remark on the air-to-ground loop and knew enough not to bother the commander again.

As Lovell worked to stabilize his ships and Lousma worked to stay out of his hair, Jerry Bostick, Chuck Deiterich, and the other off-console RETROS, FIDOS, and GUIDOS continued to work on devising a burn that would bring the crew home. The flight plans both the ground crew and the astronauts carried included a number of ready-made abort scenarios, known as block data maneuvers, that contained all of the spacecraft coordinates, throttle settings, and other information necessary for a few of the abort situations the crew was likeliest to encounter. There were block data plans for various direct aborts, block data plans for various PC+2 aborts, block data plans for aborts when the ship had left its free-return trajectory and needed merely to be nudged back in line. All of these aborts presumed a healthy command module, a healthy service module, and a LEM that would be, at best, an expendable appendage. Looking over their block data, Bostick and Deiterich did not expect to find a packaged abort that would fit the current extreme circumstances, and indeed they didn't.

Working with their respective backrooms, the controllers *were* able to cobble together the coordinates for the sometimes-considered, but almost never attempted, "docked DPS burn," a burn of the LEM's descent propulsion system engine with the command-service module docked in place. The maneuver would be an almost unprecedented one, but as near as Bostick and Deiterich could tell, it would also be a relatively simple one. From a quarter of a million miles out in space, a trajectory refinement designed to aim a spacecraft 40,000 miles closer to Earth would require the barest tweak of the vehicle's engine. With such a great stretch of interplanetary space to cover before reaching home, a change in orientation of just a fraction of a degree at one end of the journey would compound itself into a change in direction of thousands of miles at the other. Presently, Odyssey and Aquarius were traveling at about 3,000 miles per hour, or 4,400 feet per second, and the way Bostick, Deiterich, and the others figured it, they would need to accelerate the ships by just 16 feet per second to close the gap by which they would currently miss the Earth, aiming instead toward a safe ocean splashdown.

The controllers were sure the maneuver could be executed — and they, like Kraft, knew it would have to be attempted soon. The later they tried to fire the engine in the ship's earthward transit, the longer they'd have to fire it to get the same bang for the propulsive buck. But before they could attempt the burn, they'd have to sell the idea to Lunney; and before Lunney would buy it, he'd want to sell it to Kranz and Kraft. The off-duty controllers nudged the ones at the consoles, urging them to begin trying to make the sale.

"Flight, FIDO," said Bill Boone, the flight dynamics officer on Lunney's team.

"Go ahead," Lunney said.

"Let me bring you up to speed on what we're doing down here. We're looking at a maneuver that we think should get us a free return."

"Mm-hmm," Lunney said noncommittally.

"We've got the backroom working on all the vectors right now, and in about ten minutes I can have the maneuver ready, and we could execute it at about the 61:30 mark in the mission."

Lunney looked at the elapsed-time clock on the front wall of Mission Control. It was now 59 hours and 23 minutes into the flight — about three and a half hours since the accident.

"And it's a free return?" Lunney asked.

"That's affirmed," Boone assured him. "It'll be a 16-foot-per-second burn. So you can work with that number."

Lunney said nothing. Boone waited uneasily, and on the flight director's console the light for the guidance and navigation officer, which had been glowing a listen-only green, switched to a talk-and-listen amber.

"Flight, Guidance," Gary Renick said.

"Go, Guidance."

"We've got good guidance and navigation data now," Renick said, "and we do confirm that we could probably get a real good burn off now and get that free return."

"Roger."

Once again, Lunney fell silent on the loop. He did not know all the particulars of this burn yet, but he knew he didn't have to. It was the guidance guys' jobs to dope out the specifics of any maneuver, and if they said they had a burn, they probably had one. It was his job simply to give them the O.K. to try it.

During a mission like this, however, Lunney — his flight director's omnipotence notwithstanding — was not about to give that O.K. without consultation. Pushing his microphone away from his mouth, he turned toward the aisle behind his seat, where in the last ten minutes or so a small group had formed. Joining Kranz and Kraft were Space Center Director Bob Gilruth, Missions Director George Low, and chief astronaut Deke Slayton. The five men had been talking among themselves when Lunney turned around, and instantly they stepped closer, forming a tight knot around him and talking animatedly. Across the room, flight controllers strained to listen in on their headsets, but no word of the conference taking place in the aisle was audible; they craned their necks to look, but the view of the six men offered no more information than the silence on the loop did. After several minutes, Lunney signed back on.

"FIDO, Flight," he said.

"Go, Flight," Boone answered.

"Exactly how long would it take you to get that free-return maneuver? Could you get one at 61 hours instead of 61:30?"

"Uh, roger that," Boone said. "I can. It's just a question of which vector I want to do it on."

Lunney turned back around. Again, for several more minutes there was silence on the loop and animated talk behind the console. Finally, the flight director signed back on the communications channel.

"Gentlemen," Lunney said to the room at large, "we're going to proceed to do a 16-foot-per-second free-return maneuver here at 61 hours. We want to get on a free return first, and then we'll kick it at PC+2. FIDOs, get me data on this pronto for the 61-hour mark and then run a couple more sets at 15-minute increments thereafter, in case we can't get this one off."

"Roger," the FIDO said.

"Guidance, tell me what vectors we want to use for all those times."

"Roger," the GNC said.

"CONTROL, figure out where we want to pick up on the checklist for all of these maneuvers."

"Roger."

"And Capcom," Lunney said, "why don't you inform the crew about all this."

Sitting at his second-row console, Lousma reached toward his microphone switch to pass this good — or at least better — news up to the crew, but before he could, his headset suddenly filled with talk coming down from the ship. For the last several minutes, the attitude readouts on the CONTROL officer's console had indicated that Lovell was still working his thrusters this way and that, trying to gain control of his spacecraft; judging by the evidence on the air-to-ground loop, it appeared that the commander had been doing this work entirely in silence, since no communication had been coming from Aquarius during this time. Lousma knew, however, that this was probably not the case.

Like the Capcom, the astronauts had an on-off switch attached to their headset cables that they had to press to open a channel to the ground. Although flipping the switch back and forth could be a nuisance, the crew rarely objected; the microphone button provided the astronauts some degree of conversational privacy — a rare commodity in space — and, just as important, allowed them to discuss maneuvers and problems among themselves before bringing them to the attention of the ground. The only time this arrangement was changed was during especially complex procedures when the crewmen's hands were full and conversation with the ground had to be constant. In these cases, the astronauts would switch their communications system over to a "hot mike" or "vox" setting, when the sound of their voices alone would activate the microphone, transmitting every word they said directly to the Capcom. For most of the flight, the Apollo 13 crew had been using their closed-mike setting, but a minute or so ago, it seemed, they had accidentally switched over to hot mike, and the conversations they were transmitting unknowingly to the ground made it clear that if the controllers hoped to get the ship on a free-return route, the astronauts were still going to have to stabilize its attitude.

"Is there any way I can control this thing, Freddo?" Lovell could be heard saying.

"What's that?" Haise asked.

"Looks like I'm cross-coupling here. I might as well —"

"Yes you are. TTCA will give you the best —"

"I want to get out of this roll. What if I go to —"

"It doesn't matter where you go —"

"Let me get around this pitch by —"

"Do you control roll by using the —"

"O.K., try that —"

"Try what —"

"Try that —"

"Well, I'm not doing any good here —"

Lousma listened in for a few seconds, and as he said nothing to the crew, Lunney began to listen in too. Like Lousma, the flight director was concerned about what he heard.

"Jack," Lunney said, "you might let them know we're copying that vox."

Whether Lousma heard Lunney or whether he was too distracted by the troubling talk of the crew was unclear, but at first the Capcom did not respond to his flight director and continued listening in on the line.

"Why the hell are we maneuvering like this?" Lovell was asking. "Are we still venting something?"

"We're not venting," Haise was saying.

"Then why can't we null this out? What if we —"

"Every time I try to I —"

"— can't get this roll out."

"*Try* to get the roll out."

"Well what's the frappin' attitude?" Lovell asked.

"The attitude's O.K.," Haise answered back.

"Damn!" Swigert exclaimed. "I wish you guys would get to something I know."

Lunney clicked back on to the loop. "Capcom," he warned again, this time more sternly, "you might let them know we're copying their vox."

Lunney was concerned as much with the difficulty the crew was having with the ship's attitude as with the language they were using to discuss it. Now that the flight had gone from nominal to critical, the TV networks were patching directly into the air-to-ground loop, and every word Houston and the crew said would be fed out to the local affiliates. In the past, NASA's air-to-ground feed had been equipped with a seven-second delay, to give Agency's Public Affairs officers a chance to edit out any stray obscenity. Since the Apollo 1 fire, however, NASA had recognized the importance of maintaining its reputation for unvarnished honesty and eliminated the on-site censoring.

The consequences of that new candor were immediately felt. Last spring, there had been a mini-firestorm in the press when Gene Cernan, piloting the Apollo 10 lunar module with Tom Stafford, let fly an inadvertent "son of a bitch!" after accidentally engaging an abort switch that sent his ship into wild gyrations just nine miles above the moon's surface. Most of the men in NASA figured Cernan had good cause to curse and were annoyed at the media's disingenuous prissiness, but the press determined public opinion and public opinion helped determine funding, and the Agency did not want to mess around with either. As soon as the Apollo 10 crew returned, the edict was handed down that on all future lunar missions, the pilots were to remember to conduct themselves like gentlemen. Regardless of any in-flight emergencies, blue language — even light blue language like an occasional, mild "frappin'" — would not be tolerated.

"Aquarius," Lousma at last called out in response to Lunney's instructions, "just wanted to let you know we've got you on vox."

"You've got what?" Lovell called back through the static.

"We've got you all on vox," Lousma repeated, then added pointedly: "We read you loud and clear."

Swigert, who was responsible for the last expletive, caught the Capcom's meaning, looked to Lovell, and shrugged apologetically. Lovell, recalling his own recent imprecations, looked back at Swigert and waved his hand dismissively. Haise, whose side of the instrument panel controlled the spacecraft communications, reached to his vox switch and snapped it back to its normal setting.

"O.K., Jack," he said, a bit pointedly himself, "how do you read us on *normal voice* now?"

"Reading you fine."

"O.K."

"Also Aquarius," the Capcom now said, "we'd like to brief you on what our burn plan is. We're going to make a free-return maneuver of 16 feet per second at 61 hours. Then we're going to power down to conserve consumables, and at 79 hours we'll make a PC+2 burn to kick what we've got. We want to get you on the free-return course and powered down as soon as possible, so how do you feel about making a 16-foot-per-second burn in 37 minutes?"

Lovell released the controller, allowed his ships to drift, and turned to his crewmates with a questioning look. Swigert, still at sea in the alien LEM, once again shrugged. Haise, who knew the LEM better than any man on board, responded similarly. Lovell turned his palms upward.

"It's not like we have any better ideas up here," he said.

"Do you think 37 minutes is enough?" Haise asked.

"Actually, no," Lovell answered. "Jack," he now said back to the Capcom, "we'll give it a try if that's all we've got, but could you give us a little more time?"

"O.K., Jim, we can figure out a maneuver for any time you want. You give us the time, we'll shoot for it."

"Then let's shoot for an hour if we can."

"O.K., how about 61 hours and 30 minutes?"

"Roger," Lovell said. "But let's talk back and forth till then and make sure we get this burn off right."

"Roger," Lousma said.

The hour until the free-return burn would be a frantic one for the crew. In a nominal mission, the flight plan allowed at least two hours for the so-called descent activation procedure, the ritual of configuring switches and setting circuit breakers that preceded any burn of the LEM's lower-stage engine. The crew would now have barely half that time to do the same job, and do it without sacrificing the necessary precision. On top of that, there was still the elusive fine alignment to establish, something that, with all the spacecraft's wild movements, Lovell was not yet close to accomplishing. But while the hour would be a breathless one aboard the ship, on the ground it would provide a chance to *draw* a breath.

At the flight director's console, Gene Kranz removed his headset, stepped back, and glanced around the room. What was going through his mind then was not the problem of the burn — his astronauts and his flight dynamics teams would take care of that. What was going through his mind was the problem of consumables. A few minutes earlier, Kranz had started the word going around Mission Control that as soon as the preparations for the burn began, he wanted to see his entire White Team downstairs in room 210, a spare data-analysis room in the northeast corner of the Mission Operations wing. Kranz knew that the free-return and PC+2 burns were indispensable to getting the crew home, but he also knew that they wouldn't mean a damn if the water, oxygen, and power aboard the LEM couldn't somehow be stretched to last the entire journey. Now, rumor had it, Kranz was going to pull his White Team from the rotation and get them working full time on tackling the consumables problem. Borrowing a crisis-management term from the military and from industry, Kranz would rename his staff the Tiger Team. For the rest of the flight, with the exception of recovery, the Tiger Team would remain in room 210, and the Gold, Maroon, and Black teams would have to handle the console shifts.

As Kranz looked around Mission Control, he conducted a quick head count and saw that most of his team members were still at or near their consoles. Over at the EECOM console, he also saw the face of another person who had not been here at the start of the evening but who he was now happy, indeed relieved, to see in the room: John Aaron.

Anyone who had been working at the Manned Spacecraft Center for even a few weeks quickly learned that John Aaron was the stuff of folk songs. Among the men in the Canaveral blockhouse and the Houston control room, there was no greater tribute a controller could be paid than to describe him, in the rough poetry of the rocketry community, as a "steely-eyed missile man." There weren't many steely-eyed missile men in the NASA family. Von Braun was certainly one, Kraft was certainly one, Kranz was probably one too. John Aaron, a twenty-seven-year-old wunderkind from Oklahoma, had recently become one as well.

Aaron arrived at the Agency in 1964, a flight mechanics engineer straight from college, pulling down an annual salary of $6,770. Assigned originally to work in spacecraft design, Aaron showed such technical acumen that by the spring of 1965 he had already earned a slot in Mission Control, manning the EECOM console for Ed White's history-making Gemini 4 space walk. By Gemini 5 he had become a permanent part of the EECOM rotation, regularly working the start-of-mission launch shift — the most stressful, least loved shift of any flight, one which was typically assigned to the best controller any console team had. Aaron's work had always been respected, but it didn't become truly celebrated until just the previous November, during the opening moments of Pete Conrad, Dick Gordon, and Al Bean's Apollo 12 lunar landing mission.

As on nearly every manned flight since 1965, the liftoff of Apollo 12 went smoothly — but only until seventy-eight seconds after ignition when, unknown to anyone, including the astronauts on board, the booster was struck by lightning. The crew felt a capsule-rattling jolt, and as the first stage of the six-million-pound rocket was burning at full throttle, Pete Conrad radioed down the alarming news that the bottom had fallen out of nearly every reading on every electrical system aboard his ship.

Aaron looked down at his console and practically recoiled: the EECOM screen was a strobing mass of blinking lights and ratty numbers that only moments before had held not a single bad reading. Around the room, the other controllers found that their data had gone screwy too. At the flight director's console, mission boss Gerry Griffin's headset filled with the voices of men asking what the hell the problem with the rocket was and what the hell the flight director intended to do about it. In a situation like this, the flight rules dictated an abort. When six million pounds of fully fueled, freshly launched Saturn 5 begins flying out of control, you don't wait for engineering analysts to tell you what's gone wrong. You light the escape rockets at the tip of the booster, accelerate the capsule away from the Saturn, and blow up the whole wayward missile over the empty Atlantic.

In the seconds that followed Conrad's call — seconds in which the abort decision would have to be made — Aaron took another look at his screen and noticed something funny. When the electrical system in the command module crashes completely, the amp readings on the EECOM console should drop to zero; failed fuel cells yield no power, it was that simple. On Aaron's screen, however, the numbers were not at zero but were hovering at about 6 amps, well below what they should have been if the electrical system was healthy, but well above the zero that would be expected if the systems had truly blown. Aaron realized he had seen that pattern before.

It had been a few years earlier, when he was monitoring a simulated countdown of a Saturn 1B booster and the rocket had accidentally tripped a circuit breaker on its telemetry sensors. The telemetry began sending all kinds of crazy signals to the blockhouse, none of which made any electrical sense. Aaron knew enough not to trust those numbers, and guessed that if he simply pushed a reset switch and rebooted the sensors, the muddled instrumentation would disappear and normal data would return. The young technician hit the appropriate circuit breaker, and the Saturn 1B was restored to health. Four years and half a dozen launches later, Aaron suspected he might be seeing the same problem again.

"Flight, EECOM," he called into the confusion of the Apollo 12 launch loop.

"Go, EECOM," Gerry Griffin said.

"Let's get the SCE aux switch to aux," he said with a bit more authority than he truly felt. "That might restore the readings."

"Do it," said Griffin.

Aaron pushed the reset switch and instantly, as he'd predicted, the numbers were restored. Fifteen minutes later, Apollo 12 was in Earth orbit and preparing to blast away toward the moon. Before that day was over, Aaron, to both the delight and envy of his fellow controllers, was informally accorded his steely-eyed missile man designation. Now, just five months later, the man who had done so much to save the mission of Apollo 12 was back in the control room to do what he could to save the crew of Apollo 13.

Gene Kranz circulated through Mission Control, collected his newly named Tiger Team, plus Aaron, and led them downstairs to room 210. The room was a large, windowless chamber filled with conference tables and chairs. The walls and work surfaces were festooned with strip-chart recordings — telemetry readouts from the earlier, quieter hours of the mission. At some point later on, these charts were to have been read and analyzed — an unhurried look back at a presumably routine flight. Now, as the fifteen men in Kranz's group filed into the room and settled into chairs or perched on the edges of tables, the printouts were swept aside in piles and lay crumpled on the floor.

Kranz took his place at the front of the room and folded his arms. The lead flight director was known as an emotional, even combustible speaker; this evening, however, he seemed firm but restrained.

"For the rest of this mission," Kranz began, "I'm pulling you men off console. The people out in that room will be running the flight from moment to moment, but it's the people in this room who will be coming up with the protocols they're going to be executing. From now on, what I want from every one of you is simple — options, and plenty of them."

"TELMU," Kranz said, turning to Bob Heselmeyer, "I want projections from you. How long can you keep the systems in the LEM running at full power? At partial power? Where do we stand on water? What about battery power? What about oxygen? EECOM" — he turned to Aaron — "in three or four days we're going to have to use the command module again. I want to know how we can get that bird powered up and running from a cold stop to splash — including its guidance platform, thrusters, and life-support system — and do it all on just the power we've got left in the reentry batteries."

"RETRO, FIDO, GUIDO, CONTROL, GNC," he said, looking around the room, "I want options on PC+2 burns and mid-course corrections from now to entry. How much can the PC+2 speed us up? What ocean does it put us in? Can we burn after PC+2 if we need to? I also want to know how we plan to align this ship if we can't use a star alignment. Can we use sun checks? Can we use moon checks? What about Earth checks?"

"Lastly, for everybody in this room: I want someone in the computer rooms pulling more strip charts from the time of translunar injection on. Let's try to see if we can't figure out just what went wrong with this spacecraft in the first place. For the next few days we're going to be coming up with techniques and maneuvers we've never tried before. I want to make sure we know what we're doing."

Kranz stopped and glanced once more from controller to controller, waiting to see if there were any questions. As was often the case when Gene Kranz spoke, there weren't any. After a few seconds he turned around and walked wordlessly out the door, heading back toward Mission Control, where dozens of other controllers were monitoring his trio of imperiled astronauts. In the room he left behind were the fifteen men he expected to save their lives.

Up in Aquarius, Jim Lovell, Fred Haise, and Jack Swigert were not privy to Kranz's off-console address, and, for now at least, they didn't need any pep talks. The designated time for the free-return burn was only half an hour away, and the LEM was not anywhere near ready to handle the job. Off to the right side of the craft, Haise was deep into his "descent activation" checklist, and the shorthand talk between the LEM pilot and the Capcom — largely familiar to Lovell, utterly alien to Swigert — proceeded in staccato bursts.

"On panel 11," Haise was saying, "we have GASTA under flight displays and commander's FDAI. Likewise, AC bus A breaker in."

"Roger. Copy that."

"On page 3, we'll circumvent step 4, since we're sitting with descent BATs on high-voltage taps."

"Roger. And in step 5, leave inverter circuit breaker open."

Listening with one ear, Lovell followed the exchanges, waiting for the occasional procedure that would require him to throw a switch or pull a breaker Haise couldn't reach. For the most part, however, the commander had his hands full with other things. Working his attitude controller more slowly and adeptly now, he had started to get the feel of his top-heavy ship and had managed to rotate it 360 degrees in all three axes. Yet everywhere he looked out his window, the cloud of rubble that surrounded Aquarius seemed uniformly dense. Firing the jets to move straight forward, he tried to fly out of the glowing haze, but it seemed to move with him, almost as if the gravitational attraction of the ships themselves — without the moon's or the Earth's gravity to compete with them — was drawing the rubbish particles along like iron filings moving with a magnet. Now and then Lovell radioed discouraging alignment updates to the ground, but none of these reports was strictly necessary. The vertiginous angle readouts on the navigation consoles told Mission Control all it needed to know about the LEM's loopy attitude.

With time running out, Lunney had dispatched two members of the Apollo 13 backup crew — John Young, the commander, and Ken Mattingly, the grounded command module pilot — over to the fixed-base simulators to see if they couldn't come up with some maneuvers Lovell could try. Young, in turn, had phoned Charlie Duke — the backup LEM pilot whose bout with German measles had caused the shuffling of the 13 crew in the first place — rousted him out of his sickbed, and told him to get right over to the Space Center. Tom Stafford, who knew better than most the perils of piloting a LEM close to the moon, was huddled with Lousma, trying to come up with some ideas of his own. Over the last few minutes, the ground-bound astronauts and the weary Capcom had transmitted a number of suggestions to Lovell, including turning the ship so that the body of the service module blocked the sun, and shifting the LEM so that its triangular windows faced into shadow rather than toward sunlight, but all of the suggestions yielded nothing. Everywhere Lovell looked, his view of the distant stars was obliterated.

Releasing the handle of his thruster control with an exasperated flip, the commander floated back from the instrument panel. Aligning his platform by the stars, he was now convinced, would be impossible. When Houston radioed up the coordinates for the burn, Lovell would have to punch the data into his navigation computer and hope that the guidance platform was sufficiently aligned to interpret the numbers correctly and point the spacecraft in the right direction. If it was, the crew would be headed for home. If it wasn't, they'd be headed somewhere else.

"We're going to have to go with what we've got," Lovell said to Haise and Swigert. "Let's hope it's good enough."

On the ground, the flight controllers came to the same conclusion at about the same time Lovell did, and could see, by the suddenly stationary attitude readouts, that the commander agreed with them. In theory, the arithmetic Lovell had performed and the ground had checked when the guidance platform was being transferred out of Odyssey should have been sufficient to align Aquarius's platform — but theory was a thin reed to hang on to. And now it seemed that this was all they were going to have. With Deiterich, Bostick, and the rest of the guidance crew watching, Gary Renick dialed up Lunney to tell him that the time had finally come to burn.

"Flight, Guidance," the GUIDO said.

"Go."

"O.K., we've got the vectors and we're ready to have the crew load them in."

"And you've verified these are the correct loads."

"We have."

"All right," Lunney said. "Capcom, you want to have the crew get ready for the loads?"

"Roger," Lousma said. "O.K., Aquarius," he called into the air-to-ground channel, "are you ready to copy the maneuver coordinates?"

"That's affirmative," Lovell said.

"Here we go, then. The purpose is mid-course correction for free-return burn," Lousma began formally. "And the coordinates are NOUN 33, 061, 29, 4284 minus 00213. HA and HP are NA. Pitch is . . ."

Lousma droned into the loop, reading throttle settings, burn times, engine angles, and Delta V objectives, each of which Haise dutifully read back to him. According to the numbers the LEM pilot and the ground were trading back and forth, the job of actually executing the burn would proceed in several steps. Once all the data were copied down, Haise would enter the attitude coordinates in the guidance computer, telling the spacecraft, relying on its original alignment, to swing itself into the correct position for the burn.

The tests Young and Duke conducted in the simulator — with the help of suggestions phoned in by Grumman — indicated that the on-board autopilot could hold the ship in a steady attitude during the engine operation. When the ship had stabilized in the proper attitude for firing, Lovell would deploy the LEM's landing gear, extending its four spidery legs to get them out of the way of the descent engine. Next the computer, relying on other instructions Haise typed into it, would fire four of Aquarius's attitude jets for 7.5 seconds. This procedure, known as ullage, was intended to jolt the spacecraft slightly forward and force the descent engine fuel to the bottom of its tanks, eliminating bubbles and air pockets. After that, the main descent engine would ignite automatically, firing at 10 percent thrust for 5 seconds — just enough to get the ship moving. Lovell would then reach for his T-shaped throttle and ease it forward to the 40 percent position and hold it there, firing the engine at a steady 3,948 pounds of thrust for precisely 25 seconds. At the end of that period, the computer would shut down the combustion chamber, and the engine would fall silent. The crew, in theory, would then be on the proper heading to take them around the moon and back to Earth.

Haise entered the guidance platform data into the onboard computer, and as Lovell glanced out the window to the left, Haise looked out the window to the right. Swigert craned for a look over both of their shoulders, and the thrusters began firing automatically, nudging the spacecraft to the attitude the Capcom had specified. Lovell, on cue, reached toward his instrument panel and flipped the switch that controlled the LEM's landing gear.

Before the mission, the commander had looked forward to this act as a significant milestone on his planned trip down to the lunar surface. Now the stretching and limbering of the legs had no such significance, and Lovell felt a brief pang of disappointment — a pang he quickly suppressed. The legs clicked into their down and locked position, and Lovell, glancing back out the window, nodded to Haise. The commander and the lunar module pilot then settled themselves in front of their instrument panels, and Swigert retreated to the ascent engine cover behind them. Haise watched the countdown timer on the LEM's panel and then clicked on the air-to-ground loop.

"O.K.," he said, "1 plus 30 to burn."

On the ground, Lousma conveyed this information to Lunney, who quieted the men on the loop and for the next 30 seconds made one last circuit of the room.

"O.K., we're about ready to go," he said. "CONTROL, is everything all right with you?"

"O.K.," the CONTROL officer said.

"Guidance O.K.?"

"We're good, Flight."

"FIDO?"

"O.K., Flight."

"TELMU?"

"We're go, Flight."

"INCO?"

"We're good, Flight."

"GNC?"

"O.K., Flight."

"All good here at one minute," Lunney said to Lousma.

"Roger, Aquarius," Lousma relayed up to Lovell. "You're go for the burn."

Like the last time Lovell approached the moon, during the triumphant Christmas-week flight of Apollo 8, there was largely silence in the final 60 seconds preceding the lunar burn. He flipped the "master arm" switch to On and quickly glanced around to see if everything else was in order. Guidance control was set to "Primary Guidance"; thrust control was on "Auto"; engine gimbals were enabled; the propellant quantity, temperature, and pressure looked good; the ship was maintaining the correct attitude.

The computer was in charge now, and Lovell's eyes focused on the countdown display. At 30 seconds before ignition, the display flashed "06:40," telling the commander that the computer had armed the engine. Twenty-two and a half seconds later — with 7.5 seconds to go before ignition — the little jets arrayed around the outside of the ship sprang to life as ullage was initiated. Lovell, Haise, and Swigert detected a slight push as the LEM shifted subtly beneath their feet.

"We have ullage," the CONTROL officer said.

Lovell stayed focused on the computer display, and just 5 seconds before the burn, it flashed its familiar "99:40," asking the commander again if he was sure he wanted to make this maneuver. Without hesitation, Lovell pushed the Proceed button, and once again a low vibration shook the ship.

"We have ignition, low throttle point," the CONTROL officer said.

Lovell held steady at that thrust for 5 seconds, then slid the lever up another 30 percent. The vibration around him grew stronger.

"Forty percent," he called down to the ground.

"Forty percent," the CONTROL echoed. "Rates look good."

"Rates are holding good, huh?" Lunney asked uncertainly.

"Looking O.K., Flight," the CONTROL reassured him.

"O.K., Aquarius, you're looking good," Lousma said.

Lovell nodded, still holding the thruster handle as the vibration continued around him.

"Still looking good," the CONTROL repeated.

Lovell nodded again, his eyes shifting from the instrument panel in front of him to the watch on his wrist and back again. The engine burned for 10 seconds, 20 seconds, a full 30 seconds, and then alarmingly appeared to go beyond that. Then, a mere instant longer than it was planned to last — 0.72 second longer, as measured by the Mission Control computer — the burn concluded and the engine went silent.

"Shutdown," the CONTROL called.

"Auto shutdown," Lovell said.

In the spacecraft and on the ground, Lovell and the controllers glanced instantly and simultaneously at their trajectory and Delta V instruments, and smiled at what they saw. The speed of the ship had increased almost exactly as much as it had been designed to increase, and the predicted pericynthion had risen from the 60 miles that would have helped the spacecraft ease into lunar orbit, to a loftier 130 miles that would help swing it home.

Lovell waited for the order from Houston to "trim" the burn; this maneuver, a small pulse from the attitude-control jets, was usually required even after routine engine firings, to refine the trajectory further. Boone, Renick, Bostick, Deiterich, and the other navigation officers looked at their consoles to see how much trim would be required, and were stunned at the answer: none was needed at all. According to the numbers on their monitors, this burn, which violated all the rules of common sense and flight procedures, had come off perfectly, placing Apollo 13 on a path around the back of the moon and then straight toward home.

Somewhat incredulous, Lousma called up to the ship, "You're go, Aquarius. No trim required."

"You say *no* trim?" Haise asked, looking over at Lovell.

"That's affirmative. No trim required."

"Roger," Lovell said with a grin.

"O.K.," Haise echoed, smiling as well.

Pushing back from his instrument panel, Lovell rubbed his eyes with the heels of his hands. He was relieved, but only fleetingly so. While the trajectory readouts on his instrument panel were encouraging, most of the other data told another story entirely. Letting his eyes fall on his environmental and power readouts, he could not help doing some quick-and-dirty calculating. If the path the ship was now following held and its speed was not changed further, the crew should reach Earth at about the 152-hour mark in the mission, or about 91 hours from now. The $3\frac{3}{4}$-day transit time was twice as long as the LEM — with only two men aboard — was equipped to last. Though the ground had referred only in passing to a PC+2 burn, Lovell was pretty sure one was coming. However, even if he floored the descent engine when he came around the other side of the moon and burned it until the tanks ran dry, he couldn't see how this would cut much more than a day off the time of the trip. This left at least another full day of flight beyond even the outside reach of the LEM's miserly supply of consumables. It was now 2:43 in the morning on Tuesday the fourteenth. The way Lovell figured it, the earliest he could expect to be home would be sometime after midnight on the morning of Friday the seventeenth. His LEM was not up to that trip.

"If we're going to get home," Lovell said to Swigert and Haise, "we're going to have to figure out some other way to operate this ship."

In room 210 off Mission Control, Bob Heselmeyer was doing some quick-and-dirty calculating of his own. Unlike Lovell, the Tiger Team TELMU had pencil, paper, strip charts, data books, power profiles, and a support team of technical personnel to help him crunch his numbers. But like Lovell, he wasn't pleased with what those numbers told him.

Of all the consumables the crew would need for the homeward trip, it was the oxygen that caused the most obvious concern — but oxygen, it seemed, would be the least of the worries. The original flight plan had called for Lovell and Haise to spend two days on the lunar surface, venturing out of the LEM for two separate exploratory hikes. This meant completely venting and repressurizing the cabin atmosphere twice. To make this outgassing and refilling possible, Aquarius had been supplied with more O_2 than any of the LEMs that had been used on Apollos 9, 10, 11, or 12. Even with three men on board, the oxygen would be drawn from the system at just 0.23 pound per hour, a rate of consumption that the topped-off tanks could easily handle for more than a week.

The elimination of carbon dioxide would be another matter. Like the command module, the LEM was equipped with cartridges of lithium hydroxide, or LiOH, designed to trap CO_2 molecules and filter them out of the air. The ship carried two primary cartridges that could last more than a day and three secondary ones that could be snapped into place when the first two became saturated. Altogether, these five air scrubbers could work for only 53 hours — and only if there were just the two planned men in the LEM. With another passenger, the life span of the cartridges would fall to less than 36 hours. Odyssey's LiOH supply would remain untouched throughout the flight, but it could not be cannibalized to help Aquarius; the CO_2-scrubbing mechanisms of the two ships were designed differently, and the square cartridges from the command module would not fit in the receptacles of the LEM, which were round. No matter how much oxygen the lunar module was carrying, the poisonous CO_2 would soon start to crowd the life-sustaining oxygen out of the air, and the crew would asphyxiate by around three o'clock Wednesday afternoon.

Electricity was in equally short supply. A fully functioning, up-and-humming LEM required about 55 amps of current to operate. To survive for up to four days instead of the planned two, however, the spacecraft would have to lower that amp consumption to 24. Such a power-down was draconian, but it was at least manageable.

Going hand in hand with the onboard supply of electricity, however, was the onboard supply of water. All of the power-consuming hardware in the LEM generated heat, which if not dissipated properly could ultimately cause the equipment to burn out and shut down. Threading through the ship's systems was thus a web of coolant pipes carrying a solution of water and glycol. As the liquid mix rushed through the plumbing, it picked up excess heat and carried it to a sublimator; there, the water evaporated and flowed into space as steam — taking the unwanted warmth with it. The tank of pure water the LEM carried aloft was intended to satisfy both the cooling system's incessant thirst and, not insignificantly, the crew's. But neither one was supposed to be sipping from the cistern for the four days this LEM would be up and running. All told, the spacecraft carried about 338 pounds of water, which the equipment alone gulped at a rate of about 6.3 pounds per hour. To survive the trip back to Earth, that rate would have to be reduced to just over 3.5 pounds. To achieve that, electrical use would have to be cut even further, to barely 17 amps.

Agonizing over these numbers, Heselmeyer, like Lovell, pushed back and rubbed his eyes. The LEM was not meant to be flown like this.

No one, except perhaps the folks at Grumman, even knew if the LEM *could* be flown like this. Heselmeyer frowned and turned to the men sitting around him.

"If we're going to get them home," he said, "we're going to have to figure out some other way to operate this ship."

At 2:45 in the morning, just as the descent engine of the LEM was completing its burn, Tom Kelly and Howard Wright touched down at La Guardia airport. The private plane that had been promised them had indeed been waiting at Logan, and the flight from Boston to New York had taken just over an hour. Bethpage was less than half an hour's ride from the airport, but tonight it was going to take a little longer. Unlike Boston, which was experiencing seasonable temperatures for mid-April, New York was sunk in an early spring chill, and with light rain and mist in the air and the thermometer in the low thirties, the highways on Long Island were slick with ice. Kelly and Wright drove from the airport to the plant as fast as they could, but occasionally had to slow to a crawl to avoid fishtailing off the road.

When their car at last pulled up to the plant, Kelly looked out the window and was stunned at what he saw. The old red-brick airplane factory and its huge metal LEM factory were usually all but deserted at this time of night. The engineering support team that would be here to monitor the LEM during a lunar mission consisted of only a few people, and their cars would normally be lost in the asphalt prairie that surrounded the buildings.

Tonight the scene was very different. As far as Kelly could tell, there were day-shift crews here, evening crews, design crews, assembly crews, and crews that for the life of him he couldn't name. Even in an emergency, Grumman wouldn't call in this many people in the middle of the night. Clearly, these were employees who had heard on their own about the emergency in space and had come to the plant unbidden.

When Kelly entered the building, the halls were as choked as the parking lot was, and when the workers recognized the engineering manager, they waylaid him and asked what they could do to help. Kelly pushed through the group, slightly dazed, reassuring each person he passed.

"We'll get you busy," he said. "We'll get you all busy. We're going to need everybody's help in some way."

Kelly made his way to the engineering support room, where the small crew that had been on duty when the accident happened had grown many times in size. Since the moment he and Wright met at the airport in Boston, Kelly had been guessing about the same numbers Heselmeyer and the others in Houston had been totting up. Now, however, was the first time he would have any hard data to play with.

Sitting down with the men at Grumman who had been consulting with the men at Mission Control, he took his first long look at their figures — and wished he hadn't. The numbers were shocking. Kelly had never tried to operate a ship at these kinds of power levels, and hoped never to have to try. He understood that if he pushed his LEM too far, he was likely to lose the ship altogether, but if he didn't push it, he was likelier still to lose the crew.

Kelly knew only one thing with any certainty: he hadn't been kidding when he said he'd need a lot of help.

7

January 1958

WHEN JIM LOVELL arrived at the Navy's Aircraft Test Center in Patuxent River, Maryland, he was not by any measure a relaxed man. The twenty-nine-year-old lieutenant had just completed a coast-to-coast drive from northern California with a six-months-pregnant wife, a two-year-old son, a four-year-old daughter, and a five-year-old Chevy that had quit on him in practically every state between San Francisco Bay and the Chesapeake Bay. It was a dreary, wet January afternoon when the Lovell family rolled into Pax River, the kind of dingy coastal day when it's too warm to snow, too cold to rain, and the sky fills instead with a dispiriting sleet. For a man who had just driven 2,900 miles, it was not a cheering welcome. But if Jim Lovell's mood was gloomy as he slowly drove his Chevy around the unfamiliar Navy base, Marilyn Lovell's was even worse.

For the past four years, the Lovell family had made their home on the outskirts of San Francisco, in a small community near the Moffett Field Naval Air Station, and Marilyn had liked it just fine. The Milwaukee native who'd moved east to Washington to be with her Naval Academy beau had never much cared for the bitter midwestern winters or the wilting Potomac summers, and when the Navy had assigned her new husband to an air base on the temperate California coast, she could not get her bags packed quickly enough.

Arriving in Sunnyvale, Marilyn had made it her mission to find a house that would fit the idyllic image she had of West Coast living, and in short order she located one: a pretty bungalow on a street with the pretty name of Susan Way. For the first year the Lovells lived here, Marilyn had concentrated on turning the modest house into an authentic home, hanging curtains and wallpaper, buying whatever furniture she could on her husband's military salary, and filling the back and front yards of the tiny plot with lilies, tulips, geraniums, and blue hyacinths, all of which thrived in the California sun.

It was while the couple was living here with their sixteen-month-old daughter Barbara that their first son, Jay, was born. When the family was given orders to move in 1958, Marilyn was pregnant again. As she and Jim were packing up their house, they decided that if the new baby was a girl, they would name her Susan, in honor of the pleasant street they were leaving behind.

In Maryland the accommodations would not be so idyllic. Jim Lovell had been sent east with the rank of lieutenant and the job of apprentice test pilot, and neither designation carried much in the way of privileges or perks. The on-base apartments assigned to young officers and their families were in a residential complex known to the occupants as the Cinderblocks. True to their name, the units were a series of boxy, bunker-like flats, built of military-issue concrete blocks that were painted too dingy a shade to qualify as white, too bright a shade to qualify as ecru, and nowhere near subtle enough a shade to qualify as ivory.

Inside, the apartments were even more forbidding, with undersized windows, claustrophobically low ceilings, and exposed pipes poking through the floor, running up the walls, and disappearing into the apartments upstairs. The Navy provided nine hundred square feet of these unlovely quarters, a figure that was non-negotiable whether the couple had children or not. When the Lovell family Chevy pulled up in front of these blocky, Bauhaus-like structures in the steady drizzle, Marilyn's mood sank. Unloading boxes onto the wet curb in front of their new home, Lovell glanced a little edgily at his wife. "Well," he said, "I'll admit it's not quite California."

"No," Marilyn said, checking for the fifth time the address on the rain-splotched card that the clerk in the housing office had handed her, "that it's not."

"I'm afraid you won't be able to grow many flowers here," Lovell said.

"Mm-hmm."

"Do you think you can put up with it for a while?"

"I married a naval aviator. This comes with the job."

"I guess it does," Lovell said, relieved.

"But I'm telling you one thing," Marilyn said. "If we ever have a fourth child, we are *not* naming him Cinderblock."

The Navy figured it could get by with such bare-bones barracks mostly because test pilots' wives like Marilyn Lovell were well schooled in the military tradition of making do without making waves, and the test pilots themselves, who would be immersed in the business of learning to fly unproven airplanes, wouldn't be home enough to notice their surroundings anyway.

The job Lovell was taking had little appeal for the ordinary pilot; for aviators with a bit of the fighter jock in them, however, it was a true plum. But it was undeniably dangerous.

On any given day, test pilots napping in their quarters or finishing a report at their desks knew that they might hear — or feel, actually — the unmistakable concussive thump of aircraft striking turf a mile or two away, followed by the roar of rescue trucks, the scream of sirens, and the black, black plume of smoke rising from the horizon.

Often, the pilot in the cockpit was able to punch out of the flawed and falling plane in time to parachute to safety and tell the designers just what was wrong with the vehicle they'd given him to fly. Just as often, though, he wasn't able to, and another eager pilot who had volunteered for the hazardous life of a Pax River aviator would never get the chance to volunteer anywhere else again. While there were always a few fliers who relished a high-risk job like this, most of their wives — certainly military wives with a two-year-old son, a four-year-old daughter, and a five-year-old Chevy they could never hope to keep running without a man around the house — were not so enthusiastic.

In order to maximize the chance that both planes and pilots would survive their time together, the newly arrived fliers at Pax River were put through a grueling six months of test pilot school. In January 1958, when Jim Lovell and the rest of his incoming class were checking in, the military was rolling out a new generation of experimental combat aircraft, including the A3J Vigilante, the F4H Phantom, and the F8U-2N Crusader. When the fledgling test pilots were not aloft in training vehicles learning the skills that would eventually allow them to test these new jets, they were closed up in classrooms studying such aeronautical arcana as trajectory plotting, shock wave mathematics, time-to-climb rates, and dynamic longitudinal stability. At the end of the workday, when the students retreated to their tiny quarters, there were additional assignments to be done, preparing reports for their instructors on both that afternoon's flying and that morning's class-work.

Lovell immersed himself in this intensive training, stealing at least an hour or two of study time each night. He used a bedroom closet as a makeshift study, a plank of plywood as a makeshift desk, and a helicopter helmet stuffed with cotton to shut out the din of his two preschool children and newly born infant daughter. As it turned out, the self-imposed isolation paid off. When the half-year hazing was over, Lovell, it was announced, had finished first in his class, edging out even such Pax River wunderkinder as Wally Schirra and Pete Conrad.

Ordinarily, such a lofty class rank meant big things for a Pax River pilot. The various flying assignments available to new graduates were by no means equal in prestige, with the luckiest aviators getting tapped for the Flight Test Division, a squadron of men who would get first crack at a newly arrived aircraft, taking it aloft to find out how fast and nimble a machine it really was. The next group, the Service Test Division, would judge not the agility of a plane but its endurance, flying it more ploddingly through the skies to determine how far it could be pushed before it needed maintenance and repairs. Further down the ladder was the Armaments Test Division, where, as the name implied, fliers would concern themselves principally with testing the guns, bombs, and rockets a new aircraft carried. Last and least desirable was Electronics Test, where naval aviators would do little more with their training than circle lazily over military bases and nearby towns, gathering data on antenna patterns and radar trans-mitters.

All Pax River pilots lived in fear of assignment (consignment, actu-ally) to Electronics Test — all but the one who finished first in his class, that is. It was unwritten but long-established policy that this high achiever would receive any assignment he requested. What no-body in the entering class of 1958 knew, however, was that for this one year the policy had changed. The commander of the Electronics Test Division had made it clear that he was tired of being routinely denied the top finisher in the graduating class, and would like at least once to get the pick of the piloting litter. Obligingly, the school commander, Butch Satterfield, had promised that the number one man in the next group — Jim Lovell's group — would be sent packing for Electronics Test.

"Sir?" Lovell said, presenting himself in Commander Satterfield's office the afternoon the assignments were announced. "I wonder if there might be some mistake in my assignment."

"Mistake, Lieutenant?"

"Yes, sir," said Lovell. "I — I sort of assumed I was going to be assigned to Flight Test."

"What would make you assume that?" Satterfield asked.

"Well, sir, I did finish first in my class . . ."

"Lieutenant, is there anything *wrong* with Electronics Test?"

"No, sir," Lovell lied.

"Did you know the commander of Electronics Test specifically requested the best pilot in your class?"

"No, sir. I didn't."

"Well, he did. So why don't you get over there on the double. And when you do get there, remember to thank him."

"Thank him, sir?"

"For asking for you personally."

As Lovell assumed his lowly station as radar tester, events just thirty-five miles up the Potomac were conspiring to change his fortunes again. Half a year after the Soviet Union stunned the world with the launch of its Sputniks, the United States government was still struggling to overcome the blows to its technological pride. Impatient with American failures, and wary of more Soviet successes, a reluctant President Eisenhower stepped in. Since World War I, the government had operated a low-profile federal agency known as the National Advisory Committee on Aeronautics, or NACA, which was given the job of staying abreast of airplane and jet technology and helping the government determine how to spend its R&D money. What Eisenhower wanted now was to expand NACA's role to include vehicles that could fly above the atmosphere, turning the agency into something closer to a National Aeronautics and Space Administration.

One of the biggest priorities the fledgling NASA had was building a spacecraft that could put a human being into orbit. Overseeing this project was Dr. Robert Gilruth, an aeronautical engineer at the Langley Research Facility in Virginia. Though no spacecraft yet existed that could accomplish such an improbable mission, one of Gilruth's

Apollo 13 lifted off on April 11, 1970, at 1:13 P.M. Houston time, or 13:13 military time — a numerologically inauspicious start.

Photographs not otherwise credited are courtesy of NASA

Jim Lovell, an Annapolis graduate and Navy test pilot, was one of nine men NASA chose in 1962 to fly in its upcoming Gemini and Apollo programs. In the right-hand seat of a Gemini simulator in 1965, Lovell rehearses for his first trip into space.

Lovell and Commander Frank Borman set a space endurance record in December 1965, spending two weeks sealed inside Gemini 7. The tedium was broken during the twelfth day of the mission when Wally Schirra and Tom Stafford, aboard Gemini 6, met — and photographed — them in orbit, achieving the first-ever rendezvous in space.

In November 1966, Jim Lovell went back into space aboard Gemini 12, splashing down in the Atlantic after five days in orbit. For this flight, Lovell was in the left-hand seat, the commander's seat; Buzz Aldrin was in the right.

Lovell and Aldrin aboard the aircraft carrier Wasp after the splashdown of Gemini 12, the final flight in the ten-mission Gemini series.

The honor of being chosen for the first Apollo mission went to Mercury and Gemini veteran Gus Grissom, Gemini veteran Ed White, and rookie pilot Roger Chaffee. It was a mission they would never get to fly.

On January 27, 1967, a fire broke out in the Apollo 1 spacecraft during a countdown rehearsal, killing Grissom, White, and Chaffee in seconds. The temperature inside the cockpit was thought to have climbed to over 1,400 degrees. (Archive Photos)

Less than two years after the Apollo 1 fire, Frank Borman, Bill Anders, and Jim Lovell were chosen to fly Apollo 8. It was the first time human beings would orbit the moon. When the mission was over, all three astronauts were named *Time* magazine's Men of the Year.

Marilyn Lovell and three of her children watch as the Saturn 5 booster leaves the ground on December 21, 1968, launching the crew of Apollo 8 toward the moon.

On Christmas Day, while Apollo 8 orbited the moon, a Rolls-Royce pulled up to the Lovells' house to deliver a mink jacket with a card that read, "From the Man in the Moon." Later that day, Marilyn took Jeffrey, Susan, Barbara, and Jay to church, wearing her mink in the mild Houston winter.

Above: Apollo 13's lunar module waits in a maintenance building at the Kennedy Space Center prior to being loaded into its compartment in the Saturn 5 booster. The foil-covered ship was designed to sustain two men for just two days; it was eventually called upon to keep three men alive for four days.

Left top: Command-service module 109 — which would also become part of Apollo 13 — hangs in a clean room before being hoisted and attached to the Saturn 5. Inside the spacecraft was an oxygen tank with a very questionable history.

Center: Gene Kranz, the ex–Korean War aviator and NASA's lead flight director, was the unquestioned boss of Mission Control during the Apollo program.

Bottom: Chris Kraft, deputy director of the Manned Spacecraft Center and flight director during the Mercury and Gemini programs, was Kranz's one-time mentor. During Apollo 13, he would see how well his successor had absorbed his lessons.

Rookie astronaut Fred Haise, a leading candidate for one of NASA's lunar landing missions, practices retrieving core samples on a dry, moon-like patch of ground at the Kennedy Space Center in Florida.

Marking his fourth and final trip into space, Jim Lovell was chosen to command Apollo 13, the third lunar landing mission. At Ellington Air Force Base in Houston he answers reporters' questions after test-piloting the lunar landing training vehicle.

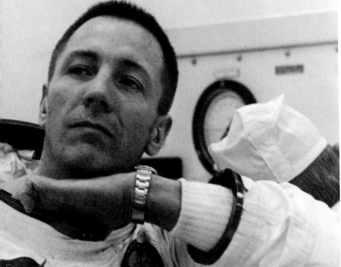

Jack Swigert, Apollo 13's backup command module pilot, was promoted to the prime crew when first-string pilot Ken Mattingly was exposed to German measles. The flight plan called for Swigert to remain in lunar orbit while his crewmates descended to the surface.

Fred Haise, who made it his business to know as much about the lunar excursion module as the men who designed it, was chosen as LEM pilot for Apollo 13.

Apollo 13's destination was the Fra Mauro highlands near the western edge of the moon. In the Lovell home in Houston, Jim Lovell familiarizes his family with the lunar landscape. (Ralph Morse/LIFE magazine © Time Warner)

At 8:24 P.M. on the evening of April 13, 1970, Apollo 13 began its last TV transmission to Earth. Forty-three minutes later, oxygen tank two exploded. In Mission Control, Flight Director Gene Kranz (foreground, back to camera) watches Fred Haise on the viewing screen.

Moments after the explosion, the crew of Apollo 14 (Stu Roosa, Ed Mitchell, and Alan Shepard) confer in Mission Control. The Apollo 14 crew was originally slated to fly Apollo 13, but after delays in Shepard, Roosa and Mitchell's training schedule, Lovell agreed to swap missions and, significantly, spacecraft.

Near the glassed-in VIP gallery in Mission Control, a team of astronauts review data coming back from the blast-damaged Apollo 13. Left to right: chief astronaut Deke Slayton; Ken Mattingly, the grounded command module pilot; Vance Brand, a rookie astronaut and capsule communicator, or Capcom; Jack Lousma, another astronaut and Capcom; John Young, Apollo 13 backup commander.

Deke Slayton holds a makeshift lithium hydroxide air filter, which the Apollo 13 crew would have to build in order to filter poisonous carbon dioxide out of their cockpit. Left to right: Flight Director Milt Windler; Slayton; Howard Tindall, deputy director of Flight Operations; Sig Sjoberg, director of Flight Operations; Chris Kraft, deputy director of the Manned Spacecraft Center; Bob Gilruth, director of the Manned Spacecraft Center.

Father Donald Raish, an Episcopal priest, leads Marilyn Lovell (in striped dress) and nine other friends and astronaut wives in a communion in the Lovell home. On the wall are mementos from the Apollo 8 lunar orbit mission. (Bill Eppridge/LIFE magazine © Time Warner)

In Jim Lovell's study, the first three men on the moon watch TV coverage of Apollo 13 with the family of the man who would have been the fifth. Seated, from left: Pete Conrad, Buzz Aldrin, Blanch Lovell (Jim Lovell's mother), Barbara Lovell, Jeffrey Lovell, Marilyn Lovell, Susan Lovell. Standing: Neil Armstrong. (Bill Eppridge/LIFE magazine © Time Warner)

Above left: Less than five hours before splashdown, the crew of Apollo 13 jettisoned their service module, seeing for the first time the devastation wrought by the exploding oxygen tank. An entire external panel was blasted away, revealing the entrails of the ship and the scorched spot where the oxygen tank should have been. (Archive Photos) *Above right:* Flotation balloons and inflatable collars kept Odyssey upright as frogmen prepared the crew's life raft. *Below:* A spacecraft that had spent six days traveling through space at translunar speeds is winched aboard the deck of the carrier. The little Mercury and Gemini ships could be plucked from the ocean with little difficulty. The Apollos required a good deal more muscle. (Capt. L E. Kirkemo, USN ret.)

One Navy man welcomes another as Lovell, Naval Academy class of '52, brings the crew of his three-man ship aboard the much larger one. (Capt. L E. Kirkemo)

Never was a homecoming prayer more in order than after Apollo 13 safely hit the ocean. (Capt. L E. Kirkemo)

Swigert, Haise, and Lovell grin and listen as they are introduced to the crew of the helicopter carrier Iwo Jima. The U.S. Navy was prepared to scramble ships nearly anywhere in the world to recover the Apollo 13 astronauts. Ultimately, the Iwo Jima got the call. (Capt. L E. Kirkemo)

The champagne was alcohol-free, but the prime rib, lobster, and "Moonfruit Melba" more than made up for it as Lovell and Swigert sat down for a celebratory dinner with the crew of the Iwo Jima. Haise, sick and feverish, remained in the ship's infirmary. (Capt. L E. Kirkemo)

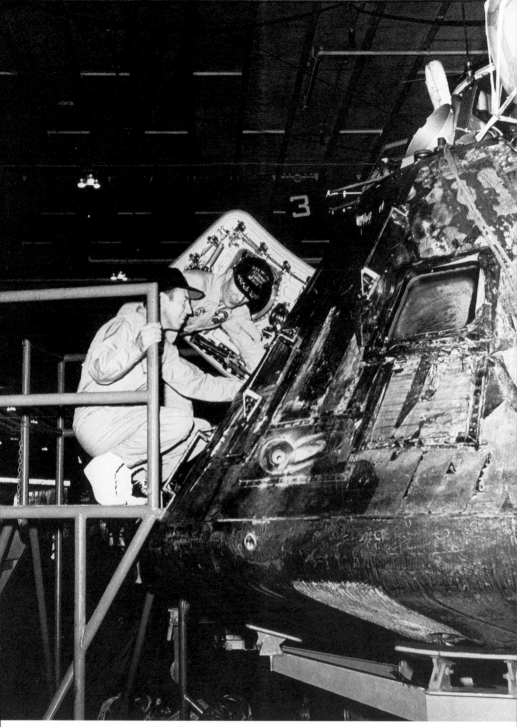

During the trip from the splashdown site to port, astronauts often found the need to return to their spacecraft, stored belowdeck, to retrieve this or that forgotten item. Safely aboard the Iwo Jima, Lovell and Haise could afford to peer into their command module with something close to fondness. Just a day earlier, the cold cockpit seemed a lot less friendly. (Capt. L E. Kirkemo)

Even on the deck of the carrier, the gaunt and stubbly Lovell showed the strain of his translunar trip. Only his crisp blue jumpsuit, which had been given to him in the helicopter that took him to the carrier, was fresh. (Capt. L E. Kirkemo)

A troika of flight directors — Gene Kranz, Glynn Lunney, and Gerald Griffin —
puff celebratory cigars in Mission Control after Apollo 13's splashdown.

Haise, Lovell, and Swigert wave to cheering sailors before going below for
medical exams. According to the original flight plan, the recovery ship should
not have expected the astronauts for another four days — after their landing at
Fra Mauro.

Following a shower — their first in nearly a week — Lovell, Haise, and Swigert receive a congratulatory call from President Nixon. The haggard-looking Haise was suffering from a kidney infection and a high fever.

In Hawaii, Lovell salutes for the national anthem after Nixon awarded the Apollo 13 crew the Medal of Freedom. The president had flown Lovell's and Haise's wives and Swigert's parents out to Hawaii on Air Force One.

Nobody planned any tickertape parades when the crew of Apollo 13 set out for the moon. When they returned, Chicago saw to it that they got one. Swigert and Lovell wave from the lead car; Haise, still ill, could not attend.

own priorities was to begin selecting the "astro-nauts" — or star sailors — who would eventually pilot whatever ship the space agency did build.

Gilruth and his staff spent several weeks determining what qualities would be needed in such fliers — height, weight, age, training — and when they were done, passed those requirements on to the Air Force and Navy. The military ran the criteria through their new, room-sized computers and came up with a list of 110 names that seemed to fit the bill. That day, telexes went out to the first thirty-four of those men, several of whom were performing their military service at the Aircraft Test Center in Patuxent River, Maryland.

The men filing out of the auditorium of the Dolley Madison House, at the corner of H Street and East Executive Avenue in Washington, D.C., were a somewhat dumbfounded group. This was going to be, so they had been led to believe, a military briefing; it would deal, so they had been certain, with military matters. The meeting that had just adjourned, however, had been like no other briefing they had ever attended.

Actually, there had been plenty of clues that the conference today would be more than a little unusual. For one thing, the pilots had been instructed not to wear their uniforms. Business attire, preferably suits, would be the clothing order of the day. For another thing, they had been instructed not to tell anyone that they were coming here at all — not their wives, not their squadron mates, not even any other men they suspected might be attending. The dispatch Jim Lovell had received had been very specific on this point.

"Report to Bureau of Personnel for CNO OP5 Special Projects Matters," it said. "CNO" was short for chief of naval operations; "OP5" meant Operations Division 5, the division that oversaw Pax River; and "Special Projects Matters" was code for "Don't ask, just show up, things will be explained to you in time."

Just as perplexing as the secrecy of the telex was the address it ordered Lovell to report to. It wasn't unheard of for a Navy officer to be summoned to Washington on official business, but when he was, he was usually instructed to present himself either at the Pentagon proper or at one of the Navy's numerous offices scattered around the District. Lovell's telex, however, ordered him to report to some place called the Dolley Madison House, a townhouse somewhere in Washington that, as the name implied, was once the home of the fourth First Lady, and had since been converted into a government office.

Jim Lovell was at his desk in the Electronics Test Division when his telex arrived. It was a Wednesday, and the dispatch directed him to be in Washington the next morning. Lovell was sorely tempted to go straight to the other men in his test flight class, show them the dispatch, and ask if any of them had received it as well and what they thought it might mean. But the novice lieutenant took military protocols seriously, and if the chief of naval operations told him to keep quiet about something, he was not about to do otherwise. Besides, by tomorrow morning he'd have his answer anyway.

Lovell awoke in the pre-dawn hours of Thursday and put on his alien business suit. As he threw his overnight bag into the back seat of his car, he saw that, sure enough, he wasn't the only Pax River flier slipping away before the sun was up. There was Pete Conrad, nodding a self-conscious hello in his own starchy civvies and heading for the parking lot. There was Wally Schirra, driving quickly off the base with a word to no one and a single wave to the guard at the gate.

All of the men leaving that morning had been careful to respect the secrecy demanded in the CNO telex, but a few hours later, as they milled about in the auditorium of the Dolley Madison House with more than thirty other Navy and Air Force pilots, they were free to speculate about why they were here at all. So far, no one had any hard answers. The smart money had it that the Department of Defense was developing some new type of rocket plane, perhaps to replace the X-15. Others offered the fanciful guess that the meeting had something to do with space. This was where Lovell was placing his bet, but it was a bet he made silently with himself. It wouldn't do to share such a foolhardy idea with the other men in the room.

After the last of the pilots arrived, the doors at the back of the room were shut, and a balding, scholarly-looking fellow, Dr. Robert Gilruth, took the podium.

"Gentlemen," he said with little preamble beyond introducing himself, "we've asked you here to discuss Project Mercury."

For the next hour Dr. Gilruth described to the group of silent fliers a plan that was, by turns, the most ambitious, the most dramatic, and the most lamebrained thing they had ever heard in their lives. What Gilruth had in mind, he said up front, was to take a man — quite possibly one of the men in this room — and have him orbiting the

Earth within three years. The ship in which the man would perform this feat in would be less a vehicle than a sort of, well, *capsule,* a titanium funnel measuring barely six feet across the base and only nine feet from bottom to top. The capsule, with the pilot sealed inside and strapped to a form-fitting couch, would be carried into orbit atop an Atlas booster, a ballistic missile with a thrust of 367,000 pounds.

Half a dozen or so men would be chosen to take these rides, each of whom would fly into orbit for a somewhat longer time than the one before him. The last man to go aloft would spend up to two days in outer space. The overall program would be administered by the civilian government, so while all of the men who volunteered for it would retain their military status and rank, they would no longer be reporting to the Department of Defense. Instead, they would be responsible to a new government agency, the National Aeronautics and Space Administration. To date, NASA had not had time to flesh out its plans much beyond those Gilruth had just described, but if anyone had any questions, he'd be happy to try to answer them.

The pilots looked at one another tentatively, teetering between genuine interest and open amusement. After a moment, a hand went up.

Didn't the Atlas booster have a reputation for, well, blowing up on the pad, one pilot wanted to know.

To be honest, Gilruth admitted, there *had* been some accidents in the past, but the engineers agreed that most of the problems had been ironed out.

Had a prototype of the, uh, capsule actually been built yet, someone else asked.

Built? No, Gilruth admitted. But some first-rate minds had already put together some first-rate blueprints.

How would the pilot control this capsule during the flight, asked another.

He wouldn't, Gilruth answered. The entire mission would be controlled automatically from the ground.

What about the landing, a fourth flier wanted to know.

No landing, Gilruth said. Splashdown. Little rockets would knock the capsule out of orbit and a parachute would carry it down into the ocean.

And if the little rockets didn't work?

That, Gilruth said, was why he wanted test pilots.

When the question-and-answer session ended, Gilruth told the men to take the night and think about what they had just heard. There would be more meetings tomorrow and the next day with doctors, psychologists, and other project officials, and any other questions would be answered at that time.

When Gilruth left the stage, the men rose and, shushing one another with their eyes, began filing out, heading toward the hotels that had been booked for them around the city. The Pax River group made their way to the Marriott Hotel on Fourteenth Street, and most of the men could not wait to get there. This Gilruth might have more meetings set up for Friday and Saturday, but what the pilots needed right now was their own meeting, a private one. After checking in at the hotel and dropping their bags, Lovell, Conrad, and Alan Shepard, a Pax River alumnus, headed down to Schirra's room, closed the door, and, as an afterthought, slipped on the chain.

"So," Lovell said. "What do you gentlemen think?"

"Well, it ain't the X-15, that's for sure," said Conrad.

"It's hazardous duty, *that's* for sure," Schirra said.

"I'd sure feel better if they were using something besides that Atlas," Lovell offered. "That thing's supposed to have walls so thin they collapse altogether if they're not pressurized."

"The lighter it is, the faster it goes," Shepard said.

"And the higher it blows," Lovell added.

"I'm not so worried about putting my ass on the line," Schirra said. "I'm worried about putting my career on the line."

The other men in the room looked at one another and nodded: Schirra had straightforwardly expressed what it was they were all thinking. While nobody in the room especially cared to strap himself on top of a booster and go the way of the unfortunate satellite that had been blown off the pad by the exploding Vanguard rocket, nobody especially feared it either. In the test pilot business, there was always a real chance that the next cockpit you climbed into would be your last. What the fliers counted on, though, was the professional payoff for taking so outrageous a risk. If they somehow stayed the course, if they came back down with their test vehicles and their bodies intact, they believed their ascent through the military chain of command would be greatly enhanced — from solo aviator to command of an eighteen-plane squadron, to command of a four-squadron air group, to a tour of duty at the Pentagon, to command of a small ship like an oiler or troop transport, and at last to command of an aircraft carrier and perhaps even flag rank. The path was long, with countless opportunities to bollix things up along the way, but it was also very clearly marked. The key was not to get sidetracked. Spend a few years doing some loony, at-the-fringes work — volunteering for some half-baked space task group, for instance — and you might never get back on course.

Wally Schirra, for one, had worked too hard getting where he was to screw around with things now. And the more he reflected on this — the more he questioned out loud how much those fellows over in the Dolley Madison really understood the sacrifice they were asking the men in the Marriott to make — the more the other people in Schirra's room began to have reservations themselves.

Or at least they did at first. But after a while, Lovell, for one, began to wonder. Might this seemingly crackpot program actually be the fastest way up the ladder? Might it be possible to vault straight over the squadron commanders and the air-group commanders and the troopship commanders and ride the Atlas rocket directly to flag rank? And might Wally, for all his good fellowship, really know this? Was he trying, in some small way, to introduce just enough doubt so that a few of his early competitors would drop out before they even got started?

There was no way of knowing. But Lovell, who had dreamed and breathed and studied rockets for twenty years, who had built his own little Atlas — right down to the explosion — more than a decade and a half ago, was not about to let a few career concerns prevent him from climbing atop a real rocket now. Within half an hour of arriving at the Marriott, every pilot in Wally's room had accepted that Project Mercury might well spell the death of his naval career. And every one had decided he'd do whatever he had to do to be a part of it.

The preliminary physical exam for Project Mercury took place at the Lovelace Clinic in Albuquerque, New Mexico. Of the elite group of men considered for inclusion in the program, thirty-two had elected to accept the invitation. This group was divided into smaller units of six or seven men and shipped, one unit at a time, to Lovelace for a week of medical tests. Of the six men who arrived at Lovelace in Jim Lovell's group, five would make it through the seven grueling days successfully.

From the moment the astronaut candidates arrived, it was clear that what this NASA had in mind would be like no physical exam they had ever experienced before. Flying willingly into the arms of the doctors were six remarkably healthy men in the prime of life, all of whom wanted desperately to pass medical muster and be accepted into the program, and none of whom, as a result, was inclined to object to whatever procedures the New Mexico hospital had planned. The doctors were almost giddy at the prospect.

On tap for the compliant pilots over the next seven days were blood studies, cardiac x-rays, electroencephalograms, electromyelograms, electrocardiograms, gastric analyses, hyperventilation tests, hydrostatic weighing tests, vestibular balance tests, whole-body radiation tests, liver function tests, bicycle stress tests, treadmill stress tests, visual perception tests, pulmonary function tests, fertility tests, urine tests, and intestinal tests. In submitting to these whole-body violations, the candidate astronauts would have their livers injected with dye, their inner ears filled with cold water, their muscles punctured by electrified needles, their intestines filled with radioactive barium, their prostate glands squeezed, their sinuses probed, their stomachs pumped, their blood drawn, their scalps and chests plastered with electrodes, and their bowels evacuated by diagnostic enemas at the rate of up to six per day.

At the end of the nightmarish week, the six men would each be handed a card, saying either that they had checked out so far and were to report to Wright Patterson Air Force Base in Dayton, Ohio, for still more tests or that they had been found wanting and were to report back to their previous billets, with the government's thanks for their time and their sacrifice. The first six days proceeded as gruelingly as the six pilots had been told they would, and on the seventh day, all but one were issued cards instructing them to report to Wright Patterson.

"Have you been ill lately, Lieutenant?" Dr. A. H. Schwichtenberg asked when Jim Lovell appeared in his office, carrying his orders to return to Maryland.

"Not that I know of, sir. Why?"

"It's your bilirubin," the doctor said, opening the folder in front of him and scanning the top sheet. "It's a little high."

"I didn't even know I had bilirubin," Lovell said.

"Well you do, Lieutenant. We all do. It's a natural liver pigment, but you have a little too much of it."

"Can that make me sick?" Lovell asked.

"Not really. Usually it means you *have* been sick."

"But if I was sick, that means I'm better now."

"That's true, Lieutenant."

"And if I'm better now, there's no reason I can't go on with the program."

"Lieutenant, I have five men out there who don't have a bilirubin problem, and twenty-six more on the way who probably don't. I have to base my decisions on something. I know you've been through a lot in the past week, and we thank you for your time."

"Couldn't we repeat the liver test?" Lovell ventured. "Maybe it was wrong."

"We already did," Schwichtenberg said. "It wasn't. But we do thank you for your time."

"You know," Lovell persisted, "if you only accept perfect specimens, sir, you'll only wind up with one kind of data. Taking someone with a little anomaly means you'll learn even more."

Schwichtenberg closed Lovell's file, pushed it aside, and looked up. "We thank you," he repeated slowly, "for your time."

Jim Lovell returned to the Cinderblocks and the Electronics Test Division of Pax River the next day. Two weeks later, Conrad followed him back. A few weeks later, both men sat glumly in front of their televisions as their Pax River colleague Wally Schirra, along with Al Shepard, Deke Slayton, John Glenn, Scott Carpenter, Gordon Cooper, and Gus Grissom, lined up in front of a swarm of reporters in the same Dolley Madison auditorium where Lovell and the others had gathered, and were proclaimed the nation's first astronauts.

Lovell watched the ceremony on his little TV set in his little family quarters, and over the course of the next three years returned to the same set to watch those men make the journeys he had been deemed unfit to make. There was Al Shepard's fifteen-minute suborbital flight atop a tiny Redstone rocket, and Gus Grissom's identical flight atop an identical missile; there was John Glenn's ride on the bigger Atlas, a ride that at last took an American into low Earth orbit. There was Scott Carpenter's later Atlas trip, duplicating Glenn's.

At the same time the Mercury astronauts were riding into test-piloting history, Lovell's own flying career was, in its modest way, improving too. Electronics Test had turned out to be not quite the backwater he had feared it would be, merging in 1960 with the far more dynamic Armaments Test and becoming the Weapons Test Division. As fighter jets grew more sophisticated, the weapons they carried did too, and it was soon apparent that a pilot who hoped to deliver his payload of bombs or rockets effectively would have to be less a mere bombardier and more a technician and electrician. The first new plane to go into development that completely integrated armaments and electronics was the F4H Phantom, an all-weather aircraft intended especially for use in night combat.

Lovell, who had trained on the carrier Shangri-La for just this kind of hair-raising piloting, was named program manager of the Weapons Test group helping to evaluate the new vehicle. The change in assignments meant both added prestige and frequent travel, mostly to the McDonnell Aircraft plant in St. Louis, where the plane was being built. Ultimately, it also meant a change in housing. When the F4H tests were completed and the time came to train the pilots who would fly the plane, Lovell was chosen for this job too. He moved his growing family out of the Cinderblocks and down to Fighter Squadron 101, at the Oceana Naval Air Station in Virginia Beach, where he would work as a flight instructor.

Toward the end of the Mercury program, in the summer of 1962, after Deke Slayton had received the devastating news that he was being grounded because of a heart fibrillation and only Wally Schirra and Gordon Cooper remained to fly into space, Lovell was in the ready room at Oceana, sipping coffee and preparing for an afternoon flight. He picked up a copy of *Aviation Week & Space Technology* and began flipping through it. With Mercury winding down, the magazine had been running frequent stories about the upcoming Gemini program and the two-man ships the men chosen for it would fly. This week's issue had no news about the spacecraft itself, but buried at the end of the news section was another, smaller item, reporting on a recent NASA press release. "NASA Will Add New Astronauts," the headline said. "Between Five to Ten Additional Astronauts for NASA's Manned Space Flight Program Will Be Selected Next Fall."

Lovell banged his coffee down on the magazine table, sloshing a bit onto his hand, read the two-sentence squib hurriedly, and, before he was even through, decided he would volunteer again. Yes, he was older now, about to turn thirty-four. But age, he figured, also meant experience. Yes, ten openings at NASA meant that even more men would submit their names than last time, but Lovell's name was already known to the folks at the Agency. And yes, there was the matter of the bilirubin. However, with four successful Mercury flights now in the books and four healthy pilots none the worse for their experience, Lovell suspected, or at least hoped, that NASA might be less concerned with finding perfect physical specimens than with looking for the best possible fliers. Quite likely, Lovell's first rejection would disqualify him the second time around. But, he decided as he sat in the ready room, he had to try again. Flying into space to help test a new spacecraft was, he figured, something more of an adventure than flying to St. Louis to help test a new jet.

"Hey, Lovell, telephone for you," someone called in the squadron office at Oceana.

Jim Lovell looked up wearily from the briefing report he had been studying for half an hour and called back across the room, "Who is it?"

"I asked him, but he wouldn't say."

Lovell dropped the report, pushed the flashing light on his phone, and picked up the receiver.

"I'm looking for Jim Lovell," the person at the other end of the line said.

The voice sounded familiar, but Lovell couldn't quite place it. It was September 13, 1962, more than two weeks since he'd returned from NASA after completing the interviews for the Gemini program, and in the time he had been at the Agency he'd met a lot of people and heard a lot of voices. If he knew this person, he wasn't sure how.

"This is he," Lovell answered.

"Jim, this is Deke Slayton."

Lovell sat up in his chair but said nothing. The NASA physicals had taken place at Brooks Air Force Base in San Antonio, Texas, and, like the last time Lovell had been through this, he had spoken mostly to doctors. Unlike the last time, he had made it past the first medical round and had been sent on for interviews at Ellington Air Force Base in Houston. Since being cut from the active astronaut roster, Deke had been given the job of director of Flight Crew Operations, overseeing the activities of all the current astronauts and the selection of all the future ones. Lovell had spent a lot of his time in Houston being interviewed by Deke, and he'd been expecting a call from him. But whether that call would bring very good news or very bad news he didn't know.

"Jim, are you there?" Slayton asked.

"Uh, yeah Deke, I'm right here."

"Well, I was just calling about the new astronaut team."

"Uh-huh," Lovell said, his throat slightly dry.

"And I was wondering," Slayton said, "if you'd like to come work for us."

"Would I?" Lovell shouted, loud enough for the other men in the office to turn and look.

"That's what I'm asking *you*." Slayton laughed.

"Yes, yes," Lovell stammered. "Of course."

"Good," Slayton said. "Glad to have you aboard."

"Glad to be aboard. Can you tell me who else made it? Did Pete make it this time?"

"You'll find out. What we need now is for all the new crewmen to come down to Houston for an announcement to the press, the day after tomorrow. We want to keep things secret till then, so what I want you to do is catch a flight down here tomorrow and then get a cab and go straight to the Rice Hotel. Have you got that?"

"Rice Hotel," Lovell repeated, reaching for a piece of scrap paper and scribbling illegibly.

"And when you get there, say that you have a reservation in the name of Max Peck."

"Ask for Max Peck," Lovell said.

"No, don't ask for Max Peck. Tell them *you're* Max Peck."

"I'm Max Peck?"

"That's right."

"Deke?"

"Hmm?"

"Who's Max Peck?"

"You'll find out."

Slayton clicked off the line. Lovell kept the receiver in his hand, pressed the cradle button, and hurriedly called Marilyn.

"We're moving," Lovell said when his wife answered the phone.

"Where?" Marilyn asked.

"Houston."

There was a pause. Marilyn, Lovell could swear, had *audibly* smiled.

"Come home," she said. "You should be the one to tell the kids."

When Lovell arrived at the William Hobby Airport in Houston the next day, the reception was a subdued one, indeed a nonexistent one.

Slayton was apparently serious about keeping things quiet, and as Lovell stepped off the plane, all that greeted him was a blast of hot, humid air. Making his way through the terminal, the overdressed Yankee did as he'd been instructed and hailed a cab.

During the ride to the hotel, Lovell told himself to pay attention; as long as he was going to be moving his family here, he figured he ought to begin learning his way around. As the cab sped along the Gulf Freeway, Lovell noticed a large billboard mounted atop a building. "While in Town, Stay at the Rice Hotel," the sign said. "Your Host in Houston!" Underneath, in smaller letters, were the words: "Max Peck, Manager."

Confused, Lovell tried to catch a second glimpse of the sign as the cab sped past, but his eye wasn't quick enough. At the hotel, he paid the driver, walked inside, and looked around. There was still no sign of Deke or anyone else who looked even remotely attached to NASA. Feeling more than a little lost, Lovell went to the desk with as much nonchalance as he could muster and nodded a hello to the clerk.

"I have a reservation for a single room," Lovell said. "I'm Max Peck."

The clerk was a girl of barely college age. "Excuse me? You're who?"

"I'm Mr. Max — I mean, Mr. Peck. I'm Max Peck."

"Uh, I don't think so," said the girl.

"No, really," Lovell said unconvincingly.

All at once, from behind the clerk, another hotel employee appeared, a large, jovial-looking man whose name tag identified him as Wes Hooper.

"I'll take care of this, Sheila," he said to the girl, and then turned to Lovell. "Glad you could make it, Mr. Peck. We've been expecting you. Here's your key, and do let us know if anything is not to your satisfaction."

Somewhat dazed, Lovell thanked Mr. Hooper and walked off in the direction he had been shown. This, he thought, was silly. Secrecy and managing the press were one thing, but all this cloak-and-dagger, cat-and-mouse stuff was ridiculous. Lovell reached his room and flipped his bag and himself onto his bed. Almost immediately his phone rang.

"Hello?" he said warily as he picked up the receiver. There was no response. "Hello," he repeated, more tersely now.

"Who's this?" a voice on the other end asked.

"Who's *this?*" Lovell demanded in return.

"This is Max Peck."

"*Who?*" Lovell now shouted.

"Max Peck."

"Do you work for this hotel?"

"Uh, no," the voice answered. "I'm just a guest. And I think you have my room."

"I don't think so," Lovell said.

"I do," the voice responded.

"Lookit," Lovell snapped, "I don't know how many Max Pecks we have here today, but for the time being you can consider me one of them. This is my room, the reservation was made in my name, and I intend to stay here. If you have a problem with that, I suggest you take it up with the manager. I understand his name is Max Peck!"

Lovell hung up the phone. Slayton may have had a reason for all this nonsense, but he couldn't fathom it. He *was* sure of one thing, though: he wasn't going to sit alone in his room waiting for someone to straighten things out. It was after 6 P.M., and Lovell was going to shower, change, go downstairs, and have some dinner. If sitting down for a drink and a meal in the hotel's restaurant was going to blow his cover, then his cover would have to be blown.

As soon as he got to the lobby, Lovell discovered that if he was unconcerned about keeping a low profile, the other men NASA had brought here today were utterly indifferent. Sitting comfortably in the middle of the hotel lounge, smoking a pipe and sipping a drink, was Pete Conrad. With him, nursing his own drink and smoking a pungent, oversized cigar, was Navy pilot John Young. Lovell could have leapt in the air: Conrad and Young, both Pax River alums. He knew the two men well, respected them both, and would be pleased to orbit the planet in any kind of spacecraft on any kind of mission with either of them. Hurrying across the lobby but avoiding Young and Conrad's notice, he sidled up on his fellow aviators and clapped them both on the shoulders.

"So the fleet has landed," Lovell announced.

"Jim!" Conrad said, turning around and peering through the cloud of tobacco smoke that shrouded his head.

"How'd you two sneak into this program?" Lovell asked, exchanging handshakes and embraces with Conrad and Young.

"Same loophole they let you in through, I suppose," said Conrad.

"Well, I guess it's a loophole they should keep," Lovell said. "So far, we seem to be an all-Navy group."

"Not quite," Young said, looking toward a chair a few feet away. Following Young's gaze, Lovell noticed for the first time another unmistakably military-looking man, sipping his own drink and reading a newspaper.

"Ed?" Young said to the man, who turned and smiled. "Like you to meet Jim Lovell. Jim, this is Ed White, Air Force."

The man rose, took a step toward Lovell, and put out his hand. Lovell studied his face for an instant. There was something vaguely familiar about it.

"Good to meet you," Lovell said, extending his own hand.

"Actually," White said, "we've met already."

I knew it, Lovell thought, and he had a dim recollection from years ago.

"But only by phone," White added.

No, that wasn't it. "Oh?" said Lovell.

"I'm the Max Peck who called your room."

"That was you? Are we all Max Peck today?" Lovell asked. Conrad and Young nodded. "Well, I can't wait to meet all the other ones they're flying in here."

None of the four men was certain who else NASA would be bringing down to the Rice Hotel today, but if the Agency wasn't going to be here to meet the new arrivals, they would. Lovell, Conrad, Young, and White stationed themselves in the lounge, ordered more drinks, then moved over to the restaurant for dinner.

Throughout the evening, they kept a group eye cocked toward the lobby, and over time, five other men showed up, all wearing the same slightly dazed expression Lovell had worn when he walked in the door. There were Frank Borman, Jim McDivitt, and Tom Stafford, all from the Air Force. There was Elliot See, a civilian test pilot for General Electric. And there was, showing up last, Neil Armstrong, another civilian — one who did much of his test-piloting work for NASA itself. Given his Agency pedigree, the odds had been good he would make this cut. The new arrivals were flagged down by the men already there, introduced around, and invited to join them for a drink.

When the ninth and last pilot had arrived, all the men sat and looked at one another, more than a little amazed. Of the hundreds of test pilots who had submitted their names to NASA earlier in the year, these nine had been selected. Every one, with the exception of Armstrong and See, had been steadily climbing the military ladder for most of his career, and every one of them had abruptly — and, it could easily be argued, recklessly — left it behind. It wasn't clear when they would get to fly in space, how well they would do when they got there, or even if, like poor Deke, they might never get there at all. But sitting in the quiet gloaming of the hotel lounge, drinking their drinks and smoking their smokes, one thing *was* clear: if the career ladder they were abandoning was preferable to the one they would now be climbing, it certainly didn't seem that way at the moment.

8

Tuesday, April 14, 1970; 7 A.M.

MARILYN LOVELL was thinking about Charlie Bassett and Elliot See when she woke up on the morning after Apollo 13's accident. Marilyn hadn't thought about Bassett and See for a long time; like a lot of people associated with NASA, she preferred to put such matters out of her mind. But on Tuesday morning, April 14, that was impossible.

Technically, Marilyn did not wake up on the fourteenth, because she did not actually go to sleep the night before. It was 7 A.M. when Marilyn began her day on Tuesday, leaving her bedroom after a restless nap she had begun only an hour or so earlier. At 6, Betty Benware and Elsa Johnson had rousted her from in front of the television, where she had spent most of the previous evening, moved her toward the staircase, and told her in the clearest terms that she was to get some sleep. Marilyn had objected, insisting that she wasn't tired, but Elsa and Betty reminded her that her children would be getting up soon, and she owed it to them, if not to herself, to take at least a quick nap. Marilyn grudgingly agreed, lay down on her bed, and after precisely an hour got back up and returned to the family room. When she did, it was with thoughts of Bassett and See in her head.

It was on February 28, 1966, that Charlie Bassett and Elliot See died. Marilyn was at home that day, tending to Jeffrey, her fourth — and, she pledged to herself, final — child, born barely seven weeks earlier. The winter just ending had been a frantic one, what with her husband leaving for his first space flight — the two-week mission of

189

Gemini 7 — during the penultimate month of her pregnancy, and reporters falling over themselves to interview the expectant mother and stoic wife. Shortly before Christmas Jim returned, and shortly after that Jeffrey was born, and Marilyn promised herself that the remaining weeks until spring would be as quiet as she could make them. She couldn't speak for her astronaut husband, but *she* intended to spend as much of her time as she could staying home and tending to her newborn, with occasional assistance from a nurse or sitter if Timber Cove cabin fever grew too great. On February 28 the nurse was on duty, and Marilyn was enjoying a tranquil moment while Jeffrey took a late morning nap. Then the phone rang.

"Marilyn," the sober voice on the other end of the line said, "it's John Young calling from the Center." Marilyn would have recognized Young's voice even if he hadn't identified himself. He had joined NASA with her husband four years earlier, and had been the first of the new recruits in space, flying with Gus Grissom the previous March in Gemini 3.

"John, how are you?" Marilyn said, genuinely happy for the phone company.

"Not well. There's been an accident," Young said. "Not involving Jim," he added quickly. "Jim's all right. But Charlie Bassett and Elliot See aren't. They were trying to land their T-38 in the fog in St. Louis, overshot the runway, and hit the parking lot outside the McDonnell Aircraft building. They were both killed instantly."

Marilyn sat down slowly. She had known the Bassetts reasonably well. Charlie and his wife lived on the other side of nearby Taylor Lake in the community of El Lago, but since Charlie was in the third group of astronauts to join the program — the group that followed Jim's — Marilyn had never had the chance to do much more than chat with the couple at NASA functions. The Sees, however, were Timber Cove folks, living just a few doors down from the Lovells. Elliot and Jim were both members of the second astronaut class, and in recent years Marilyn Lovell and Marilyn See had become close friends, often joking with each other about their common names, their near-common addresses, and their uncommon lives as astronaut wives. In the weeks Marilyn Lovell had been home with Jeffrey, Marilyn See had been a welcome visitor.

"Has anyone talked to Marilyn yet?" she now asked Young.

"No," he said. "That's what I want you to do."

"You want me to tell her that Elliot was killed?" Marilyn said, her voice rising.

"No," Young said. "I want you to do something much harder — *not* tell her. Somebody should be there with her right now, but she can't be told anything until I can come over and notify her officially. We don't want some overeager newspaperman knocking on her door. You do remember what happened when Ted Freeman was killed?"

"Yes, John," Marilyn said, recalling the horror the NASA wives had felt several years earlier when the story began circulating about the reporter who knocked on the Freemans' door, wanting a few quotes from the family before the family even knew they had something to be quoted about.

"Good," Young said. "I appreciate your help."

Marilyn hung up the phone, found the nurse upstairs, and told her she'd be going out for a while to have a cup of coffee with a friend. Then she slipped into her coat and made her way slowly down the block. Marilyn See was in her kitchen when Marilyn Lovell arrived, and when she saw her friend coming up the walk, she brightened visibly and waved her in.

"I was just getting ready to go and visit you," Mrs. See said to Mrs. Lovell. "I didn't want you to have to come out."

"That's O.K.," Marilyn said. "I can always use the break. Besides, I've got at least an hour until Jeffrey gets up."

"Is the nurse in today?"

"No," Marilyn answered absently. "I mean yes. Yes, she is."

Marilyn See looked at her oddly. "Are you all right?" she asked. "You seem distracted."

"No, no. I'm fine."

For the next twenty minutes or so the two women sat and chatted and drank coffee. Then they heard the sound of tires in the driveway, and both women turned. Out the kitchen window they could see a dark car pulling up to the house. Inside were John Young and another, unidentified man. NASA personnel did not visit astronauts' houses unannounced unless there was a reason. Usually it was a bad reason. The two wives' eyes met and held. Marilyn Lovell's gaze flickered for just a second, and in that second, Marilyn See knew what that reason was.

Without a word, Marilyn Lovell got up, answered the door, brought the visitors into Marilyn See's kitchen, and stood by her while they told her their news. Then she showed the men out, sat with her friend, embraced her, and finally did the only thing a friend and a pilot's wife could do in such a situation: began phoning other friends and other pilots' wives and explaining what had happened.

Within minutes, the first few wives arrived, and Marilyn Lovell ran back to her own house, jumped into her car, and sped off to the local elementary school to fetch the Sees' children and bring them home before they heard the news from somebody else. When she returned, the house — as she knew it would be — was filled with women and their somewhat awkward astronaut husbands, surrounding Marilyn See and speaking whatever helpful words they could. Marilyn Lovell stood back from the group and watched for a long moment. She could not help wondering what her friend and neighbor was hearing at a time like this, what she was seeing, whether she even knew they were all there. Marilyn Lovell, like the wives of all the other astronauts, knew there was only one way she could ever know exactly what her friend was experiencing. But that was a possibility she had trained herself not to think about.

Four years later, on the fourth day of the mission of Apollo 13, Marilyn was learning those answers, and she deeply, dearly wished that she weren't. The night before had been a frenzied one from the moment the Bormans, the Benwares, the Conrads, the McCulloughs and others from the NASA community arrived at the Lovells', parking on whatever patch of street, lawn, or sidewalk was available. Marilyn had no way of gauging the number of people who had been in her house at any one point, but judging by the number of full ashtrays and half-full coffee cups strewn around the family room this morning — not to mention the dozen or so people still wandering about the house or talking quietly among themselves in front of the TV — she would put the number at close to sixty.

For all of the friends and neighbors and protocol officers populating Marilyn's house, the people she had thought would need her attention the most, but who for now had demanded it the least, were her children. Jeffrey had been the first of the younger Lovells to be openly disturbed by the tumult in the family room, but Adeline Hammack had apparently satisfied his curiosity without raising his concern. The Lovell daughters had not needed any such explanations yet, and Marilyn was grateful for that. Barbara Lovell had evidently deduced her father's peril and, judging by her darkened room, her clasped Bible, and her decision to retreat into sleep, was coping with it in her own self-sufficient way. Marilyn was reluctant to disturb her with words of encouragement she was not yet seeking. Nor did she want to bother her younger daughter Susan, who, remarkably, had been sleeping through all of the early morning turmoil. Soon enough, Susan would awaken on her own and learn what the neighbors and the newsmen and much of the rest of the world already knew. Marilyn saw no reason to deprive her daughter now of what would almost certainly be the only true sleep she'd have for days.

Fourteen-year-old Jay was a somewhat different story. Phoning St. John's Military Academy at three in the morning, Marilyn had awakened one of the faculty members in Jay's dormitory, explained the crisis to him as briefly as she could, and asked him to break the news to Jay right away, before an early-rising cadet could turn on a radio and tell him first. Ideally, Marilyn would have preferred to talk to her son herself, but that, she knew, might make things harder for him. Adolescent boys are given to more bravado than is strictly good for them, and adolescent boys who also happen to be cadets are given to even more. If Jay learned the news from his mother, he would almost certainly feel obliged to put on a stronger front than was good for him. Better for him to be told by a third party and then call home for more information when he had absorbed the news. The faculty member understood this, assured Marilyn he was on his way to Jay's room, and Marilyn had been trying to keep one line clear for Jay's return call ever since.

The only other family member Marilyn had had to worry about this morning was Blanch Lovell, Jim's seventy-five-year-old mother. Blanch, who had been strong enough and self-sufficient enough to raise her only son alone, had recently suffered a stroke and moved into a nearby nursing home, Friendswood. As near as Marilyn could tell, Blanch understood that her boy was to have gone into space this week, and she appeared to understand that his destination was the moon. But whether she knew he was going there to land or simply to go flying around it again was unclear, and this, Marilyn figured, was a good thing. With the landing now canceled, it was possible that when Blanch turned on her television, she would not even notice that her son's moonwalk was not to be found. She *would* notice, however, the reporters and their stories about the misfortune that had befallen his ship. To spare her the worry the rest of the earthbound

Lovells would be experiencing today, Marilyn phoned the staff at Friendswood and instructed them that, until they were told otherwise, the television was to be removed from Blanch's room, and any questions Blanch asked about the flight were to be answered only with a smile and a hearty thumbs up.

Now, as the sun was beginning to climb in the sky, Marilyn Lovell walked into the kitchen for a cup of coffee she did not especially want, and could sense that her house was beginning to stir again. Looking through the window, she saw that so too was the street in front of her house. The sidewalk, the driveway, and the lawn were suddenly clogged with men holding notebooks, microphones, and TV cameras. Several broadcast trucks had also converged, parking wherever they could find space. Marilyn looked at the scene somewhat incredulously. Weren't these the same people who had been so conspicuously absent for the last two days? The same ones who didn't carry her husband's broadcast last night, who buried the news of his impending launch on the weather page, who gave more time to Dick Cavett's jokes than to Jules Bergman's reports?

From the study, the temporary hot line that had been set up between her house and the Space Center began ringing, and Marilyn could hear a protocol man answering it. There was a minute or so of quiet conversation, and then the man, one she didn't remember meeting last night, approached her in the kitchen.

"Mrs. Lovell," he said uncertainly, "that was the Public Affairs office. The networks have contacted them and want to know if it's O.K. with you if they put up a broadcast tower in your yard for some of the coverage they're planning."

"A broadcast tower? On my lawn?"

"Uh, yes. They're holding on the phone and I need to know what to tell them."

Marilyn thought for a moment. "Nothing," she said.

"Mrs. Lovell, I have to tell them something."

"No, *you* don't have to tell them anything, but I have to tell them plenty."

Marilyn walked back to the study with the protocol man following her and picked up the phone. "This is Marilyn Lovell. I'm told the network people want to build some kind of tower on my front lawn. Is that right?"

"Well, yes," the voice from the Public Affairs office said. "Is that O.K.?"

"Couldn't they have set their tower up yesterday or the day before if they had wanted to?"

"Well, yes," the voice answered. "But that was different."

"How so?"

"Well, the flight was going fine then. Now it's . . . you know, more of a news story."

"If landing on the moon wasn't enough of a news story for them," Marilyn said, "I don't know why *not* landing on the moon should be. You tell the networks that they're not to put one piece of equipment on my property from now through the end of this flight. And if anyone has any problem with that, tell them they can take it up with my husband. I'm expecting him home on Friday."

Marilyn Lovell hung up the phone, left the study, and walked back to the kitchen to finish her coffee. There would be no more discussions about broadcast towers for the rest of that day.

In the Public Affairs building at the Manned Spacecraft Center, reporters were a good deal more welcome, but up till now, the press, for the most part, had not been accepting the invitation. The Public Affairs department actually occupied two buildings. On one side of a gravel courtyard stood a large administration building, with offices for the employees, vaults and library space for the thousands of pages of documents and millions of feet of film that made up the NASA archives, and a small briefing room for impromptu press conferences and announcements. On the other side of the courtyard was a longer, lower building containing a several-hundred-seat auditorium, where NASA held press conferences announcing such headline-making events as the decision to send Apollo 8 to the moon, the selection of the first crew that would land on the moon, and the tentative dates, crew assignments, and landing sites for the missions that would follow. It was also here that people like Chris Kraft, Jim McDivitt, and Sig Sjoberg would be brought for midnight press conferences when something went disastrously wrong on one of those missions.

During the months-long layoffs between flights, when the auditorium building went largely unused, it was transformed into a visitors' center, with spent Mercury and Gemini capsules and showcases full of uniforms, helmets, and other artifacts displayed in the lobby surrounding the auditorium itself. During missions, this memorabilia was swept away and replaced by desks and portable typewriters for the use of reporters here to cover the flights. In July 1969, when Apollo

11 was flying, the 693 reporters who had been granted press credentials competed furiously for the limited desk space the Agency could provide. For Apollo 12, that November, the competition had eased considerably, with only 363 reporters showing up and most of them finding a place to sit. For Apollo 13, the number was down to just 250, and the press corps found they had desks to spare.

Over the past ten hours, things had changed. Beginning with the first reports of the accident, dozens of TV, radio, and print reporters, who had been patching together their coverage from stories sent over the wire services, began showing up at the Space Center's doorstep, requesting clearance and credentials and asking for access to whatever press events NASA now planned to hold. The Public Affairs officers welcomed the prodigals warmly, issued them security badges, handed them press kits, and showed them to the auditorium, where they were free to choose places to sit among the rapidly filling desks.

In Mission Control, a few hundred yards away from the auditorium building, Brian Duff was well aware of the rapidly gathering media and was pleased they were here. Duff was the Space Center's director of Public Affairs, and in the ten months he had held the job, he'd run his department according to a single, overarching rule: When things are going well, tell the media everything they want to know; when things are going badly, tell them even more. This morning, Duff was trying hard to live by the second part of that code.

Duff came by his respect for the art of public relations the hard way. Back in 1967 he was working in the Agency's Public Affairs office in Washington when NASA conducted its inquest into the deaths of Gus Grissom, Ed White, and Roger Chaffee. In the assessment of even the most partisan NASA supporters, the handling of the Apollo 1 fire was a debacle for the Agency. No one complained about the scientific investigation: the spacecraft was autopsied, the causes of the fire were tracked down, and the fixes were made in record time for so knotty an engineering problem. What the Agency botched, most people agreed, was the public relations part.

Before the Apollo 1 spacecraft had even cooled, on the night of January 27, both Cape Canaveral and the Manned Spacecraft Center had been closed down and sealed off, and reporters had been told that substantive answers and detailed information would be forthcoming only after a board of inquiry had had a chance to investigate the accident and determine its cause. NASA did quickly appoint such a board, but what escaped no one's notice was the very fact that it was NASA doing the appointing. This was an Agency crisis in which grave errors were made by Agency officials, and Agency men were being assigned to investigate it.

The media did not respond well to such self-policing. Within days, Bill Hines, the *Washington Star*'s space reporter whom NASA had come to view as a sort of windsock for the prevailing public mood, pointedly asked in one of his columns just what the Agency foxes were doing guarding their own hen house. A congressional subcommittee picked up on Hines's theme, announcing that NASA's investigation into its own mistakes would not be sufficient to set the problem to rest, and that the House of Representatives would soon conduct hearings of its own. The Senate went even further, planning yet another inquiry which would, according to Minnesota Senator Walter Mondale, investigate the possible "criminal negligence" of the nation's space agency.

Ultimately, nothing even remotely criminal was uncovered, but the episode took its toll. By the time the Apollo spacecraft was repaired and a new crew was ready to fly, the Agency found it had squandered most of the public relations capital it had accumulated over a decade. In 1969 Julian Scheer, the Public Affairs director who had helped lift the Agency to the heights of popularity it enjoyed before the fire, only to see so much of his work undone by the administrators who conducted the fire investigation, left his post, and Brian Duff was appointed to take his place.

Duff moved to fix things fast. In the event of future emergencies, the new director proposed and the Agency bosses agreed, NASA's doors would remain open, and the press's questions would be answered without delay. Within hours of an accident, a press conference would be held to announce just how much the Agency knew and when they expected to know more. Most dramatically, two more flight controller consoles would be installed in Mission Control, up in the glassed-in VIP viewing gallery at the rear of the room. During a mission, these would be manned around the clock by pool reporters of the media's own choosing who would be able to dial up the same data, the same backroom channels, the same flight director's conversations that were available to the controllers themselves, and report what they learned to the world outside.

Duff was pleased with the changes, but until last night — until the early morning hours of today, actually — he had not had a chance to see how they worked. So far, he was satisfied. The Kraft-McDivitt-Sjoberg press conference had been convened at 12:20 Houston time, just over three hours after Jack Swigert first reported that his command module had a problem. The additional media had begun arriving shortly after that and had been promptly informed of the dates and times for the announcements that would follow. Already, Glynn Lunney was preparing for the next such event, a routine change-of-shift briefing when his Black Team went off console at around eight o'clock this morning.

As day dawned in Houston and the Public Affairs auditorium was being readied for Lunney, Duff himself was in Mission Control. Like the newly plugged-in pool reporters in the VIP gallery, the Public Affairs officers had their own console with which they, too, could monitor the flight. Unlike the pool reporters, their console was down on the floor of Mission Control itself, in the left-hand corner of the fourth and last row. And unlike the pool reporters, the Public Affairs officers could use their console to do more than simply call up data and eavesdrop on conversations.

Throughout the flight, the officer on duty would patch into the air-to-ground channel and provide a running commentary of what was being discussed, translating the technical chatter in the half-whispered tones of a TV sports reporter covering a golf match. It was this communications feed, with the voices of the Capcom and the astronauts overlaid with the voice of the Public Affairs commentator, that would be sent to the networks and broadcast to the nation. The Public Affairs officers had been doing this part of their job since well before Duff's arrival — since 1961, in fact — going by the successive names Mercury Control, Gemini Control, and now Apollo Control. Today, the soothing Public Affairs voice would be more important than it had ever been, and Duff was on hand behind the console to make sure things went well.

"This is Apollo Control at 67 hours, 23 minutes," Terry White, the officer on duty, was saying this morning. "Flight Director Glynn Lunney is still in Mission Control, and we have no idea exactly when he'll be able to break away for a briefing at this time. So far, we're still leaning toward a PC+2 burn at 79 hours and 27 minutes into the mission, or about 8:40 tonight. We are about 9 hours from loss of signal, when the spacecraft will disappear behind the moon, but at the moment Apollo 13 is stable. We will keep you advised as the situation changes and as Flight Director Lunney becomes available."

Terry White clicked off and the air-to-ground babble once again filled the loop. "Aquarius, Houston," Jack Lousma could be heard to be saying. "The latest tracking data shows your eventual pericynthion to be holding somewhere around 136 miles, so the trajectory you have is still good. Over."

Lousma's message was clear and comprehensible, but the voices coming down from Apollo 13 were a different matter. When Jim Lovell — or maybe Fred Haise or maybe Jack Swigert, it was impossible to determine — answered Lousma, his voice seemed to be disintegrating into loud crackles across the expanse of space.

"Hello Houston, Aquarius," someone from the ship said. "Say again."

"We say you're holding at 136 nautical miles."

"Jack, we're getting a lot of background noise," the voice from Aquarius called. "Can you read us?"

"Jim, you're readable through the noise, but just barely," Lousma answered. "INCO is checking out what we can do about it here."

"Roger," the voice that evidently belonged to Lovell said. "Standing by."

There was a scratchy pause for several seconds, then Lousma's voice reappeared. "Aquarius, Houston," the Capcom called out. "Is it any better now?"

"This is Aquarius," Lovell said back through the static. "Negative."

A long period of hissing noise filled the line as the INCO, down in the second row, consulted with his backroom. The problem, whatever it was, was nettlesome but certainly not life threatening. At the Public Affairs console, however, Duff was uneasy. Across the country, most viewers were turning on their televisions for the first time since hearing of the accident last night, and the deterioration in communications from the sick, power-poor ship was alarming. He let the hissing play out for a minute or so, then nudged White.

"Fill," he told him. "Talk. Repeat yourself if you have to. But don't go silent. Silence sounds like we've gone dead altogether."

"Uh, this is Apollo Control," White said. "We expect communications to improve a bit after the third stage of the Saturn 5 crashes into the lunar surface. The radio frequency the stage is transmitting on is giving us a little interference, but after the impact later today that should stop."

Duff smiled with at least passing relief. It didn't really matter what explanation White was offering, as long as he was offering something. It wasn't much, but it should at least keep the nation, and more important, the press corps, from feeling as if they were being kept in the dark. An in-the-dark press corps could turn into a surly press corps, and a surly press corps could turn on you in an eye blink. Today, Duff knew, he would need the friendship of the media more than he ever had before.

In the cockpit of the distant and drifting Aquarius, Jim Lovell was at least as concerned about his air-to-ground communications as Brian Duff was, but for different reasons. For all Terry White's attempts at on-air candor, he had told only part of the story. It was true that the empty third stage of the Saturn 5 booster, hurtling toward an impact on the moon where it was supposed to jolt the seismometer left behind by Apollo 12, was playing havoc with Aquarius's radio. The Saturn stage — known within NASA as the S-4B — and the LEM did transmit on the same frequency, but since the lunar module was not intended to be fired up and flying free until after the booster had made its crash landing, radio interference between the two vehicles never seemed as if it would be a problem. Now, with Aquarius handling all the voice communications between the crew and the Earth, and the S-4B trying to shout the crew down on the same narrow broadcast band, air-to-ground conversation was being periodically garbled.

Making matters worse, the backup communications systems, which ordinarily could cut through some of the clutter, were not operating as they should. As soon as the descent engine had been shut down after the free-return burn, NASA sent the crew orders to switch off some of their nonessential equipment, in order to conserve energy until the PC+2 burn of Aquarius's descent engine later tonight. Most of the LEM's antennas and much of its secondary communications hardware were among the systems sacrificed, and as each circuit breaker controlling the selected equipment was taken off line, the air-to-ground communications deteriorated more and more. By the time all the switch-throwing was complete, Lovell found himself operating on but a single antenna at a time, continually shifting to whichever one seemed to be carrying the clearest signal and banking his spacecraft around to give it the clearest shot at Earth.

"Houston, this is Aquarius," Lovell shouted through the static in his headset shortly after White's last on-air transmission. "The comm is very, very, very noisy right now. Are you reading this?"

"Aquarius, Houston," Lousma shouted back. "Copy that. It's noisy on our end too. Stand by while we think about it."

"Houston, Aquarius," Lovell shouted again, tweaking his thrusters and shifting the ship a few degrees to port. "I'm unable to make out your transmission."

"Jim, Houston," Lousma shouted back. "We're barely able to make you out either. Stand by."

Lovell pressed his headset further into his ear and closed his eyes. "Can either of you guys figure out what he just said?" he asked, turning to his right to consult Haise.

"Barely," said Haise. "I think he's saying he can't hear you."

"Well, hell," Lovell said, "I had a pretty good idea of *that* already."

"Aquarius, Houston," Lousma suddenly crackled in the crew's headsets, causing all three to jump.

"Go ahead, Houston," Lovell said.

"At the moment it sounds like we have a slightly better downlink. How do you read us?"

"Still a lot of noise on this end."

"O.K., we've got a suggestion," Lousma said. "Recommend you push in the power-amplifier circuit breaker on panel sixteen. Over."

Lovell nodded to Haise, who pushed the appropriate button. Nothing changed in his headset.

"Houston, Aquarius," he called to the ground. "We've still got the noise."

"All right," Lousma said. "We're going to try to improve your comm and telemetry by temporarily breaking lock and reaquiring. We'll be out of contact for a few minutes and you may hear some noise in your headset."

"Couldn't be any more noise than what we're getting now," Lovell said.

Lousma clicked off the air, and the intermittent static was replaced by a loud, steady hum. Lovell pushed his earpiece an inch or two forward, allowing it to buzz away from his ear. The pause gave the commander a few moments to think, and what he was thinking about was sleep. The sun that was rising across the central time zone shone only spottily on the docked pair of Apollo 13 moonships. With the engine bell of the LEM pointing back toward Earth, sunlight flowed through the commander's window, and the astronauts found themselves awash in daylight. But when the wobbles in the spacecraft's attitude moved it a few degrees, they were plunged into darkness.

These abrupt transitions from day to night did not generally bother Lovell. On the way to the moon, the passive thermal control roll that kept the spacecraft evenly heated all around caused the sun to strobe continually in and out of the LEM's and command module's portholes. After just a day or so in translunar drift, the astronauts got accustomed to the constant flickering and went about their sleep-wake, work-rest schedules as if the sun were rising and setting outside their craft just as it did outside their homes in Houston. As long as the crew maintained that schedule, NASA's flight surgeons had learned, their circadian cycles would remain largely undisturbed.

By seven o'clock on Tuesday morning, however, those cycles had been bollixed up but good. According to the original mission plan, the most recent sleep shift for the crew was to have begun at ten o'clock last night and run until about an hour ago. Even on a routine flight, no one expected the pilots to sleep a full eight hours. The almost total lack of physical exertion in space and the almost constant output of adrenaline that accompanied the business of flying to the moon made five or six hours of sack time the most the medics could hope for. Those five or six hours, however, were absolutely essential if a crew that was flying even a nominal mission was going to make it through their day without making some serious, and perhaps disastrous, mistake. A crew that was flying a less than nominal mission would need even more rest.

By the time the free-return burn was complete, the flight surgeons had prepared a revised work-rest schedule that the crew was to begin following immediately. Haise, it was decided, would get some sleep first, retreating alone to the command module from about the 63-hour mark, or 4 A.M., to about 69 hours, or 10 A.M. Odyssey had no oxygen of its own to sustain even a sleeping man, but with the hatch open between the two ships, more than enough atmosphere would float up from the lunar module. While Haise slept, Swigert and Lovell would stay on watch, using the time to power down the backup communications system and all the other hardware NASA wanted taken offline. When Haise woke, he would eat breakfast, confer with his crewmates about any problems that had developed while he slept, and take the helm alone while Lovell and Swigert retired to the command module, from 70 hours to 76 hours. The entire crew would be back on duty by 5 P.M., in plenty of time to prep for the PC+2 burn at 8:40.

Almost as soon as Lousma radioed up the flight surgeons' instructions, it became clear that sleeping and waking according to the medics' schedule would not be an easy matter. As Haise floated up the tunnel and into Odyssey, he was stunned at what he found. The temperature in the lifeless ship had been an already chilly 58 degrees when the crew abandoned it, but in the few hours they had been gone, the thermometer had fallen even further. When he poked his head through the top of the command module's cone, he could clearly see his breath condensing in front of him.

The crew's two-piece Beta-cloth jumpsuits were not designed for warmth in the constant 72-degree atmosphere the command module was supposed to maintain, and Haise immediately wrapped his arms around himself and pushed off in the direction of his couch, to zip himself into his sleeping bag. However, the thin cloth cocoons the astronauts used during sleep periods were intended merely as restraints, to prevent a weightless arm or leg from drifting up during the night and striking a switch or circuit breaker. Haise pulled out his bag, slipped inside it, and settled as deeply into his couch as he could. But even swaddled in the extra layer of fabric, he found himself lying shivering and awake, his body pressed up against the cold metal bulkhead of the spacecraft.

As troubling as the dropping temperature in Odyssey was the noise. The open hatch between the two ships allowed not only the ambient atmosphere from the lunar module to flow into the command module, but also the ambient sounds. If the churning of the LEM's coolant systems and the bumping of its thrusters weren't hard enough for a sleeping man to tune out, the shouted conversations of Lovell and Swigert fighting to be heard over the staticky comm channels were. Haise, who had a reputation in the astronaut corps for his ability to sleep nearly anywhere under nearly any conditions, fought gamely to shut out the din from next door, but finally, at six in the morning — less than two hours after his six-hour sleep cycle had begun — he gave up, wriggled out of his sleeping bag, and floated down through the tunnel to the LEM.

"That's it?" Lovell asked, glancing at his watch as Haise appeared between him and Swigert, drifting upside down through Aquarius's roof.

"It's too cold up there," Haise muttered, reaching for one of the food packets Swigert had carried over earlier and tearing it open with only passing interest. "Too cold and too noisy. You guys can give it a try, but I wouldn't count on getting much rest."

Now, at 7 A.M., in the silence of the temporary communications blackout, Lovell closed his eyes and felt the fatigue seep in. On the ground, he knew, Gerald Griffin's Gold Team would just be replacing Glynn Lunney's Black Team, the fresh controllers coming off at least a partial night's sleep taking over the consoles from the bleary controllers coming off a full night's work. At the Capcom console, even Jack Lousma, who had worked a double shift beginning yesterday afternoon, would at last be handing off to astronaut Joe Kerwin.

Lovell was glad a new group would be coming on, but fresh as Griffin's men might be this morning, they were going to be working with a trio of astronauts who would be sleepier, and no doubt testier, than any crew they had ever worked with before. Lovell told himself he would try to keep everything percolating as steadily as possible, but the ground was going to have to make some allowances.

"Aquarius, Houston," Lousma suddenly crackled in his ear. "How do you read us now?"

Lovell jerked at the sound and opened his eyes. "We still read you with a lot of static," he said wearily. "The noise seems to indicate — "

"I didn't copy your last remark, Jim."

"I — said — we — still — have — noise," Lovell said back in a loud, slow voice.

"We do too."

"Do you want us to remain in this configuration then?" Lovell asked.

"Stay there for the next minute or two, Jim," Lousma answered. "Then we'll evaluate it."

At this point, to the surprise of no one more than Lovell, the cold and the static and the uncertain advice from the Capcom got to be too much.

"I'll tell you what we need," Lovell snapped. "We need you to try to get this squared away. See if you can't report the right procedures here, the whole works, before we get all balled up."

As chewing-outs went, it wasn't much, but in the uninflected, atonal context of the air-to-ground loop, it was as close to a harangue as Houston was ever likely to get. Lovell looked to his crewmates, who nodded at him sympathetically; Lousma looked at the men near his console, who responded to him the same way. Both he and Lovell knew that getting the right procedures up to the ship was precisely what the Capcom had been trying to do. And both he and Lovell knew that the commander appreciated that. Lovell, like his own spacecraft late last night, was merely venting, something he'd had good cause to do for the last ten hours, and something it was about time he did.

Lousma looked over his shoulder to Kerwin, who was standing by waiting to relieve him. Now, he decided, might be as good a time as any to give up the microphone. He shrugged, rose, removed his headset, and pulled his chair back for Kerwin. Kerwin plugged his own headphones into the console, sat down, and signed on the air as folksily as he could.

"Jim," he called, "how's that comm now?"

"Well," Lovell grumbled, recognizing the change in voice and softening his own in response to it, "there's still a lot of background noise."

"O.K., we'll look at it some more," Kerwin promised, "but your comm down to us is excellent now."

"Roger that," Lovell said flatly and closed his eyes again.

The commander didn't say anything more in response to Kerwin's little bucking up. If the communications channel was clear for the present, that was fine. But that fix, like every other fix the ground had come up with so far, was almost certain to be just a passing thing. Before long, Lovell thought, the comm would decay again, along with who knew what other systems.

He opened his eyes and glanced out his window at the gray-white, plastery moon that was now less than 40,000 miles away, nearly filling his triangular porthole. According to the original flight plan, today was the day that he and Fred Haise were supposed to set their lander down on the face of the giant body. That, of course, was now not to be — and for Lovell at least, it probably never would be. He had been out to this celestial neighborhood twice, and he knew he would have little chance of ever coming back. If he and Swigert and Haise did not make it home, he wondered if anyone at all would ever travel to these parts again.

"Freddo," Lovell said, turning to Haise, "I'm afraid this is going to be the last moon mission for a long time."

With Aquarius's microphones switched to vox, the commander's forlorn observation drifted 200,000 miles, into the heart of Mission Control and, from there, out into the world.

Glynn Lunney, still on duty as flight director, was barely paying attention when Jim Lovell made his prediction about the future of lunar exploration. It was a rare moment in any flight when the man who ran the mission did not have at least one ear cocked to the conversations between his astronauts and his Capcom. But with the clouds of static on the air-to-ground loop and the traffic running heavy on the flight director's loop, Lunney had to trust Kerwin to handle the ship-to-shore messages alone. Most of the other men at the other consoles had more freedom to listen in on Kerwin's loop, including Terry White, who was just minutes away from completing his shift at the Public Affairs station and heading home.

White, along with everyone else in Mission Control and the nation, heard Lovell's remark and, along with everyone else in NASA, was jolted by it. For an agency whose lifeblood was funding, and whose funding relied on good P.R., this was worse than an accidental "damn" or an inadvertent "frappin'." This was a calmly, coldly expressed statement of doubt — doubt in the mission, doubt in the program, doubt in the Agency itself. For NASA, it was profanity of the highest order.

Kerwin, a Capcom with otherwise good instincts, reacted to Lovell's inadvertently public remark in the worst possible way: he said nothing. Hoping not to draw attention to the comment, he allowed it to go wholly unacknowledged. Instead, however, it hung heavily in the air, taking on more and more meaning with each passing second. White allowed the silence to continue for several interminable moments, and then jumped into the breach.

"This is Apollo Control at 68 hours and 13 minutes," he said. "Flight Director Glynn Lunney and four of his flight controllers will soon be on their way to the Public Affairs building for the news conference. Accompanying Lunney will be Tom Weichel, the retrofire officer; Clint Burton, EECOM; Hal Loden, the CONTROL; and Merlin Merritt, the TELMU. An additional participant will be Major General David O. Jones, United States Air Force, who commands the Department of Defense recovery forces."

White's P.R. reflexes were good. The words he chose were not just soothing prattle designed to distract listeners at home. Rather, they also served as a sort of plea to the media. Bear with us, they said; work with us. We heard the same thing you did, and we'll be happy to talk to you about it. Just give us all a chance to discuss it together before you put it out in print.

Whether the media took White's meaning was unclear, and it would remain unclear until Lunney and his crew sat down with the assembled reporters. For the time being, however, Lunney remained distracted, and was likely to become only more so. Since the moment the pre-dawn free-return burn was completed, the men in the control room had been concentrating most of their energy on one thing: the PC+2 burn, scheduled for seventeen hours later. With Lunney on console and Kranz closeted with his Tiger Team, Gold Team Flight Director Gerald Griffin and Maroon Team leader Milt Windler had overseen the effort and, by any measure, managed to accomplish a remarkable amount in an impossibly short time.

For the past four hours, the two off-duty directors had been patrolling the control room almost as one, stopping at console after console, grilling the men they found there, and collecting whatever ideas they had for the long, complicated burn of a lunar module engine with a 63,000-pound command-service module attached. At most of the consoles, the Black Team member on duty was not alone but was attended by the Gold and Maroon Team members who worked the same station, and who had been showing up throughout the night. When Griffin and Windler arrived, they moved in opposite directions — Griffin toward the Gold controller, whose ideas and talents he knew best, and Windler toward the Maroon. Occasionally, the Black Team member, behind whose chair and out of whose earshot the conversations were supposed to be taking place, would overhear a scrap of the discussion, cover his microphone, and spin around in his seat to correct something the men behind him were saying or add a technical suggestion of his own. From three o'clock to seven o'clock the impromptu meetings went on, and by the time the Tuesday morning controllers were ready to relieve the Monday night crew, Griffin and Windler had sketched out three different PC+2 scenarios, none of which, they knew, was perfect, but all of which would bring the crew home faster than the trajectory they were following now.

As Brian Duff planned his early morning press conference, Glynn Lunney put in his last hour on console, and Fred Haise arose from the night's sleep he never had, Griffin and Windler sat wearily in the aisle next to the flight director's station, their elbows balanced on their knees and their heads held in their hands, hoping to suggest — if only by their poses — that for just a few minutes they'd like not to be a part of the ongoing buzz in the room. Behind them, Chris Kraft approached and laid a hand on each of their shoulders. The two men turned.

"What have we got?" Kraft asked. Griffin and Windler looked at him uncomprehendingly for an instant. "What kind of burn have we got?" Kraft clarified. "Do we know how we want to proceed with this?"

"We've got some pretty good ideas," Griffin said. "As far as we can see, we've got three options, any one of which should help a lot."

"Will they be ready to go in twelve hours?" Kraft asked.

"They should be," Griffin said.

"Would you be ready to talk about them in an hour?"

"What do you mean?" Windler asked.

"Some people are going to be getting together to discuss all this in the viewing room, and we're going to have to be able to explain things to them as best we can."

"What people, Chris?" Griffin asked.

"Gilruth, Low, McDivitt, Paine — most of the guys at that level," Kraft said. "Plus you guys, Deke, Gene, and whoever else you think we might need. Probably a couple dozen people in all."

Griffin was more than mildly surprised. Gilruth, of course, was Bob Gilruth, director of the Manned Spacecraft Center; Low was George Low, the director of Spacecraft and Flight Missions; Paine was Thomas Paine, the administrator of NASA itself. Bringing men like Deke, Kraft, McDivitt, Kranz, and the rest of the flight directors together for a meeting in Mission Control was one thing; during a mission, people on that rung of the organizational ladder met all the time in and around the control room to discuss various problems and procedures. But the Gilruths, Lows, Paines, and other men on the upper rungs were rarely part of the conferences. These were the big-picture people, the people who relied on Kranz and Kraft and the rest to run the individual missions while they ran the program as a whole. Bringing these men into Mission Control for a caucus in the soundproof, glassed-in enclosure of the VIP gallery — at once the most private and least private room in the building — was without precedent. It was a gathering of the council of Agency elders, the NASA equivalent of a joint session of Congress, and it would take place in full view of an audience of controllers who had never seen so much NASA royalty in one place before.

"This is going to happen in an hour?" Griffin asked.

"Less than an hour," Kraft said. "And before it does, I want to meet with all the flight directors to make sure we've got our ducks in a line. Pull Glynn off console and let's go find a place to talk."

"Kranz is downstairs with his Tiger Team," Windler said. "You want us to get him too?"

"Yes," Kraft said, and then reconsidered. "No. No, don't. I don't want to disturb him until it's necessary. Let him keep working on the consumables until the meeting itself. Then we'll bring him up."

Griffin and Windler nudged Lunney, told him Kraft needed him, and the Black Team flight director turned his console over to his assistant and followed the three men to a staff support room. When they got there, Kraft closed the door, sat, and inclined his head wordlessly to his controllers, inviting them to tell him what they knew. Lunney knew little more than Kraft did, so he deferred to Griffin, who began to walk Kraft through the three burns they had just finished developing. Kraft did not need the basic science explained to him; he knew the language of the FIDOs and GUIDOs and the flight directors who oversaw them. What he really wanted were the consequences of each maneuver — what the risks were, what the advantages were, how each one might affect the odds of bringing his astronauts back alive.

Griffin spoke candidly and economically and Kraft listened, nodding occasionally but saying nothing. When the flight director was done, Kraft went to work, raising questions, raising objections, poking at Griffin's projections, challenging his estimates, and on the whole trying to anticipate the grilling that would be forthcoming from the men in the VIP room. Griffin and Windler answered Kraft's concerns as best they could, and Lunney, hearing most of this for the first time, nodded in agreement. Finally, after an hour, Kraft seemed satisfied, opened the door, and prepared to lead his group over to the viewing gallery. Before he could, however, Griffin stopped him.

"You know, Chris," he said, "I'd sure feel more comfortable if we all weren't going in there alone."

"Who else do you need?" Kraft asked.

"Well, it's my FIDO and RETRO who crunched all this data."

"Who are they?"

"Chuck Deiterich and Dave Reed," said Griffin. "If I had my choice, I wouldn't go anywhere without my numbers guys."

"Go get them," Kraft said. "And get Gene too."

Kraft waited while Griffin fetched Deiterich, Reed, and Gene Kranz, and when they returned, the men made their way together over to the VIP area. When they arrived, the tableau that greeted them was an imposing one. The pool reporters manning the press consoles in the front right corner of the gallery had been cleared away, and in the front left corner, two dozen or so men were silently waiting. A few had settled into the movie theater seats, but most were standing in the aisles, perched against the backs of seats or leaning against the walls. Through the floor-to-ceiling window at the front of the gallery, the entire sweep of Mission Control was visible, and now and then a flight controller could be seen stealing a glance at the silent council sealed behind the glass. Kraft wasted little time getting things started.

"In about twelve hours," he began, "we're going to need to execute our PC+2 burn. Our objective will be to get the crew home as fast as possible while stretching our consumables as far as possible. The flight directors have come up with some possible burns, and since it's Gerry's team that worked out so many of the numbers, I'll let him explain them."

Griffin stepped forward, cleared his throat, and began to describe, slowly and deliberately, the procedures he had just gone over much more quickly with Kraft. As Griffin explained it, and as he was sure the men in the room realized, the most precious consumable Apollo 13 had to work with was not oxygen or power or lithium hydroxide; it was time. Get back to Earth quickly enough, and the problems with all the other consumables will take care of themselves. The obvious answer, then, was to burn the LEM's descent engine at full throttle for as long as the fuel supply would allow, increasing the ship's velocity until it could be increased no more.

But the obvious answer was not necessarily the best answer. Burning the engine until it ran dry would leave almost no fuel for subsequent mid-course corrections, which might well be needed: the ship would be covering more than a quarter of a million miles, and the slightest error in the initial trajectory would thus be magnified many times over. The ascent stage of the lunar module did have an engine of its own, and in an emergency it could always be fired. But in order to do so, the crew would have to jettison the descent stage first — and it was the descent stage that contained most of the lander's batteries and oxygen tanks.

The length and strength of the burn, Griffin went on, would determine not only the crew's fuel reserves and their transit time back to Earth, but what body of water they would plop into when they got here. With only some of the Earth's oceans approachable from space, and only one of those oceans, the Pacific, equipped with adequate recovery vessels, the choices were limited. The three different maneuvers Griffin and Windler had come up with would address these problems in three different ways.

The first burn, Griffin explained, would be a long one. Pushing the descent throttle all the way to the full position, Lovell would leave it there for more than six minutes before shutting the engine down. This maneuver, which for simplicity's sake Griffin called the superfast burn, would put the crew down in the Atlantic Ocean on Thursday morning, just thirty-six hours from the scheduled PC+2 time later tonight. Such an early return would be well within even the most pessimistic estimates of the LEM's projected lifetime, and for that reason alone was very attractive. But the superfast burn would also come at a price. Not only would it use an enormous amount of fuel and aim the astronauts toward a patch of ocean where the Navy had not so much as a fishing trawler presently stationed, it would also require them to make the entire trip home without a key part of their spacecraft.

In order to reduce the mass of the docked ships enough to make such a go-for-broke maneuver effective, Lovell would have to jettison the now useless service module. To be sure, Griffin explained, the flight directors did not harbor any illusions that this presumably blast-damaged part of the ship could be brought back to life, but they were nevertheless reluctant to part with it. The service module, as the room of administrators knew, fit snugly over the base of the command module, protecting the heat shield, which in turn would protect the crew during their fiery reentry through the atmosphere. Nobody had ever conducted experiments to find out what would happen to a heat shield that had spent a day and a half in the deep freeze of space, and now might not be the time to run the test. Complicating matters, even if an ordinary heat shield could survive such frigid conditions, it was possible that Apollo 13's was not ordinary. If the accident that destroyed the oxygen tanks had put even a hairline crack in the thick, epoxy-resin face of the shield, the ultra-low temperatures of sunless space could split it wide open. Still, Griffin summed up, if consumables proved to be an insurmountable problem, the superfast return might be worth considering.

The next burn would be a little slower than the superfast burn, conserving some fuel while adding only a few hours to the homeward trip. The biggest advantage of this procedure was that the added transit time would allow the Earth to make another quarter turn, presenting a different face to the reentering spacecraft and permitting it to splash down in the heavily trafficked Pacific. The biggest disadvantage was that, like the high-speed return, this one would require jettisoning the dead service module.

The final burn option was the slowest and least dramatic of all. Keeping Odyssey's service module in place, Lovell would fire Aquarius's descent engine for only four and a half minutes, and only part of that time at full throttle. Like the medium burn, this more modest maneuver would aim the crew for a splashdown in the friendly Pacific. Unlike the medium burn, however, it would get them there not midday Thursday but midday Friday — more than three days from now, or only ten hours faster than they would arrive without any PC+2 burn at all. If the heat shield and the recovery were the only considerations, Griffin concluded, this burn would certainly be the way to go. But when consumables were figured into the equation, things could get a little sticky.

Griffin finished his presentation and stepped back to allow his audience of Agency superiors to make their choice. Instantly, hands went up. What, someone wanted to know, was the likelihood that the heat shield was damaged? Probably low, Griffin answered, but if there *was* a crack, the probability of losing the crew could be 100 percent. How far, someone else asked, could the consumables be stretched? Too early to tell at the moment, Griffin admitted; Kranz, at his side, agreed. What, someone else wondered, was the precise Delta V and the precise burn time for all three maneuvers? Deiterich and Reed stepped forward and passed around their handwritten notes, explaining the meaning of each scribbled digit.

For nearly an hour, the men in the room debated their options, as Kraft and his crew of flight directors watched. Deke Slayton, as the chief astronaut and thus the chief advocate for all the other astronauts, argued strongly for the fastest burn, and several other voices soon joined his. But more numerous, and soon overwhelming, were the voices arguing for the slowest burn. Yes, consumables were going to be a problem, but weren't Kranz and the Tiger Team and the legendary John Aaron working on that? Yes, it would be difficult to explain to the media and the world at large why a crew in extremis was being kept in space for even an hour — never mind a day — longer than necessary. But wouldn't it be harder to explain why that same crew was flown most of the way home without fuel and dropped through the atmosphere on a busted heat shield into an ocean where the Navy had to scramble to put ships?

Kraft and his flight directors let the arguments play out and watched, satisfied, as the men in the room settled on the slowest alternative. It was the choice the flight directors themselves had preferred, and it was the one they had hoped the administrators would prefer. Now, as the arguments began to gel into a consensus, Chris Kraft transformed the consensus into a decision.

"So it's agreed," he summed up. "At 79 hours and 27 minutes there will be an 850-foot-per-second burn for four and a half minutes, aiming for a Pacific splash at 142 hours. If all goes well, Apollo 13 will be home by Friday afternoon."

The men in the room nodded and, almost simultaneously, rose and began to move toward the doors. As the flight controllers at the consoles peered over their shoulders for one last peek at the dispersing administrators, Gerald Griffin turned to Glynn Lunney.

"What do you say we quit talking about this thing," he said, "and go see if we can't do it."

9

WHEN GENE KRANZ walked into the VIP gallery hours after the PC+2 meeting broke up, the two reporters at the media consoles did not dream of trying to talk to him. A greener journalist would have; a greener journalist would have been crazy *not* to. When the man at the center of a maelstrom like Apollo 13 appears, unaccompanied, in your midst, with virtually no other newsman in sight to compete for his attention, you do what every reportorial instinct you have screams at you to do: you buttonhole him for a prediction or an impression or at least a space-filling quote. But the reporters at the consoles knew better. When Kranz appeared in the VIP gallery in the middle of a flight, he was not here to talk; he was here to sleep.

Ever since the Gemini program, when NASA began conducting missions that ran for four or eight or fourteen days, the Agency medics had requested, and the Agency bosses had provided, on-site sleeping quarters for flight controllers who needed to be on call around the clock. The accommodations weren't much — no more than a small, windowless room in the Mission Control building, with a shower, a sink, and two Army-style beds — but for controllers accustomed to sneaking off to an empty conference room if they needed a nap between shifts, it was unimaginable luxury.

The little bedroom was christened with great fanfare, and as soon as the next mission flew, controllers clamored for the chance to lay claim to it. Those first few who tried it out quickly wished they hadn't.

The room was directly off a hallway with lots of foot traffic passing by and lots of unceasing conversation. Most of the sound leaked through the plasterboard walls, and the little bit that didn't flooded in whenever the door was opened. The door itself had a hydraulic closing device that had apparently never been adjusted properly; when someone entered or exited, the door would groan open reluctantly and then slam shut with a sleep-shattering bang. Even the shower's pipes banged and groaned noisily.

Despite all this, on nearly every flight there were half a dozen or so overdedicated controllers — including Gene Kranz — who insisted on staying at the Center full time, so competition for the two cots was usually keen. When lunar missions started becoming almost routine, however, and fewer people worked consecutive shifts, Kranz had sworn off the noisy controllers' bedroom for good. If he needed sleep, he decided, he would retreat to the VIP gallery, select a seat in one of the far, shadowed corners, and catch as long a catnap as the schedule permitted. On Tuesday afternoon, Kranz, who had been working for more than twenty-four unbroken hours, decided to allow himself a break, and with a nod to the reporters working the VIP consoles, he settled into a cushioned chair. He knew that the nap he could steal would be a decidedly short one.

From the time Kranz turned his console over to Glynn Lunney late last night, he had been shut up in room 210 with his Tiger Team, poring over his strip charts and consumables profiles. Although the story the data told him was not a happy one, on the LEM side of the room the picture was at least somewhat promising. After quickly conducting the consumables calculations that followed Aquarius's power-up, Bob Heselmeyer, the White Team TELMU, reviewed the numbers with Kranz and then, unlike most of the members of the White Team, was sent back to the consoles.

Heselmeyer was a good TELMU, but he was also the youngest one assigned to the Apollo 13 rotation. For the LEM consumables work that had to be done, Kranz preferred Bill Peters, a TELMU on Gerry Griffin's Gold Team, who had worked every flight since Gus Grissom and John Young's Gemini 3, in 1965. The Tiger Team leader's trust in Peters turned out to be well placed. After sitting down with Kranz for half the morning — and conferring with Tom Kelly at Grumman for the other half — Bill Peters made remarkable strides in solving Aquarius's consumables crisis.

Tackling the water and power problem first, the two resources in shortest supply, Peters was able to economize even more than Kelly and Heselmeyer had thought feasible. According to the profiles Peters and his electrical specialists came up with, it appeared possible to run the LEM — which normally needed about 55 amps of current to stay alive — on a starvation ration of just 12 amps. A fully powered LEM had about 1,800 amp hours to play with, divided among four batteries in the descent stage and two in the ascent. Twelve amps wasn't much compared to this, but after factoring out that power demand over the time it would take the crew to come home and setting aside a small quantity of buffer power in case of further emergencies, Peters didn't figure he could afford to use much more. The more electricity the TELMU saved, the more water he would save as well, and Peters's strict battery diet thus conserved gallons of that scarce commodity as well.

All of the frugality he proposed came at a price, however. The partial shutdown of systems the LEM engineers had ordered between the free-return burn and the PC+2 was nothing compared to what Peters had planned for the long coast home. As soon as the speed-up maneuver was completed at 8:40 tonight, he would order the disconnection of virtually every electricity-consuming component the lunar module had, with the exception of three: the communications system and one of its antennas; the cabin fan, which would circulate the available oxygen; and the water-glycol coolant pumps, to keep the temperature down in the other two systems. Taken off line would be the computer, guidance system, cabin heater, docking radar, landing radar, instrument panel displays, and hundreds of other pieces of hardware. All of the sacrificed equipment could be powered back up in the event that it was needed for subsequent burns or other maneuvers, but to the extent possible, it would remain off for the entire return trip.

To be sure, there were gaps in Peters's draconian power-down plan. For one thing, the already uncomfortable LEM promised to become even more uncomfortable, with the loss of instruments and cabin lights plunging the cockpit into darkness and the loss of heat-generating instruments driving the chilly temperature down even more. For another thing, no one had yet solved the problem of how to scrub the cabin air free of carbon dioxide without fresh lithium hydroxide canisters to absorb the noxious gas. Perhaps most troubling, the LEM would have to provide power for more than just its own systems.

Before Lovell, Swigert, and Haise abandoned Odyssey, their dying command module had begun cannibalizing one of its three reentry batteries, automatically tapping into it for power after the three fuel cells expired. In order for the ship to be powered up again for reentry, that battery would have to be recharged, and the only place to get the juice was from the already hard-pressed electrical system of Aquarius. Even as Peters was trying to figure out how to keep his spacecraft alive for the half week that lay ahead, John Aaron came by to borrow some amps for his own sick ship.

"Bill," Aaron said, buttonholing Peters in a corner of room 210 and speaking to him in his most winning Oklahoma drawl, "you know I can't run that command module on two and a half batteries."

"I know it, John," Peters said.

"And you know I have to get it from you."

"I know that too."

"How much can you give me?"

"How much do you need?" Peters asked warily. "Those are just little-bitty batteries you've got. You don't need too much, do you?"

"We've got to top off the depleted one at about 50 amp hours," Aaron explained, "and when they abandoned ship it was down to 16. So I'm gonna have to ask you for about 34."

Peters thought a moment. "Thirty-four . . . thirty-four I could do, but you're actually asking me for a lot more than that. My chargers and umbilicals only have an efficiency of about 30 or 40 percent. To pump 34 amp hours all the way over to you is going to cost me about 100."

"I know it, Bill," Aaron said with genuine sympathy. "Can you do it anyway?"

Peters contemplated his 1,800 available amp hours and ran some quick mental math. "Yeah," he said cautiously, "I expect that I can."

On the command module side of the room, things were even more complicated, and Aaron's ability to negotiate and cajole was going to be even more essential. What was consuming most of the lead EECOM's time was not how he hoped to top off his batteries, but how in the world he expected to be able to power up Odyssey whether he had Peters's extra amps or not. Ordinarily, the process of putting an Apollo command module on line was extraordinarily costly, in terms of power and in terms of time. Before a launch, pad technicians usually needed up to a full day to accomplish the feat, using thousands of amp hours of ground power to warm up each system and check its vital signs before declaring it fit for flight. The process was painstaking, but with unlimited amps and unlimited hours at their disposal, NASA engineers preferred to be as careful as possible.

With Apollo 13, Aaron would not have this luxury. He and Kranz ran some preliminary power projections and came up with some disquieting numbers. Assuming Odyssey's third battery was indeed successfully topped off, Aaron would have merely two hours of juice to play with when the time came for reactivating the spacecraft. For an engineer schooled in NASA's hypercautious ways after Apollo 1, this seemed like recklessness of the highest order, but Aaron believed the job could be pulled off.

What concerned him most was how he was going to explain things to the flight controllers who oversaw the spacecraft's systems. In principle, every man in room 210 understood that engineering corners would have to be cut if the command module was going to make it home intact. In practice, however, nobody wanted to think that it would be *his* corner that would be affected — and Aaron did not relish breaking the news to them. With Kranz standing by, he gathered the command module controllers around the conference table and spoke with a down-home diffidence that was equal parts prairie manners and calculated salesmanship.

"Fellas," he said, "I know I'm not supposed to know about all your systems, so bear with me, and correct me when I make a mistake, but I think I have some ideas how we can get this ship on line when the time comes. Now, the way I see it, we're going to have about two hours of battery time to go from a cold stop to a full power-up."

"John," Bill Strahle, the guidance and navigation officer said, "you can't do it in that time."

"Well, now, that's what *I* thought, Bill," Aaron said, chuckling at his own wrong-headedness. "But I think if we're willing to take a few shortcuts, we just might be able to pull it off."

"Sure, you can pull it off," Strable said, "but can you pull it off safely?"

"I think just maybe we can," Aaron said. "I've got a few ideas here. Just rough stuff, nothing set in stone. But maybe if we all took a look at 'em, we could all flesh 'em out a little bit."

Almost apologetically, Aaron produced a sheaf of strip-chart paper crowded with crayon markings. The scribbles ran on for page after page, representing dozens of projections, predictions, and computations Aaron had worked out with the help of Jim Kelly, his electrical systems specialist. From even a glance, it was clear that this was not "rough stuff," these were not "a few ideas." This was a brutally realistic, exhaustively considered breakdown of exactly how much power and how much time the ship had to work with, whether the controllers wanted to hear about it or not. Aaron knew the numbers were good, and the controllers, he suspected, knew it too.

He passed his papers around, let the controllers digest them, and the first of what promised to be dozens of hours of negotiating, bartering, and dickering began. The controllers would have objections and they would have ideas, but what they wouldn't have was much time. According to the trajectory the crew was now following, Apollo 13 would collide with the Earth's atmosphere in less than seventy-two hours. Assuming the PC+2 burn went as planned later tonight, that number would be slashed to sixty-two. If Aaron didn't have a power-up checklist put together within forty-eight hours at the latest, the steely-eyed missileman was in real danger of losing his first crew.

Gerald Griffin's Gold Team was not thinking about consumables. They would eventually, Griffin knew; like all the other teams, the Gold Team had days of resource management ahead of them. But right now they didn't have to concern themselves with that.

Griffin had been in charge of this flight for more than five hours now, and so far things had been relatively quiet. It was on Kranz's White Team watch that Apollo 13's tank-blowing accident had taken place, on Lunney's Black Team watch that the power-down and free return had been executed, and on Windler's Maroon Team watch that the PC+2 would have to be tried. There was talk that Kranz's Tiger Team, formerly White Team, might come out of isolation long enough to commandeer the consoles for tonight's PC+2 maneuver, then turn them back over to Windler — and if that was what Kranz wanted, nobody would deny him. But whether it was the Tiger Team or the Maroon Team that would follow Griffin on console, the Gold Team leader's assignment was clear: keep the ship functioning, help it avoid any further technical crises, and make sure it was properly prepped for the PC+2 burn. So far, Griffin's group was performing all of its jobs well, with the distinct exception of the last one.

The earlier efforts of Lunney's Black Team to fine-tune Aquarius's platform despite the debris cloud surrounding the ship had met with failure, and when Lunney had decided to attempt the free-return burn relying solely on the alignment transferred from the command module, the men in the control room had simply shrugged and hoped for the best. That burn, they knew, would be short, and any errors in the alignment of the platform would not be magnified too much. For the PC+2, however, things would be different. Not only was the scheduled burn a sustained one — more than nine times longer than the mere engine burp that had placed the crew back on their free-return path — but it would also take place close to eighteen hours later. Guidance platforms had a tendency to drift over time, and even if the coordinates Lovell had transferred from Odyssey at 10 last night were still good at 2:43 in the morning, they would almost certainly have deteriorated by 8:10 the next night.

For much of the past several hours, Griffin and his Gold Team had thus been in constant touch with the technicians in the simulator room across the Space Center campus, where Charlie Duke and John Young were trying to come up with some alignment solutions the Black Team guidance officers hadn't. So far, the results were not encouraging. With star maps projected on the simulator's windows, and an additional light source added to represent the sun, the two pilots had cartwheeled their mock LEM through every simulated orientation they could think of, trying to move Aquarius's windows deep enough into shadow to black out the debris cloud and allow the true stars to appear. No matter which way they turned, though, the ersatz sun continued to wash around the LEM, which set the debris twinkling and made even approximate star sightings impossible. As noon gave way to afternoon and the latest gloomy report filtered back from the simulator building, Chuck Deiterich, Dave Reed, and Ken Russell — Griffin's RETRO, FIDO, and GUIDO — sat slumped at their consoles in the front row of Mission Control, utterly stuck.

"So what's our game plan here?" Reed asked his two colleagues, pushing back from his center console and looking left toward Deiterich and right toward Russell. "What do you fellows propose to try next?"

"Dave," Deiterich said, "I'm open to suggestions."

"I assume we're giving up on the stars," Russell said.

"If we can't see them," said Deiterich, "we can't fly by them."

"I suppose we could always wait until we get behind the moon,"

Russell said. "Once they're in shadow, the debris won't be lit so much."

"That cuts things pretty close, doesn't it?" Reed responded. "They've only got half an hour of shadow and only two hours after that till the burn. If something goes wrong, they won't have any time to get it right."

"Look," Russell said, "let's face it. The only thing we can see out there is the thing that's causing us all the trouble in the first place, the sun."

"Well shoot," Deiterich said, "long as it's there, why don't we just use it? It's a star, isn't it? The computer recognizes it, doesn't it? No matter how much debris you have, once you go looking for the sun, you're sure not going to mistake it for anything else."

He looked at Reed and Russell, and the two men looked skeptically back at him. Ordinarily, a fine alignment of a guidance platform was a fantastically accurate thing. With the bowl of the celestial sky stretching around the ship for 360 degrees in three dimensions, a solitary star was as close as you could get to the platonic ideal of a single geometric point: infinitely small, infinitely precise, an infinite number of them making up a single degree of arc. Take sightings on a few of these bright cosmic pinpricks and you could torque your platform to a level of accuracy that virtually eliminated any margin of navigational error.

Using the sun instead of the stars was an entirely different matter. First of all, the thing was huge. Measuring 865,400 miles in diameter and located only 93 million miles away from Earth — a mere arm's length by cosmic standards — the local star sits in the sky like a huge white blob, spreading across a full half-degree of the heavens. Within its big bright face, dozens of pinpoint stars could fit. What Deiterich was proposing, Reed and Russell instantly understood, was not to try to use this gross target to align the platform all over again, but rather simply to check the alignment the ship already had. If the astronauts instructed the guidance platform to go looking for the sun, and the platform drove the spacecraft — and, specifically, its alignment telescope — to within, say, a degree or so of where the neighborhood star actually was, they would know that Aquarius had its bearings straight and that its platform could be trusted when it came time for the burn. But no sooner did Deiterich propose his plan than he himself began to have doubts.

"Of course, we *are* talking about a pretty fat target here, aren't we?" he said.

"Very fat," Russell said.

"And what about the optics?" Deiterich asked. "You put the sun in an eyepiece where only a star's supposed to go and you're going to fry your eye right out."

"They've got filters to take care of that part," Russell said. "But I'm still not crazy about the whole idea. This is a hip-pocket procedure we're trying, fellas. It's fine in a simulator, but do you really want to rely on it in flight?"

"Not especially," said Deiterich. "But do we really have any other choice?"

Russell and Reed looked at each other.

"Not a one," Russell said.

Two rows up, at the flight director's console, Griffin was keeping an eye on his men in the front row and could see that they were deeply involved in discussing *something*. He dearly hoped that that something was an alignment plan. Like every other flight director, Griffin kept a log at his console in which he made entries when key mission milestones arrived and passed. So far, the space he had expected to fill with his fine-alignment notations remained blank, and he was growing itchy. PC+2 was seven hours away; loss of signal, when the spacecraft would disappear behind the fast-closing moon, was just over four hours away. The guidance officers were going to have to come up with at least one good idea, and come up with it right quick. In the front row, Deiterich, Reed, and Russell remained huddled for several more minutes and then suddenly stood, filed out of their aisle, and headed toward Griffin's station.

"Gerry," Russell said when they arrived at his console, "we're going to have to use the sun to check the existing alignment."

Griffin looked at his men in silence. Then he said, "That's the best we can come up with?"

"The best *we* can," Russell said. "Once we get into the shadow of the moon, we can see if a few stars pop out, and run a quick confidence check then. But that's only a fallback position."

"What's your comfort level on the sun alone?" Griffin asked.

"Pretty high," Russell said, with as much certainty as he could muster.

"*Pretty* high?"

"Yeah," said Deiterich. "But that might be as high as we're likely to get."

Griffin studied the faces of his guidance officers and then turned his palms up. "Call Charlie Duke and John Young," he said. "Get them to start trying this thing out in the simulator."

In the cockpit of Aquarius, Jim Lovell, Jack Swigert, and Fred Haise were not thinking about the sun. The body that was claiming most of their attention was four hundred times smaller — though it appeared infinitely larger — thousands of times less distant, and was getting bigger and closer by the minute. As John Young and Charlie Duke ran through their paces in the counterfeit LEM, the prime crew in the real ship was barely 12,000 miles away from the moon and whistling toward it at a speed of another 3,000 miles every hour. The closer Aquarius and Odyssey got, the more time the crewmen, despite themselves, spent stealing glances out the windows. They didn't yield to the impulse much at first, and indeed they couldn't afford to yield to it much. The communications system still demanded their constant attention, the ships themselves still required regular thermal rotation, the pre-PC+2 power-up was coming up soon, and the debris cloud still had to be monitored for chance breaks that might reveal stars. But no matter how dense the cloud became, no amount of floating rubbish could conceal the immense, plaster-gray sphere hanging in front of them.

The moon the crew was approaching was a gibbous moon — about 70 percent illuminated, with only a fat crescent on its western edge lost in darkness. At such close range, the LEM's small triangular windows could no longer contain the mammoth lunar bulk, and in order to take in the entire shape, the crewmen had to lean far forward, craning their necks as far as their limited portholes would allow. For Lovell, the proximity was starting to be a cause for concern. At the moment, his twin ships were about as far from the lunar mountaintops as an airplane leaving, say, Lisbon is from its intended destination in, say, Sydney. And Odyssey and Aquarius were moving six times faster than the jet. The commander pushed away from his window and turned to his LEM pilot uneasily.

"How do you think they're coming with that alignment business, Freddo?" he asked.

"Can't be too great or we'd have heard something," Haise said.

"Well, our margin of error is vanishing pretty fast."

"By 4,400 feet every second," Haise said, glancing at his computer's velocity display.

"What do you say we get them on the line and see if we can't hurry things along," Lovell said. Before Haise could transmit the message, however, Houston hailed the ship.

"Aquarius, Houston," the Capcom called. From the sound of the voice, it appeared that Vance Brand, another rookie astronaut, had replaced Joe Kerwin at the Capcom console.

"Go ahead, Houston," Haise said.

"O.K., we're getting a procedure ready for your alignment," Brand said, "and what we're thinking of is a sun check we'd like you to try at approximately 74 hours. We'll have the data up to you shortly, and it's our feeling that if you're within 1 degree of the target, your platform will be O.K. without subsequent alignment. Assuming the sun check *is* O.K., we will then give you a star for a confidence check on the back side when you're in darkness. Over."

Haise repeated the instructions, to make sure he'd heard correctly, then broke off the air and turned to Lovell and Swigert with a questioning expression. Of the three men on board, Haise was not necessarily the one most qualified to determine the soundness of the plan. Swigert, as navigator of this flight, and Lovell, as the first navigator of the first flight ever to come out this way, were a good deal better versed in the science of celestial steering.

"How's that sound to you?" Haise asked.

Lovell whistled to himself. "Well, it *should* confirm our alignment." He turned to Swigert. "How's it sound to you?"

"Kind of an imprecise method, don't you think?" Swigert said.

"Very imprecise," Lovell agreed. "What margin of error did they say they're giving us?"

"One degree."

"Which is two suns. It's like aiming for the side of a barn."

"The question is," Swigert said, unknowingly echoing Reed on the ground, "do you have any better ideas?"

Lovell paused. "None at all," he said. "Do you?"

"Nope."

"Call them back," Lovell said, turning to Haise. "Let's get started."

Haise brought Brand back on the line, and the Capcom began reading the LEM pilot the techniques for the sun check. As conceived by Deiterich, Russell, and Reed, and as tested by Duke and Young, the procedure would be relatively straightforward. Lovell would begin by telling the computer he wanted to look through his alignment telescope at the face of the sun. For accuracy's sake, he would specify which quadrant — or, as the guidance men liked to say, which "limb" — of the sun; in this case Reed, Russell, and Deiterich had picked the northeast limb. The guidance system was not accustomed to thinking of the sun as an alignment target, but it knew where to find it. When the computer had processed the command, Lovell would punch the Proceed button and the lunar module's sixteen jets would automatically fire, driving the spacecraft around toward the spot where the computer believed the sun to be. If the upper right limb of the giant star floated within 1 degree of the cross hairs in Lovell's highly filtered telescope, he would know his alignment was satisfactory. If it didn't, he would know he was in trouble.

Lovell listened to Brand's instructions, allowed Haise to repeat them back to the ground, and then began shooting questions down to Houston. Had Duke and Young run their simulations with the mock LEM in a docked configuration? Yes, the Capcom assured him, they had. Had the guidance system had any trouble maneuvering the ship with all the added weight? No, it hadn't. Would the docking radar, which protruded from the top of the lunar module, obstruct the alignment telescope's view of the sun? Not if the radar was retracted before the maneuver. The grilling took the better part of an hour, with Swigert and Haise throwing in questions when they could, and with such astronauts as Duke, Young, Neil Armstrong, Buzz Aldrin, and David Scott being rounded up in Mission Control to answer anything the Capcom and the guidance officers couldn't. Finally, at 2:30 in the afternoon, or 73 hours and 31 minutes into the mission, Lovell seemed satisfied.

"O.K., Houston," he said crisply to Brand, "what time is this little sun check supposed to take place?"

"At 74 hours and 29 minutes," Brand answered.

Lovell glanced at his watch. "How about if we just do it? Why don't we just do it now?"

"O.K.," Brand said. "You can start at any time."

With that go-ahead, the crewmen assumed their workstations — and for the first time since Odyssey was shut down, there was work for Swigert to do. Lovell, it was decided, would position himself at the center of the instrument panel and tend the guidance computer, typing in the data necessary to initiate the sun check and watching the attitude indicators to see if the spacecraft was moving in the correct direction. Swigert would man Haise's right-hand window, looking for the sun and alerting Lovell as it floated into view. Haise would move over to Lovell's side, where he would peer through the alignment telescope and note where the cross hairs settled on the sun.

In Houston, the ground crew assumed their stations as well. Griffin, like Lunney last night, called for quiet on the loop and asked the men standing behind the consoles to allow the men working them to focus on the job at hand. He pulled his flight log toward him, entered "73:32" in the space marked "Ground Elapsed Time," and in the blank marked "Comments" wrote, "Begin sun check." In the spacecraft, Fred Haise made a final adjustment to his communications hardware and — whether deliberately or accidentally — switched the system back over to hot mike. Instantly, the fractured voices of the astronauts, conferring among themselves, were heard by the ground crew.

"I don't have all the confidence in the world in this," Lovell was saying *sotto voce.*

"We'll get it," Haise was saying.

"Don't be so sure. I still might have screwed up my arithmetic last night."

Standing between his own station and his LEM pilot's station, Lovell now entered into Aquarius's computer the information Brand had called up to him. The computer accepted the data, processed it slowly, and then, patient as always, waited for the commander to press Proceed. With a glance first at Haise and then at Swigert, Lovell pushed the button. For a second nothing happened, then all at once, outside the windows, a fine mist of hypergolic gas appeared as the lander's jets fired. Inside, the astronauts could feel the ship begin a lazy turn. In the center of the cockpit, Lovell locked his gaze on the attitude needles.

"We've got roll," he called out. "Now yaw — roll — pitch — yaw again. Houston, are you reading all this?"

"Negative, Jim," Brand said. "We don't have high bit rate coming down from the computer."

"Roger," Lovell acknowledged, and turned to his right. "You see anything yet, Jack?"

"Nothing," Swigert answered.

"Anything over there?" he asked Haise.

"Not a thing."

In the front row of Mission Control, Russell, Reed, and Deiterich listened to the crew and said nothing. At the Capcom station, Brand held his tongue until he was called again. At the flight director's station, Griffin pulled his log toward him and scribbled the words "Sun check initiated." On the air-to-ground loop, the fractured chatter continued to flow back from the crew.

"Yaw right side," Haise could be heard saying. "Commander's FDI."

"Deadband option —" Lovell responded.

"Plus 190," Haise said. "Plus 08526."

"Give me 16 —"

"I've got HP on the FDI —"

"Two diameters out, no more than that —"

"Zero, zero, zero —"

"Give me the AOT, give me the AOT —"

For close to eight minutes, the murmuring of the crew continued as Aquarius swung its bulk around and the controllers eavesdropped in silence. Then, from off the right side of the ship, Swigert thought he saw something: a small flash, then nothing, then a flash again. All at once, unmistakably, a tiny degree of the solar arc flowed into the corner of his window. He snapped his head to the right, then turned back to the left to alert Lovell, but before he could say anything, a shard of a sunbeam fell across the instrument panel and the commander, monitoring his needles, looked up with a start.

"Call it, Jack!" he said. "What do you see?"

"We've got a sun," Swigert said.

"We've got a big one," Lovell responded with a smile. "You see anything, Freddo?"

"No," Haise said, squinting into his telescope. Then, as his eyepiece filled with light, "Yes, maybe a third of a diameter."

"It's coming in," Lovell said, glancing out the window and turning away slightly as the sun filled it. "I think it's coming in."

"Just about there," said Haise.

"We've got it," Lovell called. "I think we've got it."

"O.K.," Haise said, watching as the disk of the sun brushed the cross hairs of the telescope and slid downward. "Just about there."

"Do you have it?" Lovell asked.

"Just about there," Haise repeated.

In the telescope, the sun slid down another fraction of a degree, then a fraction of a fraction. The thrusters puffed hypergolics for another second or so, and then, silently, the little jets cut off as the ship — and the sun — came to a stop.

Lovell said, "What have you got? What have you got?"

Haise said nothing, then slowly pulled away from the telescope and turned to his crewmates with a huge grin. "Upper right corner of the sun," he announced.

"We've got it!" Lovell shouted, pumping a fist in the air.

"We're hot!" Haise said.

"Houston, Aquarius," Lovell called.

"Go ahead, Aquarius," Brand answered.

"O.K.," said Lovell, "it looks like the sun check passes."

"We understand," Brand said. "We're kind of glad to hear that."

In Mission Control, where only moments before, Gerald Griffin had called for absolute quiet, a whoop went up from the RETRO, FIDO, and GUIDO in the first row. It was taken up by the INCO and the TELMU and the surgeon in the second row. Across the room, an undisciplined, unprecedented, utterly un-NASA-like ovation slowly spread.

"Houston, Aquarius," Lovell called through the noise. "Did you copy that?"

"Copy," Brand said through his own broad grin.

"It's not quite centered," the commander reported. "It's a little bit less than a radius to one side."

"It sounds good, it sounds good."

Brand glanced over his shoulder and smiled at Griffin, who grinned back and let the tumult go on around him. Disorder was not a good thing in Mission Control, but for a few more seconds, at least, Griffin would allow it. He pulled his flight log toward him, and in the blank space under the Ground Elapsed Time column he wrote, "73:47." In the space under the Comments column, he scribbled, "Sun check complete." Looking down, the flight director discovered for the first time that his hands were shaking. Looking at the page, he discovered for the first time, too, that his last three entries were completely illegible.

To the people around her, Marilyn Lovell appeared surprisingly un-moved by the success of Aquarius's sun check. The guests gathered around the television in the Lovell family room were almost all NASA people, schooled in the ways of lunar flight and well aware of how important the event was. For those few who weren't, the network broadcasters made things unmistakably clear. The prospects for the safe return of the crew depended largely on the outcome of the PC + 2 burn, and the prospects for the burn depended almost entirely on the outcome of the sun check. The reaction in the Lovell home when Jim Lovell radioed down the success of the maneuver was thus much the same as it was in Mission Control: whoops, hugs, and handshakes. Marilyn herself, however, simply nodded and closed her eyes.

Though many of the people in the room observed Marilyn's reaction with some concern, Susan Borman sitting to her left, and Jane Conrad sitting to her right, understood. Like Marilyn — and like all of the other women who had kept similar vigils since the early days of Mercury — Susan and Jane had long since learned that one of the most important things an astronaut's wife needed to remember during the course of a flight was how to ration her reactions. Though the networks could afford to dramatize every tweak of a thruster or torque of a platform for the TV audience, the people whose fathers and husbands and sons were riding inside the spacecraft did not have that freedom. For them, the flight wasn't national news; it was, in the most literal sense, domestic news. It wasn't the future of the nation that rode on the outcome, but the future of the household. With stakes that high, the wives, at least, could not afford the luxury of a fully emotional response at each critical turning point. Whoop or weep during liftoff if you must; cry or cheer during splashdown; clasp your children's hands during ascent from the moon. But outside those moments, simply nod and move on.

The one concession Marilyn had allowed herself to displays of less-stoic emotion had been her occasional, almost dreamy lapses into reminiscences about the earlier, less newsworthy days of her husband's career. Two or three times since last night, Marilyn's face had taken on a calm, faraway expression, and, with something approximating a smile, she would turn to whoever happened to be next to her and recall some happier, safer day years back.

"Did you know Jim liked rockets as a kid?" Marilyn had said to Pete Conrad when the two of them and a handful of other guests had found themselves in Lovell's study earlier today.

"Yeah, he told me," Conrad said. "Built some exploding rocket in high school or something."

"He even wrote his Naval Academy term paper about rockets." Marilyn reached to her husband's bookshelf and withdrew an old Annapolis folder. "Read the last paragraph," she said, flipping open to a sheaf of yellowed onionskin paper stapled together at the corner.

"Marilyn . . . ," Conrad said, wondering whether reminiscing like this was a good idea at such an uncertain time.

"Please read it."

Conrad took the folder from Marilyn and read. "'The big day for rockets is still coming, the day when science will have advanced to the stage when flight into space is reality and not a dream. That will be the day when the advantage of rocket power — simplicity, high thrust, and the ability to operate in a vacuum — will be used to best advantage.'"

"Not bad for 1951, huh?" Marilyn said.

"Not bad."

"Of course, if NASA had had their way the first time Jim applied, he never would have ridden *any* rocket."

"Jim and me both," Pete said.

"You know, seven years after he flunked that first physical, the doctor in charge came by the Space Center for a visit. By that time, Jim had made his two Gemini flights, and he had all these certificates on his wall. When the doctor dropped in, Jim pointed to all the commendations and said to him, 'You guys are pretty good at measuring bilirubin, but the one thing you never thought to measure is persistence and motivation.'" Conrad smiled at this. "He loves telling that story, Pete," Marilyn said, her voice cracking. Abruptly, she looked away.

"Marilyn," Conrad had said, mustering as much conviction as he could, "he *is* coming home."

Whether it was a good or a bad idea for Marilyn to permit herself such musings no one in her house could say, but this afternoon, when her husband had finished his makeshift sun check, she apparently had no need for them. What she did instead, while her guests hugged and cheered, was rise, excuse herself, and walk toward the kitchen.

A few hours earlier, Father Donald Raish, a local Episcopal priest who had known the Lovell family for years, had phoned and offered to stop by for an impromptu Communion. Marilyn enjoyed Father Raish's company, welcomed his visit — a time when, for an hour at least, someone else would be the spiritual pillar in her family room —

and wanted to be able to offer him something other than the stale coffee she had been drinking. Before Marilyn even reached the kitchen, the doorbell rang and Dot Thompson went to answer it.

Father Raish entered and greeted Marilyn warmly, then joined the crowd in the family room. With his arrival, the atmosphere in the room changed dramatically. The volume of the television was turned down, the volume of the voices in the room dropped with it, and the house regained, if only for a short time, some of the normalcy that had prevailed before 9:30 last night.

Marilyn and the other guests had barely assembled around the coffee table where the service would be held when Betty Benware appeared at her side and whispered to her: "Marilyn, did you tell the children Father Raish was coming?"

"Of course," Marilyn said. "I mean, I think I did. Why?"

"Well, if you did tell Susan, she forgot. She just came downstairs, saw everybody talking to a priest, and now she's hysterical. She thinks you've all given up. She thinks Jim isn't coming back."

Marilyn excused herself, ran upstairs to Susan's room, and found her second-youngest child crying inconsolably. Gathering her up, Marilyn assured Susan that no, no one was giving up hope, that the people at the Space Center had nearly everything under control, and that the priest was here just to help take care of those things that were beyond even the Space Center folks.

When her daughter did not seem reassured, Marilyn took her hand, tiptoed downstairs, and signaled to Betty that the two of them would be back in a few minutes. Slipping out the kitchen door, Marilyn and Susan made their way down to Taylor Lake and sat on a patch of grass in the shadow of a nearby tree.

"Now tell me exactly what you're worried about," Marilyn said when they had settled themselves.

"What do you mean?" Susan said, confused. "I'm worried that Dad's not going to come home."

"That?" Marilyn asked with mock astonishment. "*That's* what's bothering you?"

"Uh, sure."

"Don't you know your father's too mean to die?" Marilyn said with a smile.

"He's not mean," Susan protested.

"No, of course he's not. But he's stubborn, right?" Susan nodded.

"And he's smart, right?" Susan nodded again. "And he's the best astronaut I know."

"The best one *I* know too," Susan said.

"Now, do you really think the best astronaut either one of us knows is going to forget something as simple as how to turn his spaceship around and fly it home?"

"No," Susan said, laughing hesitantly.

"No," Marilyn said, "I don't think so either. What I'm concerned about is all those other people at the house who haven't figured that out yet. Don't you think we ought to go straighten them out?"

Susan agreed, and Marilyn and her daughter walked slowly back to their home. When they arrived, the prayer service appeared to have been completed, and the first voice Marilyn heard was not Father Raish's but, she was almost certain, Jim's.

Marilyn and Susan stood in the doorway for a brief, disoriented instant before realizing that the voice was coming from the television set. In the family room, most of the guests were crowded around the screen, where Lovell was visible, looking natty in blue blazer and necktie, and sitting comfortably in an ABC studio talking to Jules Bergman. Marilyn remembered the day last month when Jim had sat for the taped interview — an interview that, he had told her later, consisted mostly of Bergman asking him repeatedly whether he'd ever experienced fear during his career as a test pilot or an astronaut. Marilyn had picked out the tie Jim wore that day, thinking it would look good on television. Now, despite everything, she couldn't help noticing that it did.

"You know, Jules," Jim was saying, "I think every pilot has known fear. I think those who say they haven't are only kidding themselves. But we have confidence in the equipment we're working with, and that overcomes any fear we have of using it."

"Is there a specific instance in an airplane emergency when you can recall fear?" Bergman asked.

"Oh, I've had an engine flame out a few times in an aircraft," Lovell said, "and I was kind of curious whether it was going to light up again — things of that nature. But they seem to work out."

"Do you believe the law of averages operates with you after all these flights? Do you worry about getting stuck on the moon, for example?"

"No, I kind of feel that every time we make these flights we count on two things. First of all, you've got to be well trained to handle emergencies. That's like money in the bank. Second, you've got to remember that every time you go, it's like a new roll of the dice. It's not like something that accumulates so that you're bound to get a seven after a while. You start out fresh every time."

"So the ascent engine failing to ignite, or things like that, don't really worry you?"

"No," Lovell said with a shake of his head. "If they worried me, I wouldn't be going."

Bergman persisted. "Let me put the equation a different way. How do the risks you take compare to the risks a combat pilot takes, say an F4 pilot in Vietnam."

Lovell drew a deep breath and thought for a moment. "Certainly we take a risk," he said at last. "Going to the moon and using the systems we use is risky. But we use the best technology we have to minimize that risk. When you go into combat, the other side is using the best technology they have to *maximize* your risk. Obviously, I think that's a very dangerous proposition."

"So you feel you have the better half of the draw in this case," Bergman said.

"I feel," Lovell said, evidently tiring of this line of questioning, "that a combat pilot in Vietnam is in a very risky position."

The interview ended, and the cameras returned live to Bergman and Frank Reynolds in the ABC studio in New York. Marilyn turned to Susan and smiled.

"See?" she said. "Dad's a lot safer than people who go into combat, and *those* people usually make it home."

Susan seemed relieved, ran out of the family room and toward the back yard. Marilyn, too, thought she felt a bit better. It was true that all over America, thousands of wives lived every day knowing their husbands were flying into combat on the other side of the world, not knowing whether they'd return. And those women didn't have Jules Bergman giving them regular updates on how things were going, Navy vessels mobilizing to pluck them out of the water, or dozens of men in a giant control room monitoring their every breath. Of course, those women's husbands also weren't a quarter of a million miles from home, surrounded by a near-absolute vacuum, flying a crippled ship, and in danger not just of failing to make it back to their carrier or their air base, but failing to make it back to the very planet where their journey began. Marilyn sat down slowly on the couch and felt her spirits sink again. All things considered, she wasn't entirely sure where she'd want her husband to be.

The sun was beginning to set on Marilyn Lovell's house in Houston at almost the same instant it was setting on Jim Lovell's spacecraft 240,000 miles away. With the exception of the two quick passes Apollo 13 had made over the nighttime side of the Earth during its parking orbit around the home planet, the sun had been a constant presence. It wasn't always directly visible, but it was always there: heating the craft during thermal rolls, illuminating the debris after the bang in the service module, glinting off the instrument panel during the alignment check. Now, at 6:30 in the evening, as the visitors in the family room settled around the TV and as Apollo 13 closed to within 1,500 miles of the moon — less than the distance of a single lunar diameter — the ship and the sun at last began to part ways.

Like all other lunar spacecraft, Odyssey and Aquarius were approaching the moon from its western edge; in the case of the gibbous moon that hung in the sky tonight, that meant its shadowed edge. The closer the spacecraft got, the deeper it moved into that darkness. Though some ambient sunshine still bathed the ships, all that reflected up from the surface and into the windows of the steadily darkening cockpit was a weak, shimmery earthshine — the reflected light of the home planet, which was itself only reflecting the light of the sun. What this deepening gloom also meant was that less and less light was reflecting off the sparkling debris cloud that still surrounded the ships. As of an hour ago, Lovell, Haise, and Swigert resumed their customary left, right, and rear stations, and as Haise pored over his engine-burn checklists, and Swigert lent a hand where he could, Lovell returned to his window.

"I've got Scorpio!" the commander announced.

"You have?" Haise asked, stopping what he was doing and turning toward his window.

"Yeah. And Antares."

"They're all coming out," Swigert said, straining for a glimpse through Lovell's window.

"You said it," Lovell said. "There's Nunki, there's Antares. We may have enough here for that confidence check."

Swigert agreed. "We've probably got more than enough."

"You want to let them know?" Haise asked.

"Yeah," Lovell said, and called into his microphone: "Houston, Aquarius."

"Go ahead, Jim," Brand said.

"Be advised I've got Antares and Nunki in the window, and I just wanted to know if you wanted to try that alignment check."

"Roger," Brand said. "We copy the stars you are seeing. Stand by for word on the confidence check."

In Mission Control, Brand clicked off the air-to-ground loop and on to the flight director's loop, to confer with his GUIDO. True to the rumors that had been circulating around Mission Control for most of the day, Kranz's group had come back on console about two hours ago and intended to stay for the next few hours at least. For much of the afternoon, Milt Windler's Maroon Team had been standing by at the edges of the Mission Control auditorium like a football taxi squad, ready to relieve Griffin's group when their shift ended shortly before sundown. But Kranz had put out the word to the room at large, and to his friend Windler in particular, that at the risk of bruising feelings, he'd just as soon send his own men out to handle the PC+2 and let Windler's team take over later. At 4:30, the Tiger Team came out of room 210 practically at a trot, fanned out around Mission Control, and with murmured "excuse me's" and shrugged apologies, commandeered the consoles they had vacated at 10:30 last night. Griffin's Gold controllers, who were minutes away from being relieved anyway, surrendered their seats and retreated to the aisles to join Windler's Maroon Team.

Now, as Brand reviewed the alignment plans with Bill Fenner, the White Team GUIDO, and Fenner reviewed them with Kranz, the first differences between the White Team's stewardship of the flight and the Gold Team's emerged. The star check that could help confirm the accuracy of the platform, Kranz announced on the loop, would be scratched. The alignment Lovell had transferred from Odyssey last night had proved itself adequate during the free-return burn and had been recertified during the makeshift sun check. Screwing around with things now, Kranz believed, was just asking for trouble, and was a sure way to squander both thruster fuel and time. He passed his decision on to Fenner, who passed it on to Brand, who called it up to the crew.

"Hey, Aquarius," the Capcom announced, "we're pretty much satisfied with our present alignment. We don't want to waste any more thruster gas trying to check it further, so why don't we just stay where we are?"

"O.K., understand," Lovell said, and then pushed his microphone away and turned to Haise with a slight eye roll. "First time in the whole flight we've got stars, and now we don't want to use them."

"They're nervous about messing things up for the burn," Haise said, trying to be diplomatic.

"*I'm* nervous about messing things up before we even get there."

The issue of the star check was fast becoming academic, since the time to conduct it was running out anyway. The spacecraft's proximity to the front side of the moon meant it was less than an hour and a half away from arcing around the back side and vanishing from radio contact. The loss of signal would be briefer than it had been the last time Lovell made the trip, because unlike the Apollo 8 crew, whose first job upon disappearing behind the lunar sphere was to apply a hypergolic brake and drop into orbit, the Apollo 13 crew would do nothing at all. Passing behind the west side of the moon at 75 hours and 8 minutes into the flight, they would come tear-assing around the east side just 25 minutes later, their speed having been gravitationally increased during the time they were out of contact with the Earth. Two hours after that, they would have to be ready to fire their engine.

"Aquarius, Houston," Brand called up. "If you're ready to copy, I can give you the maneuvering data for the PC+2, and then you might want to prepare for loss of signal."

"O.K.," Haise said, pulling out a notepad and pen, "I'm ready to copy."

Brand read up all the data, calling out vectors and yaw angles and eventual Earth landing targets, and Haise copied them down and read them back.

Lovell heard concern in the Capcom's voice, but he was pleased to find that he felt a relative sense of calm as the loss of signal and the burn itself approached. This burn, unlike the free-return burn, would be a long, strong one, with the engine firing for 5 seconds at minimum thrust, then 21 seconds at 40 percent thrust, and finally 4 minutes at full thrust. *Like* the free-return burn, it would be initiated and terminated by the computer, with Lovell handling only the throttle that would control the strength of the burn. If the engine did not fire at precisely 79:27:40.07, he would take over that function too, using two bright red, silver-dollar-sized buttons — with the words "Start" and "Stop" stenciled beneath them — on the commander's side of the spacecraft. The buttons provided a direct link between the descent engine and the batteries and, when pressed, would bypass the computer and ignite the engine on command.

Though it was only a late ignition that would require Lovell to hit the Start switch, there were many situations in which he would hit Stop. According to mission rules, the commander would be required to terminate the burn maneuver if his thruster or fuel pressure fell too low, if his oxidizer pressure climbed too high, if his attitude drifted by 10 degrees or more, or if his instrument panel flashed any one of six battery, computer, or engine gimbal alarms.

Worst of all, Lovell knew, was what would happen if he received an alarm telling him that the helium tanks in the fuel system were becoming overpressurized. Rather than using malfunction-prone pumps to force engine fuel through the lines in the descent stage of the LEM, NASA engineers relied on compressed helium fed from high-pressure tanks. Channeled into the fuel lines, the inert gas would not react with the explosive hypergolic fluid, but would instead push it along to the combustion chamber.

The system was nearly flawless, with just one exception: helium has the lowest boiling point of all the elements, so the slightest change in temperature can cause it to vaporize and expand. Compressing a gas that requires so much elbow room into a highly confining tank can be a recipe for disaster, and in order to prevent pressure explosions, NASA equipped the line that ran from the tank with a diaphragm-like "burst disk." In the event of a sudden pressure increase, the diaphragm would rupture, releasing the gas before the pressure climbed too high.

Venting the helium meant that the engine would no longer be able to fire, but in a normal lunar flight that would not be a problem. The helium system was not intended to be heated up and switched on until the descent engine was ready to burn, and the descent engine was designed to burn only once, carrying the LEM from lunar orbit to lunar landing. Any rupture of the burst disk after that would take place on the moon's surface, where the engine would already have been permanently shut down and the gas could escape harmlessly into the surrounding vacuum. What nobody had ever considered, but what the commander of Apollo 13 was now confronting, was what would happen on a mission in which the engine had to be fired and shut down and fired again and shut down again. If the burst disk in the overworked fuel line should blow now, the descent propulsion system would be lost for good.

Despite all this, Lovell felt a surprising equanimity as the burn approached, and while Haise continued taking data dictation from Brand, the commander took another moment to glance out his window. As it turned out, he picked just the right moment. At 76 hours, 42 minutes, and 7 seconds into the mission, the sun set behind the moon, and Apollo 13 moved completely into shadow. Outside the spacecraft, the sparkling debris at last disappeared, and on all sides of the ship, at all angles, and in all axes, the sky was suddenly lit up with curtains of ice-white stars.

"Houston," Lovell said, "the sun has gone down and — man — look — at — those — stars."

"Is that Nunki out there?" Haise asked, turning to the window and pointing to the star Lovell had barely spotted earlier but that now stood out like a lighthouse beacon.

"Yes," said Lovell, "and I can see Antares much better."

"What's that cloud over there?" Swigert asked, leaning over Lovell's shoulder.

"The Milky Way," Lovell answered, indicating the bright white band that bisected the sky.

"No, not the illuminated one," Swigert said. "The dark one — actually two dark ones, that look like contrails."

Lovell followed Swigert's gaze and saw a pair of eerily dark columns blotting out some of the newly visible stars. "I can't for the life of me figure out what that would be," he said. "It might be debris that was thrown out there."

"From our maneuvers?" Haise asked.

"No," Lovell said, "from our explosion."

All three astronauts looked at the clouds and fell quiet. It had been close to twenty-four hours since last night's sudden jolt and bang, and the sense memory of the experience had begun to fade. But these ghostly black fingers extending from their ship and reaching out into space brought it instantly back. It still wasn't clear what had gone wrong in the rear of their spacecraft, but lest they forget, it had made a smoking mess out of a vehicle that was supposed to be all but indestructible.

Brand's voice broke the quiet. "Aquarius, Houston."

"Go ahead, Houston."

"O.K., Jim, we have a little over two minutes until loss of signal, and everything's looking good here."

"Roger," Lovell said. "I take it you don't want us to activate any other systems or make any other preparations until we reacquire signal."

"Roger. That's correct," Brand said.

"O.K., we'll just sit tight, then. See you on the other side."

The Apollo 13 crew fell back into silence, and 120 seconds later the signal from Houston disappeared.

Slipping past earthshine and into the absolute darkness and radio silence behind the moon, the crew remained subdued. Since only a crescent on the western end of the front of the moon was in shadow, only a corresponding crescent on the diagonal end of the back side was illuminated. For most of Apollo 13's transit around the moon, therefore, there was nothing but shadow beneath the ship. The only thing that revealed there was a body down there at all was the utter absence of stars, an absence that began where the ground ought to begin and ended in the distance where the horizon ought to start.

For close to twenty minutes the astronauts coasted over this night-time nothingness, until, just five minutes before reacquisition of signal, a white-gray sickle of mottled turf appeared in the distance. Haise, at his right-hand window, saw it first and reached for his camera. Lovell, at his left-hand window, saw it next, and nodded less in rapture than recognition. Swigert, who had never seen such a thing before, grabbed his camera and glided toward Lovell's station, and the commander floated backward to let his rookie crewmate see what was unfolding below. Sliding beneath the ship, just as it had slid beneath Apollo 8 almost sixteen months earlier, was the same desolate stretch that had never been glimpsed by human beings until 1968, and that had now been seen by more than a dozen.

Swigert and Haise, like Borman, Lovell, and Anders before them, were transfixed. They scanned the mares and craters, the rills and the hills — the great sweep of the lunar terrain — in respectful silence. Unlike the crews aboard previous ships, this crew wasn't passing overhead at 60 miles but at 139, and unlike the crews of previous Apollos, they weren't here to stay. As soon as they passed to the eastern side, they would begin climbing away, and Lovell drifted to the back of the cabin and let his junior pilots look their fill. Five minutes later, at the appointed time for reacquisition of signal, he flipped his microphone switch to its Transmit setting and called back down to Earth in a considerate whisper.

"Good morning, Houston. How do you read?"

"Reading you fairly well," Brand said.

"All right," Lovell said. "We read you fairly well too." He looked over Swigert's shoulder and glanced at the formation sliding by below him. "And for your information, we're coming up on Mare Smythii now, and it looks like we're climbing away."

"We're really zooming off now," Swigert added, a little sorrowfully.

"Oh, yes, yes," Lovell responded as much to his crew member as to the ground, "we're no longer at 139 miles. We're leaving."

"Copy that, Aquarius," Brand said.

"What we're still going to need is a power-up time for the burn," Lovell reminded the ground.

"O.K. Stand by."

Brand clicked off the line, and while Haise and Swigert remained at the windows with their cameras, Lovell began moving about the cockpit, nervously fussing with his breakers in preparation for the power-up. Drifting from one section of his instrument panel to another, he found himself reaching around Haise and Swigert, occasionally muttering an "Excuse me, Freddo," or an "I'm sorry, Jack." The LEM pilot and the command module pilot would respond to their commander with a nod, absently moving out of the way to allow Lovell to reach what he needed, and then floating back into place. After two or three minutes of this, Lovell stopped, backed up onto the ascent engine cover, which until this moment he had assumed was Swigert's station, and folded his arms.

"Gentlemen!" he said in a voice deliberately too loud for the tiny cockpit. "What are your intentions?"

Startled, Haise and Swigert spun around. "Our intentions?" Swigert said.

"Yes," said Lovell. "We have a PC+2 maneuver coming up. Is it your intention to participate in it?"

"Jim," Haise said somewhat feebly, "this is our last chance to get these shots. We've come all the way out here — don't you think they're going to want us to bring back some pictures?"

"If we don't get home, you'll never get them developed," Lovell said. "Now, lookit. Let's get the cameras squared away, and let's get all set to burn. We're not going to hack it with a splashdown at 152 hours."

Haise and Swigert stowed their cameras and returned to their stations somewhat sheepishly, and for the next hour or so the crew worked purposefully. With Brand dictating the power-up instructions and the crew throwing the appropriate switches, Aquarius's systems were slowly armed and brought on line.

As with the lunar orbit insertion burn on Apollo 8, the Apollo 13 astronauts waited in silence for the final few minutes leading up to their maneuver to tick away. There would be no canvas restraints for the pilots to use this time, no couches to strap themselves safely into. Instead, they would simply stand, brace themselves against the bulkhead, absorb the sudden thrust, and feel the subtle press of acceleration through their now comfortably zero-g bodies. Lovell looked over at Haise and flashed a thumbs up, then looked back over his shoulder at Swigert and did the same.

"By the way, Aquarius," Brand announced, interrupting the quiet, "we see the results of Apollo 12's seismometer. Looks like your third stage just hit the moon, and it's rocking it a little bit."

"Well, at least something worked on this flight," Lovell said. "Sure glad we didn't have a LEM impact too."

Lovell looked down at the moon as if he could see the scatter of dust and the small crater created by the latest projectile to hit the ancient surface. What he saw instead was a tiny, perfectly triangular mountain, tucked among the craters and hills that lined the edge of the Sea of Tranquillity. It was Mount Marilyn, hailing him back as he climbed up and away, presumably forever.

"Ten minutes to burn," Haise announced. Shortly afterward he called, "Eight minutes to burn," then "Six minutes to burn," then "Four minutes to burn." Finally Brand, at his Capcom station, took up the call.

"Jim, you are go for the burn, go for the burn."

"Roger, I understand," Lovell said. "We are go for the burn."

"Two minutes and forty seconds on my mark," Brand called. "Mark."

Lovell looked at his mission timer, marked the time that remained, drew a breath, and held it. It was, he thought a little grimly, the night flight in the Sea of Japan all over again. With his cockpit blacked out and the prow of his ship pointed toward the glowing, blue algae streak of Earth, he watched the clock count down to zero and felt the LEM rumble to life beneath his feet.

10

Tuesday, April 14, 3:40 P.M. in the Pacific Ocean

MEL RICHMOND was not likely to get seasick in the South Pacific. First of all, the ship he was aboard, the helicopter carrier Iwo Jima, was too big to roll much in even the roughest waters. More important, Richmond had been out here too many times before. Indeed, it was Mel Richmond who had literally helped write the book on how to recover a returning spacecraft.

In the days leading up to the launch of a Mercury, Gemini, or Apollo spacecraft, NASA would dispatch a team of recovery experts to board the naval vessels assigned to the splashdown site and direct the retrieval of the spacecraft and its crew. It was not always a thoroughly amicable arrangement. Navy men accustomed to working only with other Navy men silently chafed at the squad of civilian engineers that appeared in their midst and, appallingly, commandeered their ship. The engineers themselves were seemingly unaware of any resentment as they blithely disrupted the ordinary routine of the vessel to perform their extraordinary rescue.

Richmond, as the second in command of the visiting NASA team, was more lost in his work than most. Well before a manned rocket ever left the pad, the former Air Force man and current trajectory expert would closet himself with the mission flight plan, maps of the potential reentry points, and worldwide weather forecasts. From this data alone, he would draw up a roster of every conceivable splashdown site the returning spacecraft could aim for, and every recovery

242

technique that might have to be used to pull the vehicle and crew out of the water. This report became the Book — the main recovery book — for that mission, and as the reentry drew closer and the likely splashdown site became clear, it would be this instruction manual that would dictate every step of the complex rescue.

Mel Richmond was not the only person who did this painstaking job. Rotating recovery teams were assigned to every second, third, or fourth space flight, and one member of the team would write that mission's manual. But Richmond had done it more than most, participating in the recoveries of Gemini 6, Gemini 7, Apollo 9, and Apollo 11, and he knew that splashdown work was not for everyone. The NASA team that set sail for these two-week tours of duty typically lived no better than the ordinary ensign, sharing small, vault-like, four-man cabins, eating in the officers' mess, and, with the exception of quick, twice-a-day phone conferences with Mission Control, losing all real contact with home.

The daily routine during these fortnights at sea alternated between periods of crushing boredom and frantic activity, depending on what exercises were planned at any given time. The hardest work came during recovery drills, held every other day, when a dummy spacecraft would be thrown overboard, the carrier would steam a few hundred yards away from it, and the entire rescue crew — frogmen, chopper pilots, deck teams, spotters — would practice retrieving it.

For several days, the recovery exercises for the return of Apollo 13 had been proceeding apace, toeing as closely as possible the line that Richmond's recovery book had drawn. But now, on the spacecraft's fourth day in space, the carefully drafted procedures and exercises prescribed in the book had been thrown into turmoil.

The Odyssey command module was, according to the original flight plan, supposed to hit the water 207 miles south of Christmas Island on Tuesday, April 21 at 3:17 P.M. — four days after blasting off from the moon's Fra Mauro foothills. Over the last several days, however, that original plan had changed, and Apollo 13, so the Houston guys now had it, would be coming home on the afternoon of April 17 — or perhaps on the evening of the seventeenth, or perhaps sometime on the eighteenth — and would be splashing down in the South Pacific — or perhaps the Indian Ocean, or perhaps the Atlantic. The exact time and place depended on the success of the PC+2 speed-up burn the guidance guys had been cobbling together. If the burn went as planned, Mel Richmond's prime recovery fleet could expect to pluck the spacecraft out of the Pacific on Friday, April 17, at about noon. If things didn't go as planned, NASA would find itself rustling up who knew what ships to meet Odyssey in who knew what ocean at who knew what time. It was not the way Richmond liked to do business.

It was 8:40 and after dark in Houston when the lunar module Aquarius was set to relight its descent engine for its four and a half minute burn, but out near Christmas Island, due south of Oahu, it was 3:40 on a still-bright afternoon. Though the entire world could listen in on Apollo 13's air-to-ground communications, courtesy of NASA's aggressive Public Affairs office, the entire recovery crew could not. One of the Iwo Jima's radio officers could pick up the conversations between Capcom and crew from a communications satellite, but the connection was fuzzy and the transmissions could not be broadcast around the carrier. As a result, it was the satellite officer alone who would be able to eavesdrop on the burn.

Elsewhere on the ship, another communications officer in another radio room was in contact with Mission Control itself. It was this officer who arranged for the regular telephone conferences between the Iwo Jima and Houston, and he would be the first to get word when the PC+2 had been completed successfully — or had not. Shortly before 3:30, Mel Richmond and a handful of other recovery team members appeared in this second radio man's office to wait for the news to arrive. Across the ship, in the satellite office, the lone officer at the air-to-ground radio listened in on the spacecraft chatter the rest of the Iwo Jima couldn't hear.

"Two minutes and forty seconds on my mark," the satellite man heard Vance Brand's voice call from Houston as the burn became imminent.

"Roger, we got you," he heard Jim Lovell respond through the storm of air-to-ground static.

There was a long silence.

"One minute," Brand announced.

"Roger," Lovell answered. Sixty more seconds of silence.

"We're burning 40 percent," the radio officer now heard Lovell call.

"Houston copies." Fifteen seconds passed.

"One hundred percent," Lovell said.

"Roger." Static roared in the background. "Aquarius, Houston. You're looking good."

"Roger," Lovell crackled back. Another sixty seconds passed.

"Aquarius, you're still looking good at two minutes."

"Roger," Lovell said. More static, more silence.

"Aquarius, you're go at three minutes."

"Roger."

"Aquarius, ten seconds to go."

"Roger," Lovell said.

"— seven, six, five, four, three, two, one," Brand ticked off.

"Shutdown!" Lovell called.

"Roger. Shutdown. Good burn, Aquarius."

"Say again," Jim Lovell shouted back through the radio hiss.

Brand raised his voice. "I — say — that — was — a — good — burn."

"Roger," Lovell said. "And now we want to power down as soon as possible."

In the satellite room of the carrier, the communications man sat back and removed his headset. He knew, if no one else aboard the Iwo Jima yet did, that Apollo 13 was in fact coming their way. Across the ship, in the second radio shed, Mel Richmond and the rest of the recovery team stood in a semicircle around the silent radio set. Finally, nearly half a minute after the burn ended, a call from Houston crackled into the radio's little speaker.

"Iwo Jima, Houston, at 79 hours, 32 minutes into the mission," the voice said. "Pericynthion-plus-two burn complete. Predicted splashdown 600 miles southeast of American Samoa, at 142 hours, 54 minutes, ground elapsed time."

"Roger," the radio man responded into his microphone. "Burn complete."

Around the room, the recovery team turned to one another and smiled.

"Well," Richmond said to the officer standing next to him, "it looks like we'll have work to do on Friday."

As soon as the PC+2 burn was complete, Gene Kranz, seated at the flight director's console, took off his headset, stood, and surveyed the room. Like Gerald Griffin's Gold Team several hours ago, Kranz's White Team responded to the successfully completed maneuver by breaking into a spontaneous, back-slapping celebration that, by Mission Control's standards anyway, qualified as pandemonium. And like

Gerald Griffin several hours ago, Gene Kranz was inclined to let the revelry run its course; he figured the team deserved its moment of self-congratulation. Besides, soon enough he would have his hands full with something else. If Kranz knew the personnel in this room — and he did — he was convinced that momentarily, three men would be converging on his station. And if he could predict what they would be coming here to say — and he could — he knew that the meeting would be stormy.

Looking one row down and to the left, he could see Deke Slayton, who had been standing behind the Capcom station, heading toward his console. Looking back to the fourth row, he saw Chris Kraft remove his headset at the Flight Operations station and walk down a level. Behind Kraft, in the glassed-in gallery, he could see Max Faget, the head of the Space Center's Engineering and Development branch and one of the first men appointed by Bob Gilruth to the Space Task Force that formed the nucleus of NASA twelve years earlier. Faget was threading through the VIP throng and making his way into the main room. Kranz sighed and stubbed out the cigarette he had lit at the beginning of the PC+2 burn, which had now burned down to his fingertips. Slayton, the closest of the three approaching men, arrived first.

"So what's our next step here, Gene?"

"Well, Deke," Kranz said, measuring his words, "we're gonna work on that."

"I'm not sure how much there is to work on," Slayton said. "We're going to put the crew to bed, right?"

"Eventually, sure."

"*Eventually* may not do it, Gene. Their last scheduled sleep period was twenty-four hours ago. They're going to need some rest."

"I know that, Deke," Kranz began, but before he could finish his thought, he heard another voice over his shoulder. It was Kraft's.

"How do we stand with that power-down plan, Gene?"

"It's coming along, Chris," Kranz answered levelly.

"We ready to execute it?"

"We're ready, but it's a long procedure and Deke thinks we ought to get the crew to sleep first."

"Sleep?" Kraft said. "A sleep period's six hours! Take the crew off stream that long before powering down and you're wasting six hours of juice you don't need to waste. Besides, Lovell agrees. Didn't you hear him on the radio?"

"But if you keep them up and have them execute a complicated power-down when they're barely awake," Slayton said, "someone's bound to screw something up. I'd rather spend a little extra power now than risk another disaster later."

From behind Slayton, Faget, who had now reached the group, nodded a greeting to Kranz.

"Max," Kranz said, "Deke and Chris were just telling me what they think our next step ought to be."

"Passive thermal control, right?" Faget said matter-of-factly.

"PTC?" Slayton sounded alarmed.

"Sure," said Faget. "That ship's had one side pointing to the sun and one side pointing out to space for hours. If we don't get some kind of barbecue roll going soon, we're going to freeze half our systems and cook the other half."

"Do you have any idea what kind of pressure it's going to put on the crew to ask them to execute a PTC roll now?" Slayton asked.

"Or what kind of pressure it's going to put on the available power?" Kraft added. "I'm not sure we can afford to try something like that at the moment."

"And I'm not sure we can afford not to," Faget countered.

For several minutes, the argument at the flight director's station played out, with Kraft, Slayton, and Faget arguing their points fiercely and the men at the nearby Capcom and INCO stations occasionally turning their heads for a sidelong look. At last Kranz, who had remained uncharacteristically quiet throughout, held up his hand and the three other men — all of them technically his superiors — stopped speaking.

"Gentlemen," he said, "I thank you for your input. The next job for this crew will be to execute a passive thermal control roll." He turned and nodded to Faget, who nodded back. "After that, they will power down their spacecraft." He nodded to Kraft. "And finally," he said, looking at Slayton with a flicker of apology, "they will get some sleep. A tired crew can get over their fatigue, but if we damage this ship any further, we're not going to get over that."

Kranz turned back to his console, and Faget and Slayton turned to leave. Kraft, however, stayed where he was. Standing behind the station he had occupied from 1961 to 1966, the man who had trained Gene Kranz to perform the job he was now performing considered objecting to the decision his one-time protégé had just made. But before uttering a word, he changed his mind and walked away. Whatever the flight director chose to do, "regardless of mission rules," was law. Kraft himself had written that rule eleven years earlier, and now he was going to have to live by it.

For the next two hours, the fatigued crew in the busted ship performed the chores the ground ordered them to perform, only afterward getting the O.K. to go to sleep. Even then, the rest periods would be stingily doled out, with Haise going off for three hours of sack time first, and Lovell and Swigert staying on watch in Aquarius until he returned.

Now, well past midnight, Haise's sleep shift was almost at an end, and the two men left behind at the helm of the lunar module found themselves nodding off for a stolen nap as well. Sleeping in the cold, noisy cabin of Aquarius was difficult, it turned out, but not impossible. The trick was to tell yourself that you weren't actually *trying* to go to sleep, that you were simply closing your eyes for a few minutes, and that — even as your mind went blank and you slipped into light slumber, floating in front of your instrument panel — you were really still awake and on watch and prepared to respond to any emergency.

"Aquarius, Houston," Jack Lousma, the graveyard-shift Capcom, called suddenly into Lovell's ear.

"Hmm, yeah," Lovell murmured, speaking with forced alertness. "Aquarius here."

"It's about time for you guys to get to bed and get Fred up," Lousma said.

"Roger," Lovell mumbled. "Looking forward to it."

"Take three hours, and return at about 85 hours, 25 minutes," Lousma said.

"Roger that."

The commander rubbed his eyes, took two steps back toward the tunnel, and jumped up into Odyssey. Approaching Haise in his right-hand couch, Lovell shook him awake. The ambient temperature in the command module was by now, Lovell guessed, down in the low forties or high thirties. Around the sleeping Haise, however, a thin layer of nearly body-temperature air had formed. In the absence of gravity, which meant an absence of convection, the warm air would not be any lighter than the cold air around it, and thus would not rise and drift away.

Helping Haise out of his couch, Lovell scattered the atmospheric blanket his junior pilot had spent the last three hours creating, and sent him down to the LEM. The commander then climbed onto his own couch, wrapped his arms about himself, and curled up against the chill that his own body heat had not had time to change. A moment later, Swigert floated onto his own couch and did the same.

At his station in Odyssey, Lovell could hear the still-bleary Haise banging around in the LEM, collecting his headset, and signing on the air to Houston. Though Haise was obviously keeping his voice low for the benefit of his crewmates, even a whisper was audible in the cramped ships, and as Lovell tried to allow his mind to carry him into sleep, he could not help listening in on the one-way conversation taking place on the other side of the tunnel.

"I left upstairs just a minute ago, Jack," Haise was saying to Lousma, "and I'm down in the LEM now. From what I see out the window, it looks like the moon is sure enough getting smaller."

There was silence from the LEM. Lousma, Lovell assumed, would be congratulating Haise on the work he'd done so far and assuring him that the moon would be getting smaller still as the hours went by.

"I'll tell you," Haise answered in response to whatever Lousma did say, "this Aquarius has really been a winner."

Silence again. Lousma, no doubt, was telling Haise that it was the crew who were the real winners.

"From the sound of all the work that's going on down there," Haise demurred modestly, "this flight is probably a lot bigger test for the guys on the ground than the guys up here."

No, no, Lousma likely said, we're just doing what we're trained to do. It's you guys who are doing the heavy lifting.

"Well, we're just trying to stay ahead of things," Haise said. "We just want to be ready for entry day on Friday."

In his commander's couch, Lovell closed his eyes tighter and turned to face the bulkhead, swirling the cocoon of air around him that had only just begun to warm. If his LEM pilot and his Capcom wanted to buck each other up with confident talk about reentry, that was fine. But Lovell, for one, did not want to hear it. The last update he had gotten from the ground indicated that he and his crew were barely 15,000 miles from the moon and moving at only 4,800 feet per second, or less than 3,000 miles per hour. Their speed, he knew, would get slower before it got faster, decreasing steadily until they had traveled another 24,000 miles or so, and the gravity of the Earth pulling them forward took over from the gravity of the moon pulling them backward. Until that happened, Lovell would not be especially comfortable. A ship that was 15,000 miles from the moon was still 225,000 miles from home — way too far to qualify as the clubhouse turn. Since Monday night, Lovell reflected as sleep began to overtake him, he'd had cause to feel a lot of emotions, but cockeyed optimism was not among them.

Ed Smylie stepped into the elevator in Building 30 of the Manned Spacecraft Center, faced forward, and watched as the silver doors shushed closed. He held a metal box awkwardly under his arm. Turning to his right, he reached toward the row of buttons and, with the faintest sense of ceremony, pressed 3, the Mission Control floor.

As chief of the Crew Systems Division, Smylie had no cause to be modest about what he did for a living. It might be Sy Liebergot and John Aaron and Bob Heselmeyer who sat at the consecrated consoles in Mission Control and kept the environmental hardware in a moon-bound LEM and command module up and running, but it was Ed Smylie and his crew who helped develop and test those life-support systems in the first place. It was important work, but it was also anonymous work. While the Liebergots, Aarons, and Heselmeyers spent their workdays in the spacious auditorium of Building 30, with the media broadcasting their every move, Smylie and his men spent their time in the warrens of labs in Buildings 7 and 4 and 45.

Today, however, was different. Today the men on the floor of Mission Control very much wanted to see Smylie — and more specifically, to see the cumbersome object he was carrying with him. Ever since Monday night, when Apollo 13 first began to bang and vent and spin, the men at the Space Center, and specifically the engineers in Crew Systems, had been fretting about the lithium hydroxide question. The problem of trying to fit the command module's square air-scrubbing cartridges into the LEM's round receptacles was a low-tech issue on a flight beset by so many high-tech malfunctions, but it was a pressing issue nonetheless. With three men living and respiring in Aquarius, the first of the lunar module's cartridges should become saturated with CO_2 by the 85-hour mark in the mission, requiring the second and last one to be snapped into place. Well before the ship reached home, that cartridge would be full as well, and the astronauts would quickly choke to death on their own waste gases.

Smylie's first move after switching on his television Monday night and learning of Apollo 13's accident was to pick up the telephone and call the Crew Systems office.

"What do we know about 13?" he asked when the desk man answered the phone.

"Not much. They're out of oxygen and they're moving over to the LEM."

"They're going to have a CO_2 problem," said Smylie.

"A big one," the desk man agreed.

"I'm coming in," Smylie said.

The Crew Systems lab in Building 7 was not a simple affair. Included in the multi-million-dollar facility was a room-sized vacuum chamber used to check out a spacecraft's environmental control system, the life-support backpacks that were used on the surface of the moon, and the spacesuits themselves. The chamber's air could be reduced from sea-level pressure down to the 5.5 pounds per square inch required in a spacecraft, or even down to the near vacuum of the moon. Like both the command module and the lunar module, the vacuum chamber also had a fully-functioning lithium hydroxide air-purification system.

As Smylie sped to Building 7, less than an hour after hearing of Apollo 13's peril, a wonderfully crude solution to Aquarius's carbon dioxide problem began to take shape in his head. The LEM's lithium hydroxide system, like the command module's, worked with the aid of a fan that drew spacecraft atmosphere through intake vents in the front of the air-scrubbing cartridge, out again through exit vents in the rear, and back into the cockpit, stripped of its unwanted CO_2. Attached to the wall of the cockpit were also two sets of hoses so that, in the event of an atmospheric leak in the spacecraft, the commander and the LEM pilot could plug their pressure suits directly into this air-purifying, life-preserving loop.

In order to make the oversized command module cartridges work in the inhospitable LEM, what Smylie envisioned doing was inserting the back half — the outflow half — of the bulky lithium hydroxide box into a plastic bag and taping the bag in place with heavy, airtight duct tape. An arched piece of cardboard taped inside the bag would hold it rigid and prevent it from collapsing against the outflow vents. Smylie would then punch a small hole in the bag and insert the loose end of one of the pressure-suit hoses into it, making this connection airtight with tape as well. With the LEM's air-purification system running, atmosphere would be drawn through the front of the square canister, out the back, into the bag, and through the hose. From there, it would run through the LEM's own air-scrubbing pipes and back into the cabin of the ship.

In essence, the CO_2-cleaning system of the LEM would work exactly as it was designed to, with the exception that the jury-rigged command module canister connected to the intake hose would take the place of the used-up LEM canister further downstream. When the new canister was itself exhausted, a fresh one could be prepared and attached in its place.

Smylie arrived at Building 7 at 11:30 Monday night and was met in the lobby by his assistant Jim Correale. The two men sprinted to their lab, fired up the environmental chamber, and, working with a dummy lithium hydroxide canister containing no actual air-scrubbing crystals, built the device Smylie had already built in his head. When the two engineers attached the makeshift device to the mockup environmental system and turned on the fan, they found that their humble invention seemed to work fine. But in order to test the system fully, they needed genuine cartridges.

The problem was, none were available in Houston. At 3 A.M. Tuesday, Smylie was on the phone to the launch center at the Cape, to see if anyone had any live cartridges on the shelf, and by 4 A.M. the technicians in Florida had managed to scare up a few — intended for installation in Apollo 14 or 15 — loaded them onto a chartered jet, and flew them to the Manned Spacecraft Center. For most of the following day, Smylie and Correale lived in their lab, pumping their LEM chamber full of carbon dioxide and then watching as the newly arrived cartridges with their cut-and-paste modifications sponged the poisonous gas out of the air, leaving only breathable oxygen behind.

Now, in the early hours of Wednesday morning, the Building 30 elevator bumped to a stop at the third floor. Smylie stepped out, carrying his strange, unwieldy invention with him. Walking down the white, windowless hall, he at last came to a pair of heavy metal doors on his left marked "Mission Operations Control Room." He opened one of the doors, stepped inside, and scanned the room uneasily. There were no humble Crew Systems engineers here, no anonymous technicians, just the glamorous EECOMS and TELMUS and FIDOS and flight directors. Smylie made his way down the aisle, looking for Deke Slayton, Chris Kraft, or Gene Kranz. With each passing minute, he knew, the three astronauts in the distant ship were coming closer to choking on their own carbon dioxide. Smylie realized that the little box he had just invented would likely save their lives. And that, he didn't need to remind himself, was something you could never accomplish with a headset or a console or a title like TELMU.

Fred Haise rather enjoyed being alone in his LEM. He liked the unaccustomed quiet, he liked the unaccustomed elbow room, and he liked, more than either, the brief chance to be in charge of his own ship. Unlike the commander of the three-man lunar crew, who enjoyed near-absolute authority over the vehicles and the men placed in his charge, and unlike the command module pilot, who would assume total command of the mother ship during the two days his crewmates were off flying their LEM, the lunar module pilot would never take the helm of either ship he was aboard. For men who, before joining NASA, made their living test-flying planes, this could rankle a bit. At three o'clock Wednesday morning, however, as Jim Lovell and Jack Swigert were entering the second hour of their sleep shift in Odyssey, Fred Haise — third in command of a crew of three — found himself drifting around his well-loved Aquarius alone.

"Houston, Aquarius," Haise radioed quietly to Jack Lousma as he floated toward Lovell's vacant station.

"Go, Fred," said Lousma.

"I'm looking back at the left-hand corner of the moon," Haise said, "and I can just barely make out the foothills of the Fra Mauro formation. We never did get to see it when we were in there close."

"O.K.," Lousma said. "It looks like you're not in so close anymore. I'm reading on my monitor here, Fred, that you're 16,214 miles away from the moon and moving at 4,500 feet per second."

"When this flight is over," Haise said, nodding to himself, "we'll really be able to figure out what a LEM can do. If it had a heat shield, I'd say bring it home."

"Well, at least you gave the folks at home a good look at the inside of the ship during that last broadcast Monday night," Lousma said. "That was a good show you guys put on."

"It would have been an even better one about ten minutes later."

"Yes," Lousma said, "things sure turned to worms in a hurry there after that."

Haise pushed away from the window and drifted backward toward Swigert's station atop the ascent engine cover. Reaching into a storage bag, he poked through a few of the food packets Swigert had carried over from Odyssey early yesterday.

"And just for your information, Jack," Haise radioed, "I'm going to pass the time by tearing into some beef and gravy and other assorted goodies."

"I presume you're doing this with the full permission of the commander," Lousma said.

"And at this very moment," Haise said with a smile, "just who do you think the commander is."

"All the same, if I was him, I'd make you sign out everything you ate, so he could keep track of it."

"Understood."

"And Fred," Lousma added, "sometime when you're not too busy chewing on that beef, how about telling us what that CO_2 reads."

Lousma's nonchalance belied the sense of urgency behind this request. Ed Smylie's visit to Mission Control had been a happy one for both the engineer and the flight controllers. The makeshift air scrubber had intrigued Slayton, Kranz, Kraft, and the knot of LEM environmental officers who crowded around the Capcom's desk, and the report of the successful test in the vacuum chamber in Building 7 had convinced them that the inelegant contraption could indeed work. Now, after Smylie had come and gone, his prototype remained atop Lousma's console, attracting controllers who would amble by and poke at it.

The fact that Smylie's box could be easily assembled in his lab was no guarantee it could be just as easily assembled in space, and the time for getting started on the job was growing short. Carbon dioxide concentrations in the command module and the LEM were tracked with a non-power-consuming instrument resembling a thermometer, which measured the pressure of the toxic gas in the overall atmosphere. In a healthy ship, the needle should climb no higher than 2 or 3 millimeters of mercury. When it rose above 7, the crew was instructed to change their lithium hydroxide canisters. If it was allowed to rise above 15, it meant that the canisters had absorbed about all they could and that before long, the first signs of CO_2 poisoning — lightheadedness, disorientation, nausea — would set in. As Fred Haise folded up his roast beef packet, left it to float near the back of the cockpit, and drifted over to the carbon dioxide gauge, what he saw brought him up short.

"O.K.," Haise said evenly, "I'm reading 13 on the gauge." He squinted at the needle a second time. "Yeah, 13."

"All right," Lousma said, "that's pretty much what we've got here, so we're going to want to get started putting together the little canister we've come up with."

"You want me to head up into Odyssey and start collecting materials?"

"Nah," Lousma answered. "We don't want to bother the skipper just yet. We'll give him a few more minutes to sleep."

As Lousma was saying this, Haise heard a rustling noise in the tunnel. He glanced up and saw Lovell, red-eyed with fatigue, floating head-first into Aquarius. The commander descended toward the ascent engine cover, flipped over, and pulled himself down to a sitting position with a thump. Bobbing at the level of his eyes was Haise's abandoned beef, which he regarded with curiosity, plucked out of the air, and tossed across the cockpit to his LEM pilot. Haise caught the packet and stowed it quickly in a waste bag.

"You're back awful early," Haise said.

Lovell yawned. "It's too cold up there, Freddo."

"You've gotta stay real still."

"I *tried* staying real still. It doesn't help anymore. If it's much above 34 degress in there, I'd be surprised."

Lovell reached forward, put his headset back on, and called down to Lousma.

"Hello, Houston, Aquarius. This is Lovell here who's got the duty again."

"Roger, Jim. Is Jack there with you?"

"No, he's still sacked out."

"O.K.," Lousma said, "as soon as he gets up, I'd suggest we go ahead and make a couple of these lithium hydroxide canisters. It's going to take all three sets of hands, I think."

"All right," Lovell said, clearing his head with a shake and moving back to his left-hand spot. "We'll make that the next project, then, getting those canisters squared away."

Though there was more than an hour left in the sleep cycle, and Swigert, unlike Lovell, had managed to fall sound asleep inside the icebox of Odyssey, the sudden chatter and bustle coming from the LEM soon roused him. Just minutes after Lovell dropped down through the tunnel, Swigert appeared as well. On the ground Joe Kerwin, who was scheduled to begin his fourth shift as Capcom in as many days, went on duty too, taking Lousma's place behind the console.

"O.K.," Lovell called down to the new man in Houston, "Jack's up with me now, and as soon as he gets on his earphone, we'll be ready to copy."

"Roger that, Jim," Kerwin said, letting his acknowledgment serve as his hello. "Whenever you're ready."

For the next hour, the work aboard Apollo 13 had little more orderliness than a scavenger hunt, and little more technical elegance. With Kerwin reading from the list of supplies Smylie had provided him, and Kraft, Slayton, Lousma, and other controllers standing behind him and consulting similar lists, the crew were dispatched around the spacecraft to gather materials that had never been intended for the uses to which they were about to be put.

Swigert swam back up into Odyssey and collected a pair of scissors, two of the command module's oversized lithium hydroxide canisters, and a roll of gray duct tape that was supposed to be used for securing bags of refuse to the ship's bulkhead in the final days of the mission. Haise dug out his book of LEM procedures and turned to the heavy cardboard pages that carried instructions for lifting off from the moon — pages he now had no use for at all — and removed them from their rings. Lovell opened the storage cabinet at the back of the LEM and pulled out the plastic-wrapped thermal undergarments he and Haise would have worn beneath their pressure suits while walking on the moon. No ordinary long johns, these one-piece suits had dozens of feet of slender tubing woven into their fabric, through which water would have circulated to keep the astronauts cool as they worked in the glare of the lunar day. Lovell cut open the plastic packaging, tossed the now useless union suits back into the cabinet, and kept the now priceless plastic with him.

When the materials had been gathered, Kerwin began reading up the assembly instructions Smylie had written. The work was, at best, slow going.

"Turn the canister so that you're looking at its vented end," Kerwin said.

"The vented end?" Swigert asked.

"The end with the strap. We'll call that the top, and the other end the bottom."

"How much tape do we want to use here?" Lovell asked.

Kerwin said, "About three feet."

"Three feet . . ." Lovell contemplated out loud.

"Make it an arm's length."

"You want that tape to go on sticky end down?" Lovell asked.

"Yes, I forgot to say that," Kerwin said. "Sticky end down."

"I slip the bag along the canister so that it's oriented along the sides of the vent arch?" Swigert asked.

"Depends what you mean by 'sides,'" Kerwin responded.

"Good point," Swigert said. "The open ends."

"Roger," Kerwin responded.

This back-and-forth went on for an hour, until finally the first canister was done. The crewmen, whose hoped-for technical accomplishment this week involved nothing less ambitious than a soft touchdown in the Fra Mauro foothills of the moon, stood back, folded their arms, and looked happily at the preposterous tape-and-paper object hanging from the pressure-suit hose.

"O.K," Swigert announced to the ground, more proudly than he intended, "our do-it-yourself lithium hydroxide canister is complete."

"Roger," Kerwin answered. "See if air is flowing through it."

With Lovell and Haise standing over him, Swigert pressed his ear against the open end of the canister. Softly, but unmistakably, he could hear air being drawn through the vent slats and, presumably, across the pristine lithium hydroxide crystals. In Houston, controllers crowded around the screen at the TELMU's console, staring at the carbon dioxide readout. In the spacecraft, Swigert, Lovell, and Haise turned to their instrument panel and did the same. Slowly, all but imperceptibly at first, the needle on the CO_2 scale began to fall, first to 12, then to 11.5, then to 11 and below. The men on the ground in Mission Control turned to one another and smiled. The men in the cockpit of Aquarius did the same.

"I think," Haise said to Lovell, "I might just finish that roast beef now."

"I think," the commander responded, "I might just join you."

As Wednesday's dawn gave way to morning, and morning gave way to afternoon, things were not as sanguine at the consoles in Houston as they were in the spacecraft speeding away from the moon.

Certainly there was *some* cause for optimism around Mission Control. At the TELMU station, where the LEM's vital environmental signs were being continually monitored, the readings of the carbon dioxide concentrations aboard Aquarius had been steadily dropping all day long. Less than six hours after Ed Smylie's ingenious air scrubber went on line, the cockpit CO_2 had dropped to a scant 0.2 percent of the overall air mass — a mere gaseous trace that could barely be detected by the onboard sensors, much less do the astronauts any harm. At the INCO station, matters seemed similarly well in hand. The tight PTC roll that Max Faget had insisted on had been successfully achieved shortly after the PC+2 burn. The ship's controlled rotation allowed the LEM to point its high-gain antenna directly toward Earth, keeping the astronauts in constant voice contact with the ground without the need for all the frantic antenna switching of the day before. Elsewhere in Mission Control, however, the numbers on the screens were not nearly as promising as the INCO's and the TELMU's. The worst data was appearing in the front row, at the FIDO, GUIDO, and RETRO consoles.

When Aquarius fired its descent engine for the PC+2 burn, the maneuver was designed not just to increase the ship's speed but to tweak its trajectory. In order to reenter Earth's atmosphere safely, Apollo 13 had to approach at an inclination no shallower than 5.3 degrees, and no steeper than 7.7 degrees. Come in at 5.2 degrees or below, and the blunt-ended command module would skip off the top of the atmosphere and boing straight back into space, entering a permanent orbit around the sun. Come in at 7.8 degrees or above, and the spacecraft would be able to reenter all right, but at so steep an angle and with such a high g force that the crew would probably be crushed well before they ever hit the water. Either way, the celebratory splashdown that the recovery forces were anticipating in the South Pacific would not be taking place.

The PC+2 burn was intended to avert both of these catastrophes, positioning Apollo 13 in the center of the narrow reentry corridor, at an approach angle of 6.5 degrees. The tracking data that appeared on the flight dynamics screens right after the burn indicated that this angle had indeed been achieved. Now, however, eighteen hours after the burn, further numbers indicated that that trajectory was becoming mysteriously shallower, falling to 6.3 degrees and below. It was Chuck Deiterich, at the RETRO station, who noticed the problem first.

"You following these trajectory numbers?" he said off loop, pushing away from his console and turning to Dave Reed, the flight dynamics officer, sitting to his right.

"Tracking them," Reed answered.

"What do you make of 'em?"

"Damned if I know," Reed said.

"We're shallow, that's for sure."

"Definitely."

"You think we did the burn right?" Deiterich asked uncertainly.

"Heck, Chuck, we *have* to have done the burn right. Those numbers were solid. The only thing I can figure is that the trajectory data itself is no good. As far away as the ship still is, we may not have a handle on all the tracking arcs."

"These numbers have been falling for a while now, Dave," Deiterich said dourly. "The data's good."

If Deiterich and Reed were both right and the numbers and the burn were both satisfactory, there weren't many things that could explain the shallowing of the trajectory. The obvious answer — the only answer, really — was that somewhere along the length of Odyssey or Aquarius something was venting, producing a tiny propulsive force that was pushing the twin ships off course.

Just where that venting was coming from was uncertain. The dead service module had long since outgased its last, and any of the systems that could spring leaks — its hydrogen tanks, for instance, or its reaction control thrusters — had been closed down. The conical command module had no such vapor-powered hardware, with the exception of its own small attitude thrusters, and those had been shut off with the rest of the ship. The LEM was just as unlikely a source of unexplained gas plumes as the command module. Nearly all of its systems had been off line since the PC+2 burn, and those that weren't were being closely monitored by the TELMU and the CONTROL officers. If any extraneous gas was escaping from any line or tank, it would almost certainly have been spotted by now.

The options for correcting the eroding trajectory were few. If something *was* found to be venting, and if the location of that leak could be pinpointed, it would be possible to roll the spacecraft stack over and allow the exhaust to blow the ships the other way. This presumably would steepen Apollo 13's angle until it crept up toward the higher end of the corridor. Finding the source of venting wasn't likely, though, and unless the mysterious shallowing abruptly stopped, the only alternative — one that the overworked FIDOS and GUIDOS and RETROS did not even want to consider — was to power the LEM back up, realign its temperamental guidance platform, and light the descent engine for yet another burn.

"If the entry angle doesn't stabilize itself," Deiterich said, "we're going to have to fire this thing again."

"Then let's hope it stabilizes itself," said Reed.

But if the GUIDOS, FIDOS, and RETROS were going to fire Aquarius's descent engine, the numbers on the screen of the CONTROL officer — the man who oversaw the LEM's non-environmental systems — would have to cooperate. At the moment, they weren't. As Milt Windler had feared before the PC+2 burn, the pressure in the supercritical helium tank, which was used to force the engine fuel into the combustion chamber, was beginning to climb.

The minus 452–degree gas was ordinarily stored at a pressure of 80 pounds per square inch, but helium expands fast, so the tanks were built to withstand many times that force. Only when the contents of the double-hulled container had boiled up above 1,800 pounds per square inch would its rounded walls begin to groan under the strain. At that point, the pressure-relieving burst disk built into the gas line would blow, venting the gas out into space.

Although this would relieve the mounting pressure, there would no longer be any way to force fuel into the combustion chamber, and thus virtually no way to burn the engine again if another maneuver was needed. The crew's only hope of relighting their descent system would depend on there being enough residual fuel left in the lines from the previous burn to support another one. It was never a sure thing how much of this so-called blow-down fuel would be left behind, though, and relying on it for any subsequent ignitions was a dubious business at best. Now, as Deiterich and Reed somewhat blithely discussed the possibility of relighting the engine for yet another mid-course adjustment, Dick Thorson, the CONTROL officer, noticed his helium indicator starting to rise.

"CONTROL," called Glenn Watkins, the propulsion officer in Thorson's backroom.

"Go, Glenn," Thorson answered.

"Don't know if you're following these readouts, but supercritical helium is going up."

"I'm following them," Thorson said. "What's your best estimate on burst pressure?"

"We don't know for sure," Watkins answered. "We're still studying it. But right now we're looking at 1,881 pounds."

"And when do we top out there?"

"Not sure on that either," Watkins said. "But we're looking for it to blow at about 105 hours."

Thorson looked at his mission timer: it was 96 hours into the mission.

"I want you guys to pull the schematics and make sure we understand what's going on," he said. "I want to know how this burst is going to happen, when it's going to happen, and which way it's going to blow when it does happen. I don't want any surprises."

For the astronauts in the powered-down spacecraft with the useless instrument panel, there was no way of tracking either the rising helium in the tank below their feet or the decaying trajectory pushing them shallower and shallower in their reentry corridor. And at one o'clock on Wednesday afternoon, the ground was almost reluctant to give them the bad news their control panel couldn't. The ten hours since the installation of the lithium hydroxide canister had been busy ones aboard Aquarius, with the crew spending most of their time monitoring their passive thermal control roll, discussing power-up procedures that would be needed two days later when Odyssey was brought back on line, and consulting with the ground about various methods for charging the command module's tapped-out battery from the LEM's four good ones. Although Haise had managed to string together a few hours of sleep before the long pre-dawn to post-noon work shift began, Lovell and Swigert hadn't, and around midday Deke Slayton and Flight Surgeon Willard Hawkins had ordered the commander and the command module pilot upstairs to Odyssey for another try. In the early hours of Wednesday afternoon — as in the early hours of Wednesday morning — the two senior men were once again asleep, and Aquarius was once again in the hands of Fred Haise.

"Aquarius, Houston," called Vance Brand, who had recently relieved Joe Kerwin at the Capcom station.

"Go, Houston."

"Just wanted to let you know that at the moment you're pretty much right in the middle of the fairway, right around 6.5 degrees," Brand reported encouragingly, and then paused a bit. "We are getting a little drift, though, and if we don't correct it, you're going to shallow out of the corridor."

"All right," said the commander pro tem. "What do we want to do about that?"

"What we're thinking about," Brand said, "is a mid-course burn at about 104 hours. Just a little one, about 7 feet per second."

"O.K.," Haise said, "that sounds good."

"The only complication," Brand added, "is that we're also looking at your supercritical helium tank pressure, and we do expect it to blow. We don't know exactly what time it will happen — maybe about 105 hours or so. Even if it goes early, we figure we've got plenty of blow-down capability, so we'll probably be all right."

Haise said, "That sounds O.K. too."

Whether any of this was indeed O.K. with Haise was unclear from the dispassionate voice that came across on the air-to-ground channel. A change in trajectory serious enough to require a burn was by no means "a little drift." Moreover, the idea of another uncontrolled venting from one of Apollo 13's gas tanks — this one inside the descent stage of Haise's beloved lunar module — could not have sat well with the LEM pilot.

But if Haise, temporary heir to the skipper's station, was disturbed by these developments, he was not about to reveal it. That wasn't the way Lovell would have done it, or Conrad or Armstrong or any of the other men who had commanded ships out around these parts before, and it wasn't the way Haise was about to do it now. Those men would have accepted whatever the latest development was and moved on to the next piece of business.

Floating at the vacant left-hand station of the LEM, Haise allowed the air-to-ground loop to fall silent and drifted back to the storage cabinet at the rear of the cockpit. Among the few personal items the crew had brought aboard was a small tape player and a handful of cassettes containing songs chosen by the astronauts. Nobody had expected to have much time to listen to music on the way out to the moon, but at the end of the week, when the LEM had been jettisoned and the crew was heading home with their cargo of Fra Mauro moon rocks, they'd planned to break out the tapes and enjoy them. Now, of course, Aquarius was still attached to Odyssey and the storage space set aside for rocks was empty, but Apollo 13 was indisputably heading home, and Haise was going to have his music. As Vance Brand listened to the air-to-ground static at his Capcom station, what broke the silence from the spacecraft was not a worried question from the stand-in commander, but the opening chords of "The Age of Aquarius," one of the first songs the astronauts had requested when they made up their playlist. Around the room, controllers listening in turned to one another and smiled. Fred Haise, it appeared, did not rattle easily.

"Hey Fred, you got a woman up there or something?" Brand called out.

"No way I could handle that," Haise laughed back.

"Well, since you're in such a good mood," Brand said, "let me make it better. Somebody just handed me the latest consumables status report, and it looks like you're only using between 11 and 12 amps an hour. That's a couple amps below what the TELMU guys projected, so you look real good."

"Roger," Haise said, the music tinkling behind him.

"And in addition, according to our little tracking plot here, you're now about 44,000 miles out from the moon. FIDO tells me that means we're in the Earth's sphere of influence and starting to accelerate."

"I thought it was about time we crossed," Haise said.

"Roger," Brand said.

"We're on our way back home."

"That you are."

Haise turned the volume on his tape player down a little, left it to float in the air behind him, and drifted forward toward his window. If he had indeed crossed that invisible gravitational line between the Earth and the moon, he wanted to take a long, final look back. With the feet of the LEM pointed toward the moon, and the windows angled down in the same direction, the lunar view should be unobstructed. And with his crewmates asleep and the cockpit silent except for the tinny tune coming from the tape player, the atmosphere for a farewell gaze would be a good one. But suddenly this atmosphere changed.

Just as Haise approached the right-hand window, a chillingly familiar bang-whump-shudder shook the ship. He shot his hand out, braced himself against the bulkhead, and froze in mid-float. The sound was essentially the same as Monday night's bang, though it was unquestionably quieter; the sensation was essentially the same as Monday night's shudder, though it was unquestionably less violent. The locus of the event, however, was utterly different. Unless Haise was mistaken — and he knew he wasn't — this disturbance had not come from the service module, at the other end of the Aquarius-Odyssey stack, but from the LEM descent stage below his feet.

Haise swallowed hard. This should be the helium burst disk blowing; if the ground has told you to expect a venting, and a moment later your ship bangs and rocks, chances are the two are connected. But viscerally, Haise — the man who understood Aquarius better than anyone else on board — knew this wasn't true. Burst disks didn't sound this way, they didn't feel this way, and, floating cautiously up to his porthole and peering out, he also saw that they didn't look this way. Just as Jim Lovell had discovered vented gas streaming past his window more than forty hours ago, Fred Haise, the LEM pilot, was alarmed to see much the same thing outside his window now. Drifting up from the Aquarius's descent stage was a thick white cloud of icy snowflakes, looking nothing at all like misty helium streaming from a burst disk.

"O.K., Vance," Haise said as levelly as he could, "I just heard a little thump, sounded like down in the descent stage, and I saw a new shower of snowflakes come up that looked like they were emitted from down that way. I wonder," he said somewhat hopefully, "what the supercritical helium pressure looked like now."

Brand froze in his seat. "O.K.," he said. "Understand you got a thump and a few snowflakes. We'll take a look at it down here."

The effect of this exchange on the men in Mission Control was electric.

"You copy that call?" Dick Thorson, at the CONTROL console, asked Glenn Watkins, his backroom propulsion officer.

"Copied it."

"How's that supercrit look?"

"No change, Dick," Watkins said.

"None?"

"None. It's still climbing. That wasn't it."

"CONTROL, Flight," Gerry Griffin called from the flight director's station.

"Go, Flight," Thorson answered.

"Got an explanation for that bang?"

"Negative, Flight."

"Flight, Capcom," Brand called.

"Go, Capcom," Griffin answered.

"Anyone know what that bang was about?"

"Not yet," Griffin said.

"Anything at all we can tell him, then?" Brand asked.

"Just tell him it wasn't his helium."

As Brand clicked back on to the air-to-ground loop and Griffin began polling his controllers on the flight director's loop, Bob Heselmeyer at the TELMU station began scanning his console. Looking past the oxygen readouts, past the lithium hydroxide readouts, past the CO_2 and H_2O readouts, he noticed the battery readouts, the four precious power sources in Aquarius's descent stage that, working together, were barely providing enough energy for the exhausted, overtaxed ship. Gradually, the readout for battery two — just like the too easily recalled readout for Odyssey's O_2 tank two — had slipped below what it should be and was falling steadily.

If the data were right, something had arced or shorted in the lunar module's battery, just as it had arced or shorted in the service module's tank on Monday night. And if there had been a short, the battery, like the tank, would soon go off line, killing fully one quarter of a power supply that Houston and Grumman were rationing down to the last fraction of an amp. The numbers on the screen were too preliminary to be conclusive — too preliminary even for Heselmeyer to pass them on to Griffin. And if Heselmeyer didn't pass them on to Griffin, Griffin couldn't pass them on to Brand, and Brand couldn't pass them on to Haise.

At the moment, that was probably just as well. Standing at his window and looking out at the growing cloud of flakes surrounding the bottom of his LEM, Fred Haise had more than enough burdens of command.

11

DON ARABIAN was in Building 45 when battery two in Aquarius blew. Though Arabian's offices were a good quarter mile from Mission Control — tucked away in one of the bland, blockhouse-like structures where people like Ed Smylie worked — Arabian himself was hardly at the periphery of things. He and his staff were equipped with much the same monitor screens as the men who worked in Mission Control proper; they listened in on the same air-to-ground loops; and they tracked the same data streaming back from the spacecraft. The only difference was, each man at each console in Mission Control was expected to keep track of only his small part of the command module or LEM. Arabian was expected to keep track of everything. When battery two in Aquarius went down, he knew his phone would start ringing.

The part of Building 45 where Don Arabian worked was known among Space Center employees as the Mission Evaluation Room, or MER. Arabian himself was known as Mad Don. To the men who worked in the MER, the moniker fit. In a community of scientists in which the prevailing accent was Texan, the prevailing cadence was sleepy, and questions were answered with a nod as often as with a word, Arabian was a verbal tornado. And what he liked to talk about most were his systems. To Arabian, as well as to the fifty or sixty other men who worked in the Mission Evaluation Room, every nut or bulb or piece of hardware in a spacecraft could be defined in terms of systems. A fuel cell was an energy system; the LEM was a landing system; a single warning light — with its filament, its threaded base, and its brittle little glass shell — was an illumination system. Even the astronauts themselves, whose job it was to push the buttons that made all the other hardware go, were, in their own big and clumsy ways, systems too.

Altogether, in the command module there were 5.6 million such systems; in the LEM there were several million more. When something went wrong with any one of them, it was Don Arabian's job to figure out why. Somewhere in every accident was a piece of hardware that had been pushed beyond the job it was designed to do, and while the men in Mission Control worked to fix the busted part, Arabian worked to find out why it had failed in the first place. When Fred Haise reported a bang in his descent stage, and the data on the LEM screen in the Mission Evaluation Room showed battery two starting to falter, Arabian went to work. Only a few minutes after he got under way, the phone at his console rang.

"Mission Evaluation," Arabian said.

"Don? It's Jim McDivitt."

Arabian had been expecting to hear from McDivitt. The one-time commander of Gemini 4 and Apollo 9 and the current head of the Apollo program office would be monitoring Apollo 13 from the back row of consoles in Mission Control. If something else was going wrong with either Aquarius or Odyssey, McDivitt would be the first one to press Arabian for answers.

"I see you've had a problem over there," Arabian said.

"You're tracking battery two?" McDivitt asked.

"Tracking it."

"And what do you think?"

"I think you've got a problem." There was worried silence from McDivitt's end of the phone. "Jim," Arabian said with something close to a laugh, "have you had lunch yet?"

"Uh, no."

"Well, why don't you come on over and have it with me. I'll order us a pizza and we'll get this thing sorted out."

Arabian's nonchalance was born less of hubris than of confidence. In even the brief time he'd had to investigate Aquarius's problem, he was reasonably certain he'd found the source. Each of the LEM's four batteries consisted of a series of silver-zinc plates immersed in an electrolyte solution. As the plates and the fluid worked together to produce electricity, they also gave off, as byproducts, hydrogen and oxygen. Typically, the two waste gases were generated in such small quantities that they could barely be detected. But occasionally a battery would overproduce the vapors, and a few stray wisps would collect in a nook in the battery's lid. Arabian had always been a bit skittish about that nook. Combine oxygen and hydrogen in a small enough space, and pressure begins to build; and when pressure begins to build, all you need is a spark to get yourself a tidy little explosion. The inside of a battery, of course, is just the place you'd most likely find a spark, and when Haise reported his bang and flakes, Arabian figured that the little bomb that had been waiting to go off in every battery of every LEM that ever flew had finally blown.

The diagnosis was not all bad, however. After sitting down with an on-site representative of the Eagle Picher company, the contractor that manufactured the batteries, Arabian concluded that the injury the LEM had sustained was easily survivable. The explosion clearly had been a small one, given the fact that battery two was still working. More important, to the degree that the battery *had* been damaged, the rest of the electrical system seemed to be compensating. The LEM's power grid was designed in such a way that if any one of the spacecraft's four batteries was unable to perform its job fully, the other three would pick up at least part of the slack. As Arabian and the on-site technician studied the numbers, they could see that batteries one, three, and four had already increased their electrical output, allowing battery two to stabilize. On later flights, Arabian knew, the system would have to be redesigned. No more LEMs could be allowed to fly with miniature grenades built into their bodies. For now, though, Apollo 13's batteries looked stable.

Arabian, along with the Eagle Picher man and a MER electrical engineer, made their way to the Building 45 conference room. Within minutes, Jim McDivitt, accompanied by two representatives from Grumman, the manufacturer of the LEM, showed up as well. Soon after that, Arabian's pizza arrived.

"Fellows," the MER chief said, tearing off a pizza slice and pushing the box across the table toward McDivitt, "we've been looking at the numbers, and the good news is, this is no big deal." He turned to the Eagle Picher engineer. "You agree?"

"No big deal," the engineer said.

"So the battery will stay on line?" McDivitt asked.

"It should," said Arabian.

"And we can make it back on the power we've got?"

"We should," Arabian said. "We were pulling fewer amps than we thought we would anyway, so we should stay within our margin of error."

"Then there wasn't an explosion?" the Grumman man asked.

"Oh, there was an explosion," Arabian said.

"But nothing actually . . . blew up," the Grumman man amended.

"Sure it did," Arabian said, chewing pizza. "The battery blew up."

"But do we have to actually use that term? I mean, the battery's still operating. People get awfully excited when you say something blew up."

"What term would you suggest?"

The Grumman man said nothing.

"Look," Arabian said after a pause, "you know this is no problem and I know this is no problem. But if the battery screws up, I'm going to say so. And if a tank screws up, I'm going to say so. And if the crew screws up, I'm going to say so. Fellows, these are just systems, and if you're not honest with yourself about what went wrong, you ain't gonna be able to fix anything."

Arabian finished his slice of pizza, fished another one out of the box, and glanced fleetingly at his wristwatch. There were seven or eight million other systems aboard Apollo 13 that might require his attention today, and a few more minutes was all he could afford to waste on a working lunch.

Jim Lovell was surprised by what had become of his LEM in the time he was asleep. It had been a little after ten o'clock Wednesday morning when he floated up the tunnel to Odyssey to begin his sleep cycle, and it wasn't until close to three in the afternoon that he prepared to swim back down. The four and a half hours of sack time was far and away the most rest he'd had since the accident, and with splashdown now less than forty-eight hours away, the sleep couldn't have come at a better time.

As always on this trip, Lovell stirred himself from slumber well before his scheduled wake-up call from the ground. Rising from his couch in the frosty command module, he looked around bleary-eyed and drifted through the lower equipment bay toward the tunnel.

Before swimming down to the LEM, however, he stopped and considered something. Off and on, Lovell had been entertaining the idea of breaking what was ordinarily an ironclad rule on any space mission, and now, almost impulsively, he decided to do it. Opening the top two or three buttons of his flight suit, he reached beneath his thermal undershirt, felt around for the biomedical sensors that had been glued to his chest since before liftoff Saturday, and began painfully removing them.

There were a lot of reasons, Lovell decided, that the electrodes had to go. First of all, they itched. The adhesive used to hold the sensors in place was supposedly hypoallergenic, but after four days, even the most skin-friendly glue was going to become annoying, and this glue most assuredly had. More important, pulling off the sensors would save power. The biomedical monitoring system that beamed the astronauts' vital signs to Earth drew its juice from the same four batteries that powered everything else aboard the LEM, and although the electrodes were hardly power gobblers, they still consumed their share of amps. Finally, there was also the question of privacy. Like any test pilot, Jim Lovell had long prided himself on his ability to keep his emotions out of his voice — whether he was flying over the Sea of Japan in a blacked-out Banshee or flying over the far side of the moon in a blacked-out LEM. But while the voluntary nervous system responds to such exertions of will, the involuntary one doesn't, and nobody could control the accelerated respiration and triphammer heartbeat that even the most imperturbable pilot could experience in an emergency. Lovell did not know how high his cardiac rate climbed after the explosion that aborted his mission on Monday night, but it rankled him to know that everyone from the flight surgeon to the FIDOs to the pool reporters did. In the event of another crisis in the next two days, he saw no reason that his heart rate should be broadcast to the world. Peeling off the electrodes, he balled them up, stuffed them into a pocket, and pushed off toward the LEM.

"Morning," Haise said as Lovell's head poked through the tunnel. "Looks like you finally got some rest."

Lovell glanced at his watch. "Wow," he said, "looks like I did."

"Jack coming down?" Haise asked.

"Nope." Lovell floated all the way into the cockpit. "Still sawing wood. What's the status of things down here?"

"Well," Haise said, "they've definitely decided on a mid-course burn sometime tonight, probably around 105 hours. We're shallowing pretty bad now."

"Mm-hmm," Lovell said.

"And they're pretty sure we'll try to get it in before the helium bursts."

"Makes sense."

"Also," Haise said. "Looks like we had a bit of an event in the descent stage."

"An . . . event?"

"A bang. And some venting."

The commander looked at his LEM pilot for a long moment, reached for his headset, and pressed his push-to-talk switch.

"Houston, Aquarius," Lovell called.

"Roger, Jim," Vance Brand said from Houston. "Good morning."

"Say, Vance, what is our status on descent-stage venting? Was it venting? Is it still venting?"

Brand, who had not yet gotten a report from Arabian and McDivitt in Building 45, hedged. "Fred reported it. Does he still see it?"

Lovell turned to Haise with an inquiring expression. Haise shook his head.

"No," Lovell said. "Fred hasn't seen anything else."

"O.K.," Brand said, without elaboration.

Lovell waited to see if his Capcom had anything to add, but Brand said nothing. In the clipped code of the air-to-ground loop, Lovell knew, this silence said a lot. Brand didn't *know* what the bang was yet, and he would almost certainly prefer it if the commander didn't pursue it. It was one thing for the omnipresent press to hear the Capcom explain a problem to the crew, quite another for them to hear the commander ask for an explanation of something and the Capcom come up empty. Lovell let another instant go by and moved on to other things.

"Also," he said to Brand, "I understand we can expect the super-critical helium to relieve itself at about 105 hours."

"Closer to 106 or 107," Brand said.

"And we'll be doing a mid-course a little before that?"

"Roger," Brand said. "Not only will that guarantee you fuel pressure, but it means you'll still have your thrusters powered up from the burn when the helium blows. That way, if the venting tosses you around a little, you can regain control."

"Roger," Lovell repeated skeptically. "I can regain control."

He clicked off the air, pursed his lips, and decided that he didn't like what he was hearing one bit. These newest problems might have surfaced on Haise's watch, but they would have to be resolved on Lovell's. He felt his jaw clench once with an unexpected flutter of tension. Suddenly Brand's voice reappeared in his ear.

"And there's only one more item for now, Jim. Could you switch your biomedical switch to the position opposite wherever it is now? We're getting a signal but no data."

Lovell paused. Brand paused. Three seconds went by, and the man on the ground, sitting impassively at his console, outwaited the man in the spacecraft.

"Now you know, Houston," the commander said at last, "I don't have the biomed on."

Lovell listened on the air-to-ground channel and braced for what he assumed would be a reprimand from Houston. Instead, he heard another few seconds of silence. Finally Brand, an astronaut himself who, like Lovell, had learned his craft test-flying jets and who, like Lovell, might one day find himself in a busted spacecraft far from home, clicked back on the air.

"O.K." was the only thing the Capcom said.

Lovell smiled to himself. When this flight was over, he'd have to remember to buy Brand a beer.

"Marilyn!" Betty Benware called from the master bedroom of the Lovell home in Timber Cove. There was no answer.

"Marilyn!" she called again. Again there was no response.

Marilyn, as far as Betty knew, was in the family room. It was only about a dozen steps from there to the bedroom, where Betty stood with the telephone in her hand. The call for Marilyn was, by any measure, an urgent one. But if Marilyn heard her friend's voice, she gave no indication of it.

Betty looked at her watch and immediately saw why. It was just after 6:30 on Wednesday, and 6:30 meant the evening news was on. As always when Jim was in space, Marilyn cherished this time. This was the half hour when she would position herself in front of the television set, turn to CBS, and lose herself in Walter Cronkite's reports on the progress of her husband's mission.

For astronaut wives who wanted to get the straight dope on the status of a spacecraft and the astronauts flying it, the man to turn to was usually Jules Bergman. The ABC correspondent generally made it his business to offer his audience only the darkest, least varnished truth, whether the audience wanted to hear it or not. It wasn't always easy to accept what Bergman had to say, but the up side was that once you'd listened to one of his reports, you knew you'd heard the worst. If he wasn't concerned about the status of the mission at a given moment, you could be pretty sure there was nothing to be concerned *about*. The down side was that a little Jules Bergman went a long way. After a day or so of following his brutally frank reports, the family of a crewman could find themselves fairly wrung out. When that happened, it was time to switch to Walter Cronkite.

Cronkite's reports were no less reliable that Bergman's, no less honest; but they were, on the whole, a lot more palatable. News that came from Walter Cronkite just seemed to go down easier. At the end of the day, Marilyn Lovell and most of the other astronaut wives would thus make a point of tuning in to the fatherly newsman. Tonight was no different, and as Betty Benware stood in the master bedroom, looking nervously at the receiver in her hand and questioning whether she dared tell the caller to hold, Marilyn perched on the edge of the family room couch, leaned forward, and shut out the rest of the world.

"Good evening," Cronkite began, sitting behind his broadcast desk and in front of a projected image of the Earth and the moon. "The Apollo 13 spacecraft is slightly off course as it limps home tonight. It's now about a quarter of the way back from the moon, but its present course would not return it to Earth as you see here. Instead, it would miss the atmosphere and the crew would perish. That's why a critical burn to correct its course is scheduled for 11:43 eastern time tonight."

"Earlier this evening, White House news secretary Ron Ziegler said there is no need for other nations to assist in recovering the Apollo 13 crew, 'although we do appreciate the offer,' he said. Nevertheless, the Soviet Union sent six naval vessels toward the Pacific splashdown site, and Britain sent six naval vessels toward the alternate site in the Indian Ocean. France, the Netherlands, Italy, Spain, West Germany, South Africa, Brazil, and Uruguay have put their navies on alert. President Nixon originally scheduled a report to the nation on the Vietnam War for tomorrow evening, as a sort of public relations counterattack to the antiwar rallies taking place around the country. But this morning the president postponed that speech until early next week, saying he didn't want to do anything to detract from concern over the astronauts. CBS White House correspondent Dan Rather has more."

Whatever it was that Dan Rather had to say, Marilyn Lovell would not get to hear, because just as the correspondent appeared on her family room television screen, Betty Benware appeared in the family room doorway.

"Marilyn!" Betty said in an urgent whisper. "Didn't you hear me calling you?"

"What?" Marilyn said distractedly. "No, no. I was watching the news."

"Well *stop* watching it. You have a phone call from President Nixon."

"Who?!"

Marilyn leapt from the couch and ran to the bedroom. She was flattered to get a call from the president but, even under these circumstances, was also surprised. While nobody around Houston questioned Nixon's genuine interest in the welfare of the Apollo 13 crew, no one harbored illusions that any space flight was at the top of his daily priorities.

It was John Kennedy — never a Nixon favorite — who had committed the nation to landing on the moon before the end of the 1960s, and it was Lyndon Johnson who had pushed the program doggedly along. Though Apollo 11's historic moon landing last July had occurred during Nixon's tenure, the president felt, quite rightly, that the public gave him little credit for the achievement, tipping their hats instead to the recently retired Johnson and the less recently martyred Kennedy. Now, as Apollo 13 headed for home, Marilyn Lovell had no reason to believe that the president had the time or inclination to devote more concern to this crisis than to the numerous other crises he was facing in his first year in office.

In fact, Nixon was profoundly concerned. Ever since Apollo 8's successful lunar orbit, just one month before his inauguration, Nixon had developed a fascination with moon flight and a special admiration for the crew of that first circumlunar trip. After their return from the moon, Frank Borman, Jim Lovell, and Bill Anders had been invited to attend the new president's inauguration, and then later, when he had moved into the White House, to join him for dinner, not in one of the formal dining rooms on the first floor of the mansion, but in the family quarters upstairs. Marilyn had remembered being charmed during the chatty tour Nixon gave his guests of his new home when, on several occasions, he would stumble on a room even he didn't know existed, and fall silent, gesturing toward it with an abashed smile and a your-guess-is-as-good-as-mine shrug.

While Nixon must have known that the Apollo 8 crew appreciated the presidential attention, like many powerful men he felt the highest compliment he could pay someone he admired was to put the man to work for him. After Apollo 8, Jim Lovell had made it clear that he intended to stay in the space program at least until he got the chance to land on the moon, and Nixon was not inclined to question that decision. Frank Borman and Bill Anders, however, left the space agency shortly after their return from the moon, and the president quickly pounced. Borman, generally dubious about politics, declined an offer to join the White House staff in a nonspecific "policy-making" role. Anders was not so chary; he accepted an appointment as executive secretary of the National Aeronautics and Space Council, an advisory body traditionally chaired by the vice president — in this case, Spiro Agnew.

Last Saturday, when Anders's old Apollo 8 crewmate boarded Apollo 13, it was the new executive secretary's responsibility to accompany the vice president down to Florida for the launch. After the crew was safely on its way to the moon, Agnew flew off for a political event in Iowa, and Anders was free to go on his way. On Monday all that changed. When Apollo 13 began to bang and vent, both Agnew and Nixon made it clear that they wanted to be kept posted on events, and the job fell to the National Aeronautics and Space Council.

Anders himself was not ordered to Washington right away, but his assistant Chuck Friedlander was, receiving instructions to fly promptly from Florida to deliver twice-hourly updates in the cabinet room of the White House. Friedlander arrived at National Airport early the next morning, but when he got there, he found there wasn't a cab in sight. Instead, he climbed on a city bus idling at the terminal, showed the driver his credentials, hastily explained why he was in town, and asked if the bus was going anywhere near 1600 Pennsylvania Avenue. The driver responded even better than Friedlander had hoped, abandoning his planned route and driving this passenger — as well as the handful of others aboard — directly to the White House gate. Within minutes, Friedlander was inside and beginning the first of his briefings. The next day Anders arrived, and both he and Friedlander were called into the Oval Office to consult with the president personally. When the two men presented themselves, Nixon had just one question.

"Bill, I want to know what the odds are that this crew is going to come back."

"The odds, Mr. President?" Anders asked.

"Yes, the statistical likelihood."

"Well, sir, if I had to give odds at this point, I'd say 60–40."

The president snorted disapprovingly. "I've already talked to Frank Borman. He said 65–35."

Anders and Friedlander looked at each other.

"Well, Mr. President," Anders said accommodatingly, "I suppose Frank knows best."

The two men spent much of Tuesday and Wednesday in a small office adjoining Nixon's, watching the TV coverage of the mission with Apollo 11 veteran Mike Collins, drafting statements with a presidential speechwriter, and preparing themselves to provide the president with running odds as he requested them. Now, at the end of the day on Wednesday, Nixon seemed satisfied that the percentages had turned in favor of the Apollo 13 crew, and he decided that it was time to call their families and offer his own words of encouragement. He started with the wife of the commander whose achievements he had respected so since 1968.

"Mrs. Lovell?" a White House operator's voice said.

"Yes?" Marilyn was nearly out of breath after running to the master bedroom.

"Hold for the president, please."

Marilyn waited through a few seconds of silence and then heard a click and a receiver being lifted.

"Marilyn?" said a familiar, growly voice. "This is the president."

"Yes, Mr. President. How are you?"

"I'm just fine, Marilyn. More important, how are you?"

"Well, Mr. President, we're holding up the best we can."

"And how are . . . Barbara and Jay and Susan and Jeffrey?"

"About as well as can be expected, Mr. President. I'm not so sure Jeffrey understands quite what's going on, but the other three are following everything on TV."

"Well, I just wanted you to know, Marilyn, that your president and the entire nation are watching your husband's progress with concern. Everything is being done to bring Jim home. Bill Anders, an old friend of yours, has been briefing me."

"That's nice to hear, Mr. President. Please give Bill my best."

"I certainly will, Marilyn. And Mrs. Nixon wants you to know that her prayers are with you. Hang on for just another couple days and maybe we'll all get a chance to have dinner together again at the White House."

"I would enjoy that very much, Mr. President," Marilyn said.

"Well then, we'll see you soon," the president said, and the line went dead.

Marilyn hung up the phone somewhat dazedly, smiled at Betty, and made her way back to the family room. She was grateful for the call, but also eager to return to the TV. Richard Nixon might have good wishes, but Walter Cronkite had hard news. When she resumed her place in front of the television, CBS was still on the topic of Apollo 13, and the face that now filled the screen belonged to another space correspondent, David Schumacher.

"At 179,000 miles out," Schumacher began, "Apollo 13 has passed the last hour without even a minor problem developing. The astronauts are relaxing right now before the course correction they need to make to get inside the reentry corridor. Once again, that burn will come at 11:43 tonight. The crew actually would have all day tomorrow to make the burn, but they would very much like to go to bed tonight knowing they're back in their corridor. And just for historical purposes, we note that in the original flight plan, Aquarius would have landed on the moon, with Lovell and Haise aboard, nine minutes ago. In all the excitement, we've also forgotten that this is the day that Ken Mattingly was supposed to get the measles. He has not."

Marilyn reached forward, turned down the volume, and frowned slightly. Having watched dozens of these reports over dozens of evenings during the four trips her husband had made into space, she was never quite clear how the broadcasters picked what they were going to report. But with the president ringing her bedroom phone and TV trucks ringing her block, Ken Mattingly's rubella status and Apollo 13's original flight plan seemed like minor matters indeed.

The crew did not have time for chin-up phone calls from the president. As the Wednesday evening news hour passed and full night fell in

Houston, Lovell, Swigert, and Haise had a lot more on their minds than the mid-course correction coming up in a few hours. Mission Control had just decided that for one brief period, the command module Odyssey, sunk in slumber since Monday night, would be powered back up and brought on line.

For the nearly forty-eight hours since the three astronauts abandoned ship and crawled over to Aquarius, Odyssey had been in a state of almost constant cold soak. Bad as this was for the men in the relatively insulated cocoon of the cockpit, it was even worse for all the electronic equipment lying just beneath the thin skin of the spacecraft. With temperatures outside the ship down to minus 280, even the best passive thermal control roll was not always enough to keep the electrical entrails of the ship warm. Instead of relying on the PTC roll alone, the most sensitive hardware was also equipped with heaters that would switch on when the ship rotated away from the glare of the sun and switch off when it rotated back into it. But when Odyssey was shut down, the heaters went off line too, and the protection they offered vanished.

Of all of the millions of command module systems, few were more sensitive to the cold — or more essential for reentry — than the attitude-control jets and the guidance platform. The jets in the command module, like the jets in the LEM, ran on a liquid fuel that flashed into gas when vented into space. Like any liquid, this one could only be chilled for so long before turning to ice or thick slush, making it impossible to feed through the fuel lines and into the thrusters.

The guidance platform was even more cold sensitive. If the temperature of the apparatus fell too far, the lubricant that kept its three gyroscopes spinning on their posts would become viscous, causing the platform to become sluggish and imprecise. At the same time, the system's finely milled beryllium components would begin to contract, throwing the balance of the carefully calibrated instrument even further off. On Wednesday night, with the command module facing at least forty more hours of coasting through the deep freeze of space, Gary Coen, the Gold Team's guidance, navigation, and control officer, or GNC, decided to ask around and find out how much more cold his systems could take. The first person he spoke to was the on-site representative of the subcontractor that manufactured the guidance platform.

"I need you to do something for me," Coen said to the visiting technician, hurrying into the GNC backroom where the company reps camped out. "I need you to consult your manufacturing records and see what experience you have powering up an inertial maneuvering unit from a completely cold state to a fully operational state."

"A *completely* cold state?" the engineer asked.

"Completely," Coen said. "No heaters."

"That's easy. We don't have any experience with that."

"None?" Coen asked.

"None. Why would we? That unit's supposed to be heated. We already know that if you fly without the heaters, the thing's not going to work."

"So you've got no data on this at all?" Coen asked.

"Well," the engineer said after a pause, "one of our people up in Boston did take a guidance unit home with him one night and accidentally left it in his station wagon till morning. It got down to about 30 degrees, but the next day the thing started right up with no problem."

Coen looked at the man. "That's *it?*"

The man shrugged. "Sorry."

With so little useful data available, the GNC, as well as the FIDOS, GUIDOS, and EECOMS, knew there was only one answer. At some point well before reentry, the command module's heat sensors and telemetry would have to be powered up for a short time, to allow the controllers to check on the state of the ship's innards. If the systems were found to be too cold, the heaters would have to be considered.

Powering up the command module at all — even just long enough to take the ship's temperature — would draw precious energy from the reentry batteries. But with the LEM available to help top off the batteries, an amp or two could probably be spared. It was at 7 P.M. Wednesday that Jack Swigert was given the word to bring his command module briefly to life.

"Aquarius, Houston," Vance Brand called up from the Capcom console.

"Go, Houston," Lovell answered for the crew.

"While we're getting ready for the mid-course burn, we've got a procedure here we'd like you to copy down for powering up the command module and turning on the instrumentation so we can check telemetry."

"This is to power up the *command module?*"

"That's affirm," Brand said.

Lovell cut his connection to the ground and looked over his shoulder to Swigert, who had been sorting through food packets and taking inventory of the crew's remaining supplies, and who now looked up, surprised.

"You following this?" the commander asked.

"Sure," Swigert said. "I'm just assuming it's a mistake."

"Go figure," Lovell said, then switched back on the air. "O.K., Houston. I'll have Jack get some paper and he'll copy down whatever procedure you've got for him."

Swigert snatched up a nearby flight plan, plucked his pen from the sleeve pocket of his jumpsuit, and signed on the air himself.

"Vance, this is the third officer of the LEM crew here, ready to copy," he said.

"All right, Jack, this is a lengthy procedure. It'll probably take two or three pages."

Swigert turned to the blank back sides of his flight plan sheets. As Brand dictated, Swigert started furiously copying, and both men could see it would be slow going at best. There were batteries to engage, buses to connect, inverters to switch, sensors to activate, antennas to maneuver, telemetry to turn on. Worse, unlike any other activation procedure Swigert had ever rehearsed, this one was wholly improvised, a mere partial power-up he would never have dreamed of attempting before. Nevertheless, just half an hour after Swigert had begun scribbling the procedure down, he finished, pulled off his headset, and sprang up the tunnel into Odyssey to put what Brand had read him into practice.

Down in Aquarius, Lovell and Haise had no real evidence of the work Swigert was doing, beyond the sound of the occasional snapped switch or thrown breaker, but on the ground things were different. At seven o'clock on Wednesday evening, the Gold Team was on watch, which meant that Buck Willoughby was at the GNC console, Chuck Deiterich was at the RETRO console, Dave Reed was the FIDO, and Sy Liebergot — who had swapped shifts with Tiger Team member John Aaron — was the EECOM. On Liebergot's screen, which for the past two days had been blinking nothing but zeros, pixels began to flicker. In moments, the flickering turned to numbers, and the numbers to hard, healthy data.

"You picking up these readings?" Liebergot asked Dick Brown in the EECOM backroom.

"Affirmative."

"Looks mighty nice," Liebergot said.

"Mighty nice," Brown agreed.

Elsewhere around the room, similar readouts from thrusters and fuel lines and guidance hardware began to appear on other screens. At their consoles, the controllers, who had grown to accept the absence of Odyssey as a given on this mission, were as transfixed as the EECOM. In the spacecraft, Swigert, who had brought about the resuscitative magic, finished his work, swooped back down the tunnel into the LEM, and donned his headset.

"All right, Vance," he called out, "I've gone all the way through the procedure. How are you reading?"

"O.K., we're indeed getting data from you, Jack," Brand said.

"And how does the telemetry look on old Odyssey?"

Brand scanned the readouts on his screen and listened to the reports coming in from other controllers on the flight director's loop.

"It doesn't look too cold," he said after a moment. "Looks pretty good. You're ranging from 85 degrees down to about 21 degrees, depending on sun angle, so it looks like no sweat."

"Roger. Thank you very much," Swigert said.

"Now we're going to want you to get back up there, use the back-out procedure, and shut down again."

"Roger," Swigert said, already beginning to remove his headset. "I'm on my way."

As Jack Swigert disappeared back up the tunnel, Jim Lovell floated backward and leaned against the wall behind him. He was relieved at the state of his command module — but only a little. It was indisputably good news to get such temperate temperature readings from the guts of the ship, but 21 degrees above zero was still 11 degrees below freezing, and for cold-sensitive equipment, that was still less than optimal. Besides, even if his command module was temporarily healthy, his LEM evidently wasn't.

A short time before the Odyssey power-up began, Brand had at last come on the line to tell him that the earlier bang and snowflakes from the descent stage had been an explosion in battery two, and while the Capcom hastened to pass along Don Arabian's diagnosis that the problem was minor, the commander felt uneasy. The infirm battery kept triggering a master alarm light on the instrument panel, and since the engineers had failed to predict that the battery would blow in the first place, their prognosis for its continued good health was somewhat suspect.

More troubling for Lovell was the upcoming mid-course burn. Even if his LEM battery did stay stable enough to keep putting out juice, and even if his command module did stay warm enough to function when the time came, it would all be pointless if the spacecraft didn't get back into the middle of its reentry corridor and get there soon. Lovell reached for his press-to-talk switch, to call Brand and ask him just when the crew in Houston expected the crew in the spacecraft to get started on the pre-burn procedures. But before he could sign on the air, Brand hailed the ship. The Capcom evidently was thinking the same thing.

"O.K., Jim, we'd like you to turn to page 24 in your systems book and get ready for the power-up at 105 hours."

"O.K., Vance," Lovell said, reaching gratefully for the book. "Midcourse at 105. I'm going to page 24."

"Now the basic situation at the moment," Brand said, "is that we're presently a little bit shallow, and a 14-second burn at 10 percent thrust will put us more in the center of the corridor."

"Roger. Understand." Lovell withdrew a pen from the pocket in his sleeve and wrote this down.

"We don't want to power up the spacecraft completely, so that means no computer or mission timer. We'll simply go with a manual burn, with you controlling the engine with the Start and Stop switches."

"Roger," Lovell said, still writing.

"And for attitude, what we're going to want to do is manually orient the spacecraft to place the Earth in the center of your window. Put the horizontal line of the optical sight's cross hairs parallel with the Earth's terminator. If you hold it there throughout the burn, the attitude will be correct. Got that?"

"I think so."

Lovell started to write this instruction down as well, but after realizing what he had heard, he stopped himself. When the LEM was powered down after the PC+2 burn, the guidance system was powered down with it. When that happened, the alignment that Lovell had transferred so painstakingly from the command module on Monday night and checked so painstakingly against the sun on Tuesday vanished completely. Catastrophic as this would have been before the long free-return burn or the even longer PC+2, it did not present much of a problem for the mere 14-second engine sneeze Lovell would be asked to perform now. For a burn this brief, only an approximate alignment was needed, one with a margin of error of as much as 5 degrees.

Coincidentally, Lovell knew how to pull off just such a maneuver. Sixteen months ago, during Apollo 8, the FIDOS and GUIDOS in Houston had wondered what would happen if a lunar ship on its way back from the moon suddenly lost its platform and could no longer align itself by the stars. Would it be possible to point the optical sight toward Earth, line up the horizontal line on the planet's terminator — the twilight line that separates the nighttime side of the globe from the daytime side — and burn the engine accurately enough to get the crew home? With Jim Lovell serving as navigator, the crew ran a few quick experiments, and sure enough, it seemed, for a short burn at least, such cosmic eyeballing might actually do the trick. The procedure, a decidedly last-ditch one, was tucked away in the contingency flight plan files and shortly forgotten altogether. Now, as Lovell copied down Brand's instructions, he saw that the procedure he had helped improvise the first time he came out here might help save his life the second.

"Hey," he said to Brand, "sounds just like what we came up with on Apollo 8."

"Yes, everybody wondered if you would remember that, and by golly you did," Brand said. "And Fred, when Jim has the Earth centered in his window, you should also be able to see the sun in the alignment telescope. It'll be at the very top of the field of view, just splitting the cursor. That will confirm that your attitude is correct."

"I understand, Vance," Haise said.

"Freddo," Lovell said, turning to Haise, "what do you say we stop this PTC spin and see if we can't go hunting for the Earth."

"Whenever you're ready."

Lovell took a few minutes to race through the power-up checklist on page 24, engaging all of the instruments he would need for the burn, including the circuit breakers for his thrusters. When he was done, he reached forward, grabbed his attitude controller, eased it slightly to the right, and vented a plume of propellant through the nozzles in the direction opposite the spacecraft's spin. With surprising responsiveness, Aquarius bumped to a stop. On the opposite side of the tunnel, Swigert felt the rumble, surmised what his crewmates were up to, and, throwing the last few switches needed to put Odyssey back to sleep, swam down to the LEM and resumed his spot atop the engine cover. As Lovell began shifting the spacecraft around in pursuit of the home planet, Haise leaned toward his own triangular window.

"Whoa!" he called to Lovell. "I've got the Earth."

"So have I," Lovell responded.

"You're getting good at this maneuvering, Jim."

Lovell jockeyed to keep the Earth in his optical sight, and Haise glanced into his telescope. As Houston had promised, the sun was splitting the cursor and holding steady.

"Houston," he called, "Jim has the Earth aligned, and you're right, the sun is in the AOT."

"Roger. Good going, 13." the Capcom answered. Haise could hear that in the last few minutes Brand had gone off console and been replaced by Jack Lousma. "If the attitude looks O.K. to you, why, I guess it's your choice when you want to burn."

Lovell looked at his watch. The time for the burn had not nearly arrived.

"We're counting down, aren't we?" he asked. "Or do you want us just to start anytime?"

"Your choice," Lousma answered.

"You guys are getting easy."

"It's not time critical, Jim."

"I understand." Lovell turned to his crewmates. "You guys ready to try this?"

Haise and Swigert nodded.

"All right," the commander said. "Jack, since we don't have any countdown clock, you time the burn with your watch. We're firing for 14 seconds at 10 percent. Freddo, since we don't have an autopilot, you grab your attitude controller and keep us from yawing too much. I'll handle pitch and roll with my controller and also take care of ignition and shutdown. Got it?"

Haise and Swigert nodded again.

"I hope the guys in the backroom who thought this up knew what they were doing," Lovell muttered. "Houston," he then called, "let's say we'll make this burn in two minutes."

"Roger. Two minutes. We got it."

At his station, Lovell set his throttle to 10 percent and positioned one hand over his Start and Stop buttons and the other around his attitude controller. At his station to the right, Haise centered the Earth in his window and kept his right hand on his own controller. Behind them, Swigert fixed his gaze on his watch.

"Two minutes on my mark," he said. "Mark."

Sixty seconds of silence went by.

"One minute," Swigert announced to Lovell and Haise.

"One minute," Haise announced to the ground.

"Roger," the ground answered.

"Forty-five seconds," Swigert said.

"Thirty seconds."

Then: "Ten, nine, eight, seven, six, five, four, three, two, one."

With a gentle flick, Lovell pressed the big red engine button set in the bulkhead and once again felt the vibration below his feet.

"Ignition," the commander said to his crewmates.

Swigert looked at the second hand of his watch. "Two seconds, three seconds."

Haise, at his window, kept staring at the distant Earth. The planet began to slide left, and the LEM pilot finessed the thrusters, bringing it back to the center. "Holding steady in yaw," he murmured.

"Five seconds, six seconds," Swigert said.

"Pitch and roll O.K.," Lovell said as the planet jittered in his optical sight.

"Eight, nine seconds," Swigert called.

"Hang on," Lovell said. The planet jumped up slightly but the commander pitched up and caught it.

"Hanging on," said Haise.

"Ten, eleven," Swigert counted.

"Almost there, Fred," Lovell said, his index finger hovering over the Stop button.

"Twelve, thirteen."

The planet fluttered.

"Fourteen seconds."

Lovell mashed the button hard, far harder than he needed to.

"Shutdown!" he called.

"Shutdown!" Haise echoed.

Instantly, the lunar module fell silent and the vibration shaking the crewmen stopped. In the optical sight, the crescent shape of Earth came to rest directly atop the horizontal line of the cross hairs.

"Houston, burn complete," Lovell said.

"O.K., guys," Lousma said. "Nice work."

Lovell took another look through his reticle, then at the blacked-out dashboard, then back through the sight at the distant, dime-sized Earth.

"Well," he said to Lousma, "let's hope it was."

*

"I want everyone in this room to finish what they're doing and go home."

Gene Kranz stood at the front of room 210 and spoke in what he assumed was a loud enough voice to cut through the babble of the two dozen controllers hunched over their strip charts and power profiles. As far as he could tell, nobody heard him.

"I want everyone in this room to finish what they're doing and go home," he repeated, somewhat louder this time. Still no response.

"Hey!" the former Air Force man barked. This time his controllers stopped what they were doing and turned toward him. "The Tiger Team is shut down for the night. I expect every one of you to knock off for six hours, and I don't want to see you back here till morning."

There was a brief silence in the room, then a few controllers made a move to object. Looking at Kranz, however, they thought again. The lead flight director had already returned to his own strip charts, and it was clear he was not interested in listening to any dissent. It was shortly after midnight, early Thursday morning, thirty-six hours before splashdown, and with the exception of an occasional stolen hour or two, the Tiger Team had not set foot out of room 210 since Monday night. Then, as now, their mission had been to figure out a way to power up and operate the command module on the two hours of juice its three reentry batteries could provide. The difference tonight was that it appeared they had at last worked the problem out.

The task of rationing Odyssey's electricity had, of course, fallen to John Aaron. Many of the controllers in the room, who could easily imagine someone else's subsystems running at only partial current, but could never imagine their own, did not believe Aaron could pull off such a power-stretching feat, but as the hours wore on, the lead EECOM's crayoned charts suggested that he had.

But Aaron's work was only half of what was going on in room 210. Just as important as determining how much juice each switch in the command module would draw when it was turned on was determining the order in which those switches should be flipped. On a normal mission, the command module power-up followed an established sequence, and for good reason. Ground engineers could hardly turn on, say, the spacecraft's guidance system before turning on the heaters that would warm it up; they could hardly engage the buses before connecting the batteries that would be feeding them juice. But Apollo 13 had long since departed from the norm, and with so many of the spacecraft's systems being sacrificed for this power-up, a whole new checklist would have to be developed. That job fell to Arnie Aldrich.

Aldrich was one of the Space Center's leading command module engineers, and as well as John Aaron understood Odyssey's electrical constraints, Aldrich understood the checklist's constraints. As soon as Aaron could work out a power budget for a particular system or subsystem, he would pass it on to Aldrich, who would figure out a switch-throwing sequence that stayed within those limits.

Aldrich, in turn, would forward this plan to the INCO or EECOM or GNC who oversaw that part of the spacecraft, and who, often as not, would at first express disbelief at what he was seeing, insisting that such a half-baked power-up would kill his subsystem, and then, after closer examination, concede that, well, maybe it could work. The INCO or EECOM or GNC would then pass the procedure on to Kranz, who would scan it, O.K. it, and have a courier run it over to the crew training building, where Ken Mattingly, whose feared case of measles had still not surfaced, was sealed up in the command module simulator. Mattingly would run through the procedure he had been handed and then radio back to room 210 that yes, the new method Aldrich and Aaron had come up with was a good one, or that no, they would have to try again. Now, shortly after the mid-course correction and a day and a half before splashdown, the entire checklist — running tens of pages and consisting of hundreds of steps — was virtually complete, and Kranz was at last willing to dismiss his team for the night.

Shortly before he made his announcement, however, there had been one more piece of business to attend to, which Aaron and Aldrich knew would spark a firestorm. The way the power profiles broke down, it looked as if there would indeed be just enough electricity to get the command module up and running, provided that the one system that would let the controllers and the astronauts know they were doing the job right — the telemetry — was not turned on.

Powering up a spacecraft without the temperature, pressure, power, and attitude readouts that would let you monitor the equipment as it went on was a bit like trying to paint a portrait in a dark room. No matter how good your artistic instincts, when the lights went on you'd almost certainly be disappointed by the results. The catch was, the telemetry in the spacecraft, like the lamps in the artist's studio, used up juice, and it was just not juice Apollo 13 could afford to spare. As the final pages of the checklist were being assembled, Aaron and

Aldrich called the other members of the Tiger Team together to explain this conundrum.

"Gentlemen," Aaron said, taking a place at the head of the conference table in room 210. "Arnie and Gene and I have been crunching the numbers every way we can, and while the checklist looks pretty good to us, there's one small glitch." He paused for a moment. "From the amp profiles we've got so far, it looks like we're going to have to perform this power-up blind."

"And that means?" someone asked.

"No telemetry," Aaron said flatly.

The voices of protest that suddenly called out from around the table in room 210 jolted Aaron but did not surprise him.

"John, this is just asking for trouble," someone objected.

"Doing it any other way is just asking for more," Aaron said.

"But no one's ever tried this kind of thing before. No one's even *thought* of trying it."

"It wouldn't be the first thing about this flight that's been irregular," Aaron said.

"This isn't just irregular, John," another voice objected, "this is downright dangerous. Suppose something starts to overheat or blow. We won't know until it's too late."

"And suppose we use up all our juice monitoring the systems and don't have enough left to bring them on line?" Aaron asked. "Then where are we?"

The grumbling continued around the table, and Aaron, it was clear, had not made his case. Unfolding his power profiles, he looked at them slowly, then all at once seemed to notice something. A tiny flicker — part inspiration, part surrender — crossed his face.

"Wait a minute," he said, looking up with an abashed, how-could-I-have-missed-this smile, "how about we try this. How about we set aside a few amps so that when we get all powered up, we switch the telemetry on for just a few minutes and take a good scan then. I admit it's not as good as monitoring everything as we go along, but at least we'll have a chance to spot any problems and catch 'em before they do any damage. How would that be?"

The men at the table looked at Aaron and then at one another. They had no way of knowing if this was a stroke of Aaron inspiration or if he had been planning this concession all along. There was no denying, however, that it *was* a concession, and gradually the members of the Tiger Team nodded their assent. If John Aaron, the steely-eyed missile man, thought he could power up a crippled command module without a single data point of telemetry to help him along, who were a few garden-variety controllers to disagree? Besides, in a few minutes Gene Kranz might let them go to sleep, and that was something none of them had had a chance to do in two days.

Fred Haise noticed the fever beginning at about three o'clock in the morning. It started the way most fevers do: lightheadedness, ashen skin, tingly nerve endings. While the sensations were unpleasant, they didn't take Haise by surprise. The first clue he'd had that he might be falling ill came yesterday morning when he tried — for one of the few times in the past day — to urinate, and had noticed that this most ordinary of acts was all at once accompanied by a most extraordinary pain.

To be sure, nobody aboard Apollo 13 had been doing much urinating lately, and the reason was simple: they hadn't been doing much drinking. As the TELMUS had told the astronauts during the earliest hours of the mission's crisis, water was one of the most precious consumables the crew had. Since the water supply in Odyssey would quickly freeze, the supply in Aquarius would be the only usable one. But since the water for drinking and the water for cooling equipment were drawn from the same tank, the crew would have to think carefully about taking so much as a sip. If they drank too freely from the central supply, they could easily wind up quenching their thirst at the expense of the spaceship that was keeping them alive.

But even if there had been plenty of water aboard, there were other reasons for the crew to pass it up. Like the command module, the LEM was equipped with a venting system that would allow the crew to dump urine and other wastewater overboard. The problem was, expelling this fluid, like any other liquid or gas streaming from a spacecraft, created a tiny propulsive force that could change the trajectory of the ship. With Odyssey and Aquarius having so much trouble maintaining attitude, and with the crew having labored so hard to get themselves back into the center of their reentry corridor, it seemed dangerous, and indeed ludicrous, to urinate themselves back out of it. Instead, what urine the astronauts did produce in the past forty-eight hours they had been instructed to store in plastic bags gathered from around the ship.

Over the course of two days, three nervous men — even three nervous, water-deprived men — can produce an inconvenient amount of urine, and the interior of the spacecraft had begun to grow cluttered with the plastic-sealed specimens. Rather than accumulate even more such mementos, the astronauts had decided to quit drinking almost entirely, limiting themselves to about six ounces of water per day, less than a sixth of the average adult intake.

The crew members were well aware that the results of such deprivation could be serious. Time and again during training, the flight surgeons had cautioned all astronauts that if they did not consume and pass enough water in space, their bodies could not excrete toxins. And if they did not excrete toxins, the noxious substances would accumulate in their kidneys, leading to an infection, which could be recognized first by burning during urination, then by a high fever. At ten o'clock Wednesday morning, Haise had experienced the first symptom, and now, at three o'clock Thursday morning — just thirty-three hours before he would be expected to participate in perhaps the most perilous reentry in the history of lunar spaceflight — he noticed the second.

Jim Lovell glanced over at his pale crewmate. "Hey, Freddo. You all right?"

"Yeah, sure," Haise mumbled. "I'm fine. Why?"

"You sure don't look fine, that's why."

"Well, I am."

"You want me to get the thermometer, Fred?" Swigert asked. "It's right upstairs in the first aid kit."

"No, don't bother."

"You sure?" Swigert asked.

"I'm sure."

"It's no trouble."

"I said," the LEM pilot repeated firmly, "I'm fine."

"O.K.," Swigert said, exchanging a look with Lovell. "O.K."

Lovell regarded both his crewmates and reflected on what he ought to do next, but before he could reach any conclusions, his thoughts were interrupted. From beneath the LEM's floor came a dull pop, then a hiss, then another thump and vibration rattled through the cabin. Lovell leapt forward toward his window. Below the cluster of thrusters far to the left of his field of vision, he could see a far too familiar cloud of icy crystals floating upward. For an instant Lovell was startled, and then just as quickly he knew what the sound and the vent were.

"That," he said, turning to his crewmates, "was the end of our helium problem."

"About time," Haise said, looking at his watch.

"I'd almost forgotten about it," Swigert admitted.

"Aquarius, Houston," Jack Lousma called. "You notice anything in the last second or so?"

"Yes, Jack," Lovell answered. "I was just about to call you. Underneath quad four I noticed a lot of sparklies going out. I'm assuming that was the helium."

"Roger," Lousma said. "Readings here indicate that your pressure was up to 1,921 pounds, and it's now down to about 600 and falling fast."

"Well that's good to hear," Lovell said, "but this probably means we're going to have to worry about reestablishing thermal roll." As the commander looked out his window at the spreading helium cloud, he could see that the Earth and the moon, which had been passing by the approximate center of his window as the spacecraft spun through the PTC rotation he had set up after the last engine burn, had moved noticeably, with the Earth rising higher and the moon falling lower, both threatening to move out of his field of vision altogether. "It looks like the burst reversed my yaw completely and put in a little pitch. Is this what they call a non-propulsive vent?"

"Right," Lousma said. "I'd hate to see a propulsive one."

"You and me both."

"Well, the pressure's going down through 50 pounds now, so are you seeing fewer sparklies?"

Lovell looked out the window. "Yes," he said, "far fewer."

"O.K.," Lousma said. "Then for the time being, why don't you just monitor the spacecraft's position, see how the pitch and yaw are going, and keep us apprised. We'll make a recommendation later as to whether we think you ought to reestablish PTC."

"Roger. Monitoring."

Lovell settled down in front of his window, folded his arms against the chill in the spacecraft, and began to watch the Earth and moon slide by. The movement of the bodies was almost hypnotic, and in the quiet pre-dawn hours of Thursday, Lovell found a curious tranquility come over him. He knew that within the next hour or two he might have to reengage his attitude-control jets and once again go through the whole tedious routine of establishing the PTC roll, but right now that caused him little concern.

As the commander gazed out the left porthole, his crew apparently became affected by the same strange serenity and decided to bunk down for an unscheduled rest period. The feverish Haise, avoiding the frigid command module, backed halfway up the tunnel and, with his head hovering over the ascent engine cover, fell instantly asleep. Swigert, claiming the LEM pilot's spot Haise had abandoned, curled up on the floor on the starboard side and wrapped a wire restraint around his arm, to hold himself in place. Lovell watched them both settle in, and after a time he hailed the ground.

"Houston," he quietly called into his microphone.

"Houston here," Lousma answered, unconsciously matching Lovell's modulated tone. "How you doing, Jim?"

"Not bad. Not bad at all."

"You up alone or are Jack and Fred with you?"

"Jack and Fred are both asleep, and at the moment," Lovell said, looking out at the now-stabilizing Earth and moon, "it looks like we're in no real trouble as far as the PTC goes."

"Good. Everything looks pretty smooth from here too. We'll keep monitoring and we'll let you know if there are any other procedures."

"Roger," Lovell said.

"Actually," Lousma added, "there is one procedure we can go over if you've got the time. I've just been handed some notes from the guidance officers that they want you to start thinking about." The Capcom paused. "How would you like to discuss a few ideas about reentry and splashdown?"

Lovell did not answer right away, but simply cast his eyes around the cockpit. Scanning back and forth, he glanced first at his darkened instrument panel, then at his unconscious crew, then at the off-center Earth and moon moving by his off-center LEM, and then at the remaining snowflakes venting into space from his now all-but-dead descent engine.

Yes, he decided. Splashdown was exactly what he'd like to discuss.

12

IT WAS BARELY the beginning of the morning work shift, and Jerry Bostick, the Maroon Team flight dynamics officer, was already having a lousy day. He suspected it would soon get even worse.

"Damnit," Bostick muttered softly, standing behind his first-row console and reading the screen in disgust. Leaning over the shoulder of Dave Reed, the FIDO on duty, he gave the phosphorous numbers a second look.

"Damnit!" he repeated, this time loud enough for Reed to turn around in his chair.

"What's the problem, Jerry?" Reed asked.

"You don't want to know," Bostick answered.

"Try me."

Bostick reached around Reed, ran his index finger down a column of numbers on the screen, and came to rest on a single data point. Reed leaned forward and squinted. The column Bostick was pointing to was headed "Trajectory." The number he was pointing to read "6.15."

"Oh no," Reed groaned, dropping his head in his hands.

Since ten o'clock last night, after Apollo 13's mid-course correction was executed, the number on the screen had been one of the most encouraging bits of telemetry streaming back from the ship. Earlier in the evening, before the burn of the descent stage, the trajectory of Aquarius and Odyssey had deteriorated to 5.9 degrees, just a little more than half a degree away from the shallow end of the reentry corridor — the end that would prevent the crew from descending through the atmosphere and instead send them bouncing off it and back into space. After the mid-course burn, things looked dramatically better, with Apollo 13 climbing up to a comfortable 6.24, cozily close to the 6.5 that marked a perfect bull's-eye reentry. Now, however, at eight o'clock Thursday morning, twenty-eight hours before splashdown, the trajectory appeared to be decaying again.

"Jerry, what the hell is going on here?" Reed asked, moving aside so Bostick could get closer to the screen.

"I don't have any idea."

"Well, it wasn't the helium vent."

"Nah, that's not nearly enough to cause this."

"Maybe the tracking arcs are bad."

"The arcs are good, Dave."

"Maybe there's some noise in the data."

Bostick looked at the steady 6.15 shining unblinkingly on the screen. "Does this look like ratty data to you?"

If the helium and the tracking data weren't the problem, and the spacecraft was truly dropping to the bottom of its corridor, it meant that the LEM's descent engine would have to be fired again to straighten things out. But with the helium that pressurized the fuel tanks gone, it was unlikely the engine *could* be fired. Before Bostick could contemplate this new development, Glynn Lunney, the Black Team flight director, approached him from behind.

"Jerry," Lunney said, "I need to talk to you. We've got a problem."

"I've got a problem here, Glynn," Bostick said. "It looks like we're shallowing again."

"Are your tracking arcs good?" Lunney asked.

"They appear to be," Bostick said.

"Are you venting anything?"

"Not that we can see," Bostick said.

"Well, make that your priority," Lunney said, "but start working on this, too: I just got a call from the Atomic Energy Commission; they're worried about the LEM."

Bostick had been afraid of this. During Aquarius's brief planned stay on the lunar surface, Jim Lovell and Fred Haise were intended not just to retrieve rocks but to leave behind a number of automated scientific instruments, including a seismograph, a solar-wind collector, and a laser reflector. Since the experiments were intended to operate for well over a year, and since fuel cells or batteries could not keep them running that long, the equipment was instead powered by a miniature nuclear reactor, fueled by spent uranium taken from nuclear power plants.

On the surface of the moon, the tiny generator posed no danger to anybody. But what, some people worried when the system was first proposed, would happen if the little rod of nuclear fuel never made it to the moon? What if the Saturn 5 rocket blew up before the spacecraft even reached Earth orbit, dropping the uranium who knows where? To prevent such accidental contamination, the LEM's designers agreed to seal the craft's nuclear material in a heavy, heat-resistant ceramic cask that would allow it to survive an explosion, a fiery atmospheric reentry, and even a violent collision with the planet's surface without any leakage of radiation. Once a LEM was out of Earth orbit and on its way to the moon, the protective cask became superfluous and nobody gave it another thought. But now Apollo 13's LEM was on its way home, heading for just the fiery reentry the doomsayers had feared, and Jerry Bostick had been suspecting that the Atomic Energy Commission might soon come poking around, fretting about the radioactive rod and its ceramic protection.

"When did you hear from them, Glynn?" Bostick now asked Lunney.

"Just a little while ago. They're pretty jumpy about that fuel rod."

"Did you tell them we've tested the cask repeatedly?"

"I did."

"And did you tell them we have no reason to think it won't survive reentry?"

"I did."

"And they didn't believe you?"

"Oh, they believed me, but they still want insurance. They want to make sure that when the LEM comes down, we don't just dump it in any old ocean, but in the deepest water we can find. Can you handle that for them?"

Bostick, in his own restrained way, went through the roof. "Oh crap, Glynn, that's ridiculous. We built the damn ceramic cask just so we wouldn't have to worry about this kind of thing. As long as we bring that LEM down someplace it doesn't bonk anybody on the head, it's not going to do anyone a bit of harm."

Glynn Lunney may have agreed with Jerry Bostick — indeed, he probably did agree with him — but to the extent that he did, he kept it to himself. The AEC was an arm of the government, the government paid NASA's bills, and if the people who controlled the Agency's purse strings wanted a flight director to address this problem, then the flight director would have no choice but to comply. For the next several minutes Lunney listened sympathetically while his FIDO blew off steam, shrugged along with him at the thinking of Washington bureaucrats, and then suggested that maybe, just maybe, the AEC had a point. Certainly, the first order of business had to be fixing Apollo 13's shallowing trajectory, but once that was taken care of, wouldn't it be a simple matter to humor the AEC, pick an especially bottomless patch of ocean and aim the LEM to come down there?

"We'll take care of it, Glynn," Bostick finally said. "No problem. I think there might be a spot off New Zealand that would be just what you're looking for."

Lunney nodded gratefully and went off to tend to other things, and Bostick returned to his own business. Turning back to his console, he could see Reed — looking even more worried than he had minutes ago — in fretful consultation with the Black Team FIDO. Leaning over the two men and squinting at the screen, Bostick could see that the flight path, which had been deteriorating before, seemed to be collapsing altogether: the number in the trajectory column was just a fraction above 6.0 and continuing to fall. His lousy day was indeed getting worse.

Jim Lovell was eating a hot dog when Joe Kerwin called to tell him about the trajectory. Actually, Jim Lovell was *trying* to eat a hot dog, but he wasn't having much luck. It was early Thursday morning when the astronauts in Aquarius and the Maroon Team on the ground began their workday, and while Lovell could not vouch for the crew back in Houston, the crew in his ship seemed at least marginally refreshed. When Fred Haise and Jack Swigert drifted off for their impromptu three-hour rest period at 3:30 in the morning, Lovell figured it was best not to disturb them, and the decision turned out to be a good one.

Swigert, who had appeared almost surrealistically exhilarated by the chance to go to work in his command module yesterday, looked downright chipper this morning. And Haise, whose face had been a sickly gray yesterday, looked something close to flushed now. Lovell was not sure if his LEM pilot's rosiness was a sign of renewed health or a symptom of an even higher temperature bringing feverish blood to his cheeks. But Haise had already made it clear that he did not welcome inquiries into the matter, and Lovell told himself to respect that preference. For the first hour or two the entire crew was up on their last full day in space, they rattled around the cockpit, attending to their various chores without speaking, like three just-awakened men in a lakeside cabin preparing for a pre-dawn fishing trip. Now, at 8:30, as Jerry Bostick, Glynn Lunney, and Dave Reed discussed shallowing trajectory and nuclear fuel, Lovell figured it was time he got his crew fed.

"Say, Jack," the commander said over his shoulder. Swigert, as always, was atop his engine cover, flipping through a systems book. "How do we stand on supplies back there?"

"Let me check," Swigert said. He released his book, let it hover in the air next to him, and opened the large storage bag in which he had stashed his cache of food packets.

"Not too great, Jim," he said, sifting through the clear plastic pouches. "Cold soup, more cold soup, and . . . it looks like some desserts."

"How about running up to the bedroom and bringing back some rations?"

"No problem."

"You want anything, Freddo?" Lovell asked.

"Sure," Haise said. "How about a few of those hot dogs."

Swigert bounced up into the frigid command module, floated over to the food locker, and dug through the remaining packets. At the bottom of the bin were the sealed pouches containing the hot dogs. Each hot dog was individually wrapped, color-coded with the red, white, or blue Velcro indicating which crew member it was intended for, and each, Swigert was amazed to find, was frozen solid. Removing one from the bin, he regarded it curiously and then, collecting two more, swam back through the tunnel, laughing.

"Well, gentlemen," he announced as he reappeared, "I got what you asked for, but I'm not sure you want them."

Lovell reached out, took the frost-covered packet Swigert offered, and then, laughing himself, knocked it against the bulkhead. It made a resounding *bonk*.

"Sounds delicious," Lovell said.

"*Looks* delicious," Haise said.

"Eat hearty," Swigert said.

Before Lovell could discard the frozen frank, Joe Kerwin's voice came through his headset.

"Aquarius, Houston."

"Go, Houston," Swigert answered for the crew.

"Say, fellows, just wanted you all to know the plot shows you 130,000 miles out now, which is about, gee, 10,000 miles closer than you were when I came on a couple of hours ago. And your smiling FIDO tells me you're making 3,427 miles per hour in a 3,000-mile zone."

"Real good," Swigert said.

"There is just one other thing," Kerwin said. "The good FIDO does give us a slightly shallowing flight path, and he's kind of . . . tossing around the idea of doing another mid-course maneuver at about five hours before entry. If we do it, it won't be more than two feet per second."

Lovell, Swigert, and Haise gave one another dubious looks.

"That FIDO is really cooking today," Swigert said with exasperation.

"Oh, he's having a ball," Kerwin answered, and then quickly clicked off the line.

Lovell didn't like this at all. If his engine was indeed out of commission following the helium eruption, the attitude-control jets could probably handle the job, but while a two-foot burn would take just a few seconds of low-throttle power from the big, roaring descent system, it would take a good half minute of full-bore fire from the little thrusters, running them to near exhaustion.

"I don't like the sound of this," Lovell said to Haise, tossing his no-longer-amusing hot dog aside.

"I'm with you," Haise agreed.

The commander pushed away from his station and prepared to move up the tunnel to find a more palatable breakfast, but before he could, Kerwin came back on the air.

"Jim, the next action item we want you and Jack to pursue is transferring some LEM power up to the command module so we can get that reentry battery charged."

"O.K.," Lovell answered, signaling to Swigert, "I'll let you have Jack."

Swigert signed on and Lovell removed his headset so he could make it up the tunnel untethered, but as soon as Kerwin began explaining the procedure to Swigert and the command module pilot's "uh-huh"s and "mm-hmm"s began filtering up to him, Lovell started to worry.

"Are they sure they want to be screwing around with the juice now?" he called to Swigert, popping his head back into the LEM. "We've still got to run this LEM for twenty-four more hours."

Swigert relayed this to the ground. "One question here. If we transfer power now, we aren't going to cut short what we have left in the LEM to get us back to entry interface, are we?"

"That's a negative, Jack. According to the latest update, we've got amp-hours out to 203 hours, and splashdown is set for 142."

"No problem," Swigert called to Lovell. "They've got us projected to 203 hours."

"Have they tested this thing," Lovell called back, "or are we going to cook both sets of batteries trying to pump power upstream to the command module?"

"Say, Houston," Swigert said, "Jim would like to know whether this procedure has been tried and found to be O.K. There's no danger of shorting out the batteries or anything, is there?"

"O.K., Jack, the procedure has not been tried out as such, but given the hardware paths through which the current flows, there's not a problem with shorting out a battery. But remember, the reason for all this is that your entry battery is 20 amp-hours short, and we've got no choice but to juice it up to get you home."

Swigert turned back to Lovell. "No, they haven't tested the procedure. No, they don't think it's a problem. And they remind us that no, we can't get home without it."

Lovell grumbled his assent. Swigert got back on the line and spent much of the morning copying the power-up procedure and swimming back and forth between the two ships, throwing the necessary switches to execute it, and monitoring the current as it was transfused from one spacecraft to the other. While he busied himself with this assignment, the Capcom — Vance Brand by now — came back on line with an assignment for Lovell and Haise.

In the same way that the FIDOs needed to know the precise weight of cargo and crew in Aquarius before burning the descent engine, so too did the guidance and navigation officers need to know how much ballast Odyssey carried before aligning the platform and aiming the ship for reentry. The computers in an Apollo spacecraft were programmed to expect a command module returning from the moon to weigh one hundred pounds more on its earthward journey than on its moonward journey — that hundred pounds representing the rocks and soil samples the crew went out there to get in the first place. But this Apollo was returning from the moon rockless, and before it could reenter the atmosphere, the astronauts would have to transfer a few armloads of equipment from the LEM up to the command module, pack them away in the storage areas that were to hold the priceless bits of the moon, and hope that the weight was right and the computer was fooled.

"O.K., Jim," Brand called up while Swigert worked, "when you have some time to copy, we've got the entry stowage list, which specifies the equipment you're going to have to move before splashdown."

"I'm able to copy now," Lovell said, removing his pen from his sleeve pocket and signaling to Haise to toss him a scrap-paper flight plan.

"All right, you want to carry over the two 70-millimeter Hasselblad cameras, the black-and-white TV camera, all 16-millimeter and 70-millimeter exposed film, the LEM data recorder, extra oxygen hoses, extra oxygen screen caps, the waste management system chute, and the LEM flight-data file. Copy that?"

"Copy."

Lovell showed the equipment list to Haise, and the two men began scavenging up the cargo the Capcom had specified. Opening one storage compartment, Haise gathered the two still cameras and left them to drift behind him; opening another, Lovell retrieved the oxygen hoses and let them snake nearby. Opening a third, Haise spotted something curious and stopped what he was doing. Stacked on top of one another in the compartment were the astronauts' personal preference kits, or PPKs, Beta-cloth pouches in which each crew member was permitted to carry a few mementos or good-luck charms, which contributed nothing to the mission technically but a lot to the men emotionally. Some astronauts carried the odd bit of sentimental jewelry; some brought a coin or a miniature flag; Lovell himself took along a small gold brooch with a diamond-studded "13," which he'd had made before the mission and that he intended to give to Marilyn upon his return.

As Fred Haise looked at his own PPK, he noticed a sealed envelope taped to the top of it, with the words "To Fred" written on it. The handwriting was instantly familiar to him. Looking around to see that his commander wasn't watching, Haise removed the envelope and tore open the flap. As soon he did, a flock of pictures drifted out. The first one he saw was of his wife, Mary; the second was of his oldest son, Fred; the third was of his other son, Stephen, and his daughter, Margaret. Haise snatched up the floating faces and peered into the envelope. Inside was a piece of notepaper covered in the same neat script.

Dear Fred, the note said. *By the time you read this, you will already have landed on the moon and, hopefully, be on your way back to Earth. This is to let you know how much we love you, how proud we are of you, and how very much we miss you. Hurry home! Love, Mary.*

Haise read the letter quickly, folded it back into the envelope with the pictures, and tucked them into his jumpsuit.

"From Mary?" Lovell asked quietly from over his shoulder. Haise looked startled.

"Mm-hmm," he said. "She must've slipped it to whoever was stowing the PPKs last week."

"Nice," Lovell said with a knowing smile. He had earlier found a letter from Marilyn tucked into his own kit.

"Mm-hmm."

By mutual, unspoken agreement, the two men said no more about the note and finished assembling their equipment in silence. Though Lovell could not know what Haise was thinking, he suspected it was the same thing he was. This mission, he decided with sudden exasperation, was getting old. He'd had enough poignant reminders of the lunar landing that was not to be: the backward peeks at receding Fra Mauro, the longing looks at his unused lunar suit, the sorrowful glances at his useless lunar landing checklist. If the touchdown on the moon that he and Haise had trained so long for was not going to happen, fine; but it was time, then, to get things stowed, get geared up, and get the rest of this ill-fated trip over with.

"Freddo," he said, "what do you say we pack this stuff up, call the ground, and see how they're coming with that damned reentry checklist."

"This is Apollo Control at 119 hours, 17 minutes ground elapsed time," Terry White said into his microphone at the Public Affairs console just after lunch hour. "Spacecraft position is 112,224 nautical miles out from Earth. Velocity continuing to build up, now 3,725 miles per hour. Looking now at an entry interface time of 142 hours, 40 minutes, and 42 seconds, which is 23 hours, 22 minutes from now. A mid-course correction burn of something less than two feet per second will probably be performed at about 5 hours before reentry.

"At 3 P.M. today in the main auditorium of Mission Control, Neil Armstrong, the commander of Apollo 11, will hold a press conference to discuss various technical aspects of Apollo 13. Also, the president of the Chicago Board of Trade has forwarded the following message to Mission Control: 'The Chicago Board of Trade suspended trading at 11 A.M. today for a moment of tribute to the courage and gallantry of America's astronauts and a prayer for their safe return to Earth.' This is Apollo Control."

Chuck Deiterich stood in front of the blackboard in the staff support room just off Mission Control. Everywhere he looked, it seemed, was a FIDO, a RETRO, or a GUIDO. There was Jerry Bostick, Bobby Spencer, Dave Reed, and others, all trained in the black art of steering a spacecraft across 250,000 miles of void and bringing it back home again. An EECOM or INCO or TELMU who wandered into this room would barely understand the tongue that was spoken here, but the RETROS and FIDOS and GUIDOS were fluent in it.

In the past twenty-four hours, Deiterich had had a lot of luck working with this council of navigational elders, and he was hoping for a little more this afternoon. While Bostick, Reed, and Bill Peters had been busy figuring out why Apollo 13's trajectory was continuing to shallow and whether it was possible to drop its lunar module in an ocean that would make the Atomic Energy Commission happy, Deiterich had been occupying himself with other problems.

The biggest issue he'd been addressing was how the crew could safely jettison their dead service module and their quite alive LEM when the time came to position the conical command module for reentry through the atmosphere. Had the mission of Apollo 13 gone as planned, the service module's thrusters would have handled much of this work, moving Odyssey a safe distance away from Aquarius when the lander was cast adrift in lunar orbit, and moving the service module itself away from the command module when the time came to expose the heat shield and begin the reentry. But the mission had long since stopped going as planned, and the thrusters that would have been relied on to perform these maneuvers had long since stopped working.

Deiterich and his colleagues had come up with some elegant solutions. When the time came to jettison the service module, they decided, Jim Lovell and Fred Haise would stay in the LEM while Jack Swigert would scramble up into the command module. Moments before separation, Lovell would fire the LEM's thrusters for a single pulse, pushing the whole spacecraft stack forward. Swigert would then press the button that fired the service module's pyrotechnic bolts, cutting the huge, useless portion of the ship loose. As soon as he did, Lovell would light his thrusters again, this time in the opposite direction, backing the LEM and its attached command module — with Swigert aboard — away from the drifting service module.

Easier, but no less elegant, was the procedure for jettisoning the LEM. Before a lunar module was released on a normal mission, the astronauts would close the hatch in both the lander itself and the command module, sealing off the tunnel from the cockpits of either ship. The commander would then open a vent in the tunnel, bleeding its atmosphere into space and lowering its pressure to a near vacuum. This would allow the twin vehicles to separate without an eruption of air blowing them uncontrollably apart.

During the flight of Apollo 10 last spring, the controllers had experimented with the idea of leaving the tunnel partially pressurized, so that when the clamps that held the vehicles together were released, the LEM would pop free of the mother ship, but in a slower, more controlled way than it would if the passageway between the two spacecraft was fully pressurized. This method, the controllers figured, would come in handy if a service module ever lost its thrusters. Now, a year later, a service module had done just that, and the flight dynamics officers were glad they had the maneuver tucked away in the contingency flight-plan books. Yesterday, the procedure had been explained to Jack Lousma, and the Capcom had proudly relayed it up to Lovell.

"When we jettison the LEM," he had reported, "we're going to do it like we did in Apollo 10 — just let the beauty go."

Lovell had radioed back a far more skeptical "O.K."

Now, in mid-afternoon on Thursday, Deiterich had one more new procedure to clear with his fellow FIDOS, GUIDOS, and RETROS. This one concerned Apollo 13's guidance systems. Before the command module could reenter the atmosphere, its guidance system would have to be reactivated and then — with the aid of telescope sightings of the sun and the moon — realigned. The job could be painstaking, and would probably be made all the more painstaking by the condensation that was now leaching into the spacecraft's optics. Nevertheless, Deiterich and the other flight dynamics officers felt confident the crew could pull it off without too much difficulty.

To make sure they did, the reentry alignment, once established, would have to be checked. The customary method for doing this required the command module pilot to watch the horizon of the Earth as it moved across his window. If the alignment of the ship was true, the arc of the planet would cross specific hash marks etched on the window frame at specific times. As long as the planet moved as planned, the computer could control the reentry. If it didn't, the crew would know that the guidance platform was somehow skewed and the man in the commander's seat might have to take over the reentry, manually steering the spacecraft to splashdown. The problem on Apollo 13 was, just before reentry there wouldn't *be* any horizon to take a sighting on. With the hurry-up route the spacecraft was following home, Odyssey would wind up approaching the Earth from its nighttime side, meaning there would be nothing below in the critical moments before reentry but a dim mass where the planet ought to be.

But Chuck Deiterich, the Gold Team RETRO, had an idea. "Fellows," he said to the other flight dynamics men in the staff support room, "tomorrow around lunchtime we're going to have a problem — specifically, we're going to be trying to check our attitude against a horizon that isn't there."

He turned to the blackboard and drew a large downward arc representing the edge of the Earth. "Now while the Earth will be invisible, the stars will always be there" — he tapped a few chalk dots onto the board above his horizon — "but as fast as the ship will be moving, there might not be time to determine which ones we're looking at." He eliminated his stars with a sweep of his eraser.

"Of course, what we'll also have out there," Deiterich said, "will be the moon." He drew a neat little moon above his ragged Earth. "As the spacecraft arcs around the planet and gets closer and closer to the atmosphere, the moon will appear to set." Deiterich drew another moon below his first one, then another and another and another, each moving closer to the chalk horizon, until the last one vanished partially behind it.

"At some point," he said, "the moon will set behind the Earth and disappear. It will disappear at the same time whether it's daytime below or nighttime, whether we can see the horizon or can't see it." The RETRO touched the corner of his eraser to the blackboard and carefully erased only the long arc that represented the horizon, leaving all his moons behind. He pointed to the one moon that was half obscured by the horizon that was no longer there.

"If we know the exact second the moon is supposed to disappear, and if our command module pilot tells us it indeed disappears, then gentlemen, our entry attitude is on the mark."

Deiterich put his chalk and eraser down on the blackboard's ledge, turned and faced his audience, and waited for questions. There weren't any. The Gold Team RETRO wasn't immodest, but he knew a good idea when he heard one, and the men in the room, he suspected, did too.

It had been more than a day since the crew of Apollo 13 had been able to see out the windows of their command module. Since Monday, of course, the view from the lunar module had been somewhat obstructed, what with the astronauts' constant respiration introducing moisture into the air and the spacecraft's low temperature causing condensation to form on the two triangular windows that were supposed to provide a clear line of sight into space. But for most of this period, the command module had been spared this problem, largely because the astronauts had been spending the majority of their time — and doing the majority of their breathing — downstairs in Aquarius.

Now, as Apollo 13 began its final evening in space, the temperature in the command module had dropped lower than it had been the entire trip, and the water in even its considerably drier air had at last made itself visible. The crew noticed with alarm that every window, wall, and instrument panel in the clammy cockpit had become covered with pearl-sized beads of water. In zero g, the droplets couldn't drip, but in a gravity environment they almost certainly would have, and had Odyssey been resting on Earth, it would have soon developed the eerie, *plink-plink* ambience of a limestone cave.

For Jim Lovell, this boded trouble. If the windows and walls and the front of the instrument panel were so well soaked, it was a safe bet that the back of the instrument panel, where the spacecraft's wires, bulbs, and soldered joints lived, was too. Engineers at North American Rockwell had taken care to waterproof every one of the millions of electrical connections that ran through the ship, but the protective sealant was sufficient only to guard against moisture in ordinarily humid cabin air. Nobody had ever thought it would be necessary to seal the electronics against free-running water condensed enough to *plink*. When the ship was turned on tomorrow and power once again began to flow through its instruments, there was a very real chance that a single raw wire or porous seal could blow the whole system right back out.

As the dinner hour came and went in the marginally warmer LEM, Lovell desultorily sipped a bag of cold soup, then gave up and pushed off toward the command module to check on the status of his ship.

"What are you doing?" Haise asked, looking — and even sounding — more feverish to Lovell than he had yesterday.

"Checking on the condensation upstairs," Lovell said.

"I'll go with you," Haise offered.

"Why don't you stay put. You look lousy, Freddo, and it's freezing up there."

"I'm fine," Haise said.

Lovell sprang up the tunnel. Haise followed him, and the two men floated straight toward the commander's left-hand window, through which Lovell had first spotted the venting seventy-two hours ago. Now nothing at all was visible through the drenched pane, and as Lovell ran a finger across it, a few droplets broke free and floated in the air.

"This is a mess," he said, shaking his head.

"A mess," Haise repeated.

"Well, we're not going to know a thing until we actually power up."

"And we're not going to power up until they get around to reading us the checklist."

Ever since Lovell and Haise had finished carrying their Aquarius gear over into Odyssey, Lovell had been prodding Houston for the whereabouts of the list John Aaron and Arnie Aldrich had been working on for so long. The read-up, they knew, could take hours, what with Swigert having to copy down each step in longhand and read them all back to make sure he got them right. And that was assuming no glitches or gremlins were found hiding in the list. If a problem cropped up and Aldrich and Aaron had to return to room 210, who knew how much longer things could take? The first time the commander asked the Capcom on duty — Joe Kerwin at the time — how the list was coming, he had responded evasively.

"It exists," Kerwin said.

"'It exists'?" Lovell had mouthed to Haise, and then radioed back "O.K., that's good," to the ground.

The last time Lovell inquired — reminding the new Capcom, Vance Brand, that today was Thursday, tomorrow was Friday, and splashdown was Friday at noon — Brand had tried to jolly him out of his pique. "We're almost ready," he said with an ingratiating laugh. "We'll have it for you Saturday or Sunday at the latest." The commander was not amused.

Now, at 6:30 on Thursday, eighteen hours before splashdown, Lovell had had enough. Swimming back through the tunnel with Haise following him, he called to Swigert.

"Hey, Jack, you ready to get to work here?"

"Do I look busy?" Swigert said.

"Then let's get those guys on the horn and get those procedures up here. I'm through waiting." Lovell pressed his push-to-talk switch. "Houston, Aquarius."

"Go, Jim," Brand answered.

"Just one more reminder that I'm waiting for the power-up procedures you're working on, so I can run through them with the crew and make sure we've got our signals straight."

"Jim, we really are going to get it up to you," Brand said.

"O.K. . . . ." Lovell's voice betrayed obvious annoyance.

"We just about have them in hand."

"O.K. . . . ."

"We should have them . . . within the hour."

"I'll be standing by," Lovell said, and signed off with a loud click.

Though he did not believe Brand's promise — and probably Brand himself didn't either — as it turned out, the Capcom was unknowingly telling the truth. Almost as soon as Lovell clicked off the line, the doors opened at the back of Mission Control and Aaron, Aldrich, and Gene Kranz appeared. With the exception of the hour before and the hour after the PC+2 burn on Tuesday night, none of these men had been seen in Mission Control since the accident on Monday, and when they entered the room, the men at the consoles could not resist turning around for a quick, respectful look.

Aaron, they could see, was carrying a thick sheaf of papers. It was evident from the protective way he held it against his body and the flying wedge Aldrich and Kranz formed on either side of him that it was the power-up checklist the lead EECOM was bearing. The three men walked down two tiers of consoles, stopped at the Capcom station, and huddled briefly with Brand. Aaron handed Brand what appeared to be one copy of the checklist, turned to Kranz and handed him another, then turned to Aldrich and handed him a third. The fourth and last he kept for himself. Brand turned happily around to face his console, and on the air-to-ground loop the controllers heard him hail the ship.

"Houston, Aquarius."

"Go, Houston," Lovell answered.

"O.K., we *are* ready to read you the first checklist installment."

"All right, Vance. I'm going to get Jack on the line, so stand by."

In the spacecraft, Lovell motioned to Swigert to don his headset, collected two or three of the obsolete flight plans, and handed these and his pen to his command module pilot. "You're on, Jack, and you're going to need these." Swigert took the pen and paper, adjusted his earpiece and microphone, and prepared to sign on the air.

As Brand waited for his call, more people suddenly converged on the Capcom station. From the flight director's console came Gerry Griffin and Glynn Lunney, the Gold and Black Team flight directors. From the EECOM station came Sy Liebergot.

"O.K., Vance," Swigert now said on the loop. "I'm ready to copy."

"O.K., Jack," Brand said, "but we have to ask you to wait one minute again. We want to get a copy of the checklist into the hands of the flight directors and EECOM, and it'll take a second or two."

"Roger, Houston," Swigert said, his voice carrying a bit of the same edge as Lovell's.

Aaron picked up the telephone at the Capcom's station to call out for extra copies, and a full two minutes of silence elapsed on the loop as the men at the Capcom console paced, the crew in the spacecraft waited, and the crowd in Mission Control glanced periodically at the back door, waiting for the copies to arrive. Kranz, looking impatient, mimed a keep-talking gesture to Brand.

"Say, Jack," the Capcom said to Swigert. "How you doing on command module water? You guys have any of that bagged water left?"

"Negative. I went up and tried to repressurize the potable tank, but nothing came out."

"Ah," Brand said. "We understood there isn't any more in the potable tank, but we just wondered if any of the bags were left."

"No."

"O.K."

As Brand searched his mind for some other conversational gambit, the back door of Mission Control burst open. Expecting to see an engineer rushing in with a stack of stapled flight plans, the men at the Capcom console groaned when instead half a dozen flight controllers, all from the White-Tiger Team, headed for the communicator's station. Like Kranz, Aaron, and Aldrich, all of these men would want to be present when their masterpiece was read up to the ship, and all of them would want their own copies of the mimeographed sheets.

"Jack, we're probably going to have to hold on for about five more minutes. We have some more people coming on to listen to this. It took a lot of people to devise this procedure, and a few of them have been testing it out, so we'd like to have them all on hand while we give you the rest."

Brand waited for an answer, but all he got was a frosty five-second silence. All at once, on the air-to-ground loop a new voice broke in. It belonged to Deke Slayton, and Brand welcomed it. As an astronaut — albeit an astronaut who had not yet flown — Brand recognized the well-nigh mutinous tone coming down from the spacecraft, and he knew he had only so much authority with this crew. As the chief astronaut — albeit a chief astronaut who had never flown either — Slayton would have a good deal more.

"How's the temperature up there, Jack?" Slayton asked casually. "You guys chopping wood to keep warm?"

The change in Swigert's voice was instantly recognizable. "Deke, it's now about, I think, 50 in the LEM," he said with new conviviality, "and less in the command module."

"A nice fall day, huh?"

"Absolutely. And just so you know, the command module has been stowed per your earlier checklist, with the exception of the Hasselblads, which we'll use to photograph the service module when we let it go."

"Roger. Got that, Jack."

"The LEM is pretty well stowed too, with the exception of a few things we have yet to bring over."

"Roger. Got that too."

Slayton's presence on the loop brought about just the change in Swigert's demeanor that the chief astronaut had hoped it would. But the first-time astronaut was only the second in command aboard

Apollo 13. First in command was Lovell, a veteran of three other space flights, and he was less easily placated by the voice of Deke Slayton.

"Look, Vance," the commander snapped, sidestepping Slayton and speaking, as protocol dictated, directly to his spacecraft communicator, "you've got to realize that we've got to establish a work-rest cycle up here. We can't wait around for you to read up procedures all the time. We've got to get them up here, look at them, and we've got to get our people to sleep. So take that into consideration, and get ready to send up that checklist."

For four and a half minutes there were virtually no exchanges between the air and the ground. Then the back door of Mission Control banged open once more and a breathless engineer arrived with a large stack of checklists. From 7:30 Houston time until just past 9:15, the endless list was read up to the ship, and Swigert copied it down. Finally, fifteen hours before the scheduled splashdown, and just over twelve hours before the power-up was set to begin, the last command was sent, and Swigert stowed his pen and shut his book.

"O.K., Jack," the Capcom said. "Amazingly enough, it looks like we've closed up the loose ends here."

"All right," Swigert said. "If we have any questions, we'll be coming back at you."

"O.K. We did run simulations on all of this, so we do think we got all the little surprises ironed out."

"I hope so," Swigert said, "because tomorrow is examination time."

The laughter started at one end of the lunar module control room at the Grumman plant in Bethpage and gradually spread to the other. Tom Kelly, who had been wedded to his console in a corner of the room since he and Howard Wright had flown down from Boston in the early hours of Tuesday morning, had not heard much merriment in the three days he had been here, and he didn't have a clue as to what this outburst was about. Several consoles away, he noticed that a thin sheet of yellow paper was being passed from controller to controller, each of whom looked it over and emitted a loud bark of laughter.

Kelly waited for the paper to arrive at his station. Scanning the sheet, he recognized what it was immediately, and with a mixture of surprise and amusement, read on.

The thin yellow sheet Kelly had been passed was a copy of an invoice that Grumman would send to another company when Grumman supplied it a part or service. In this case, the company being billed was North American Rockwell, the manufacturer of the command module Odyssey.

On the first line of the form, underneath the column headed "Description of Services Provided," someone had typed: "Towing, $4.00 first mile, $1.00 each additional mile. Total charge, $400,001.00." On the second line, the entry read: "Battery charge, road call. Customer's jumper cables. Total $4.05." The entry on the third line: "Oxygen at $10.00/lb. Total, $500.00." The fourth line said: "Sleeping accommodations for 2, no TV, air conditioned, with radio. Modified American Plan with view. Prepaid. (Additional guest in room at $8.00/night.)"

The subsequent lines included incidental charges for water, baggage handling, and gratuities, all of which, after a 20 percent government discount, came to $312,421.24.

Kelly looked at the controller who had handed him the form, then looked back at the paper and smiled, despite himself. The men at Grumman would love to send this out, and the men at Rockwell would hate to receive it. For that reason, as much as any other, Kelly guessed that someone was actually going to put this thing in an envelope and mail it to Downey, California.

He figured there was nothing wrong with taking advantage of any opportunity to tweak the boys at Rockwell — provided, of course, that the tweaking took place well after splashdown. The form that was amusing the room at Grumman seemed funny enough now, but it wouldn't seem nearly so funny if something else went wrong with either Rockwell's Odyssey or Grumman's Aquarius between now and then. Kelly was about to pass the paper on, but before he did, he looked at it once more. This time he noticed a line typed near the bottom of the page that he had overlooked before.

"Lunar module checkout no later than noon, Friday," the line read. "Accommodations not guaranteed beyond that time."

Kelly, for one, was a little surprised the crew's outlandish "accommodations" had lasted this long.

Jack Swigert could not get the image out of his head, and it was driving him mad. In the nightmare scenario he kept imagining, he was up in Odyssey, setting his switches and arming his pyros in preparation for jettisoning the service module — just as he would be doing a few hours from now — while down in the LEM, Lovell and Haise were poised in front of their windows, hoping to see the cylindrical back end of Odyssey as it popped free and floated by — just as *they* would be in just a few hours. Swigert could see himself in his center seat, counting down to the moment of jettison, and moving his hand with slow-motion, dream-like grace toward the switch marked "SM JETT." At the last second, however, just as his fingers grazed the control, he would lose his focus or become distracted, and his hand would drift a fraction of an inch to the left, toward another switch, the one marked "LEM JETT."

In Swigert's black musings, he could hear the dull *chunk* as the twelve latches that held Aquarius in place released their grip, sense the slight shaking as the lander popped free, feel the tornado of wind as the 5.5 pounds of atmospheric pressure in the command module rushed out through the tunnel and into space. Glancing down that newly opened hole, Swigert could see through the roof of the LEM, where Lovell and Haise, set adrift in the ship that was supposed to be their lifeboat, glanced back at him in sudden horror and confusion. The final thing Swigert could see before the last of the atmosphere in Odyssey rushed into space and the last of the atmosphere in Aquarius did the same was the rapidly shrinking lunar module drifting balletically away, its crinkled foil skin catching glints of sunlight and flashing them back to the dying command module pilot.

The terrible fantasy had begun playing itself out in Swigert's mind late Thursday night, perhaps triggered by a playful comment his Capcom made earlier in the day when reviewing the procedures for closing out and jettisoning the LEM: "Don't forget to transfer the commander over to the command module first." The man at the console in Houston laughed.

"Roger," the man in the spacecraft answered, with no hint of a chuckle.

Now, in the earliest hours of Friday morning, Swigert could take it no more. Climbing off his ascent engine cover, he jumped up into the command module and scrounged around until he found a piece of paper and a small strip of silver duct tape. Removing his pen from his sleeve, he leaned against the bulkhead, wrote a large, block-letter "NO" on the slip of paper, and taped it to the LEM JETT switch. He lifted the paper to make doubly sure it was the LEM JETT switch and not the SM JETT switch he had taped it to. Then he checked again.

Then he called Haise, who drifted up the tunnel and, at Swigert's request, looked at the slip of paper. A bit confused, Haise confirmed that yes, it was taped where it ought to be.

Now, back down in the lunar module, Swigert had at last gotten some peace of mind. But with all his fretful imaginings, what he hadn't gotten was rest. In this sense, at least, he was not alone. For all of the sleep periods Houston scheduled for the crew, nobody was doing much actual sleeping. Each time one of the astronauts returned to the mike after three or four hours in the rack, the Capcom would casually ask him how much real sleep he had managed to get. And nearly every time, the answer was the same: an hour, maybe a little bit more; often a good deal less.

In the second row of Mission Control, the flight surgeon had been copying down the answers the men gave, and the totals had begun to alarm him. Since Monday night, the crew had been averaging about three hours of sleep apiece per day. It was 2:30 Friday morning, ten hours before splashdown, and Swigert had not added to that average, and from the restless thrashings of Lovell and Haise, it appeared that they hadn't either.

"Fred," Jack Lousma called up to the one astronaut who was supposed to be awake. "Are you sleeping?"

"Go ahead," Haise grumbled, opening his eyes and repositioning his headset.

"I've got a few minutes work for you guys, a few changes in the switch configurations in the checklist."

"O.K.," Haise said, "I'll get Jack."

Swigert, hearing this, signed on the air. "O.K. Houston, Aquarius," he said wearily.

"How much rest did you get, Jack?" Lousma asked.

"Oh, I guess maybe two or three hours," Swigert lied. "It was awful cold and it wasn't very good sleep."

"Roger. The way we're looking at it here, it looks like you would have a couple hours to try again before we have to get busy with the final mid-course burn."

"Well," Swigert said, "maybe we'll try, but it really is awful cold."

Swigert nudged Lovell, who didn't really need to be awakened.

"Work to do," he said.

"Swell," Lovell responded.

All three astronauts gathered themselves up and moved sluggishly back to their stations. On the ground, the controllers exchanged worried looks. At the Flight Crew Operations console, Deke Slayton cut in on the line.

"Hey, Jim. While you're up and things are nice and quiet, let me give you one or two other things to think about. One specifically. I know none of you are sleeping worth a damn, and you might want to dig out the medical kit and pull out a couple of those Dexedrine tablets apiece."

"Well . . . I hadn't brought that up," Lovell said. "We might . . . we might consider it."

"O.K." Slayton paused. "Wish we could figure a way to get a hot cup of coffee up to you. It'd probably taste pretty good about now, wouldn't it?"

"Yes, it sure would. You don't know how cold this thing becomes, especially when it's in a thermal roll that's slowing down. Right now the sun's on the engine bell of the service module and not getting down to the LEM at all."

"Hang in there," Slayton said unconvincingly. "It won't be long now."

"Not long," as Slayton well knew, was a relative term. With a final mid-course correction not scheduled for another four hours, the lunar module would not be powered up, and warmed up, for at least another three. Three hours was not much time for the thirty men working the graveyard shift in the climate-controlled Mission Control, but for the men in the icebox of Apollo 13, it was an eternity.

Slayton, like every other man in the room, had been monitoring Aquarius's power consumption since Monday and had been growing increasingly confident about what he was seeing. With the ship pulling a mere 12 amps from its four batteries, a surplus of electricity — albeit a small one — had been created. Logging on to the controllers' loop, Slayton called the flight director's console to ask Milt Windler if it might be possible to use a bit of that additional power to bring the LEM on line early.

Windler called Jack Knight at the TELMU console, who in turn contacted his backroom. Knight's assistants put him on hold, conducted some quick-and-dirty amp projections, and came back with the good word: the crew was free to switch on their ship.

"Jack, they're go for power-up," the backroom called to the TELMU.

"Flight, he can power up if he wants," the TELMU called to Windler.

Windler relayed this to Lousma: "Capcom, tell him to turn on the lights."

"Aquarius, Houston," Lousma called.

"Go ahead, Houston," Lovell answered.

"O.K., skipper. We figured out a way for you to keep warm. We decided to start powering up the LEM now. Just the LEM, though, not the command module. So open your LEM prep checklist and turn to the thirty-minute activation. You copy?"

"Uh, copy," said Lovell. "And you're sure we have plenty of electrical power to do this?"

Slayton cut in. "Jim, you've got 100 percent margins on everything from here on in."

"That sounds encouraging."

The commander turned to his crewmates, gestured to the instrument panel, and with the help of Haise, went into a frenzy of switch-throwing, completing the half-hour power-up in just twenty-one minutes. As soon as Aquarius's systems came on line, the crew could feel the temperature in the frigid cockpit begin to climb. And no sooner did the temperature start to climb than Lovell took a step to make sure it climbed even further. Grabbing his attitude controller, now active again, he spun his ships in a half somersault, so that the sun, which had been falling uselessly on the rump of the service module, fell across the face of the LEM. Almost at once, a yellow-white slash of light flowed into the ship. Lovell turned his face up to it, closed his eyes, and smiled.

"Houston, the sun feels wonderful," he said. "It's shining straight in the windows, and it's getting a lot warmer in here already. Thank you very much."

"Duck blinds are always warmer when the ducks are flying," the Capcom said.

"Right." Lovell opened his eyes. "And looking out the window, Jack, the Earth is whistling in like a high-speed freight train. I don't think many LEMs have seen the Earth from this perspective. I'm still looking for Fra Mauro."

"Well, if you are," Lousma said, "you're going the wrong way, son."

As the sun came up on Friday, and the street in front of the Lovells' house began to fill again with reporters and cameras, the family room began to fill again with friends and astronauts and family members.

Among the first to arrive, courtesy of a driver from the Friendswood Nursing Home, was Blanch Lovell, the mother of the Apollo 13 commander, prettily dressed and cheerfully upbeat, anticipating her son's return from his trip to the moon with the same optimism she had felt the previous times he had flown into space.

Marilyn had still not informed her mother-in-law that there was any reason for her to feel any different about this return, and for the rest of the morning she would have to do what she could to maintain that happy fiction. To make this as easy as possible, Marilyn decided that Blanch would watch the recovery not on the television set in the family room, where most of the other company would be, but off in the den, where she would be protected from any casual comments any of the dozens of other guests in the house might make. As for similarly disturbing remarks by newsmen on TV, Marilyn figured that somebody would simply have to remain with her mother-in-law to distract her attention or explain things away if the on-screen commentary became ominous. Before Blanch arrived, no one had yet been assigned this job, but when she was at last helped through the front door, Neil Armstrong and Buzz Aldrin stepped forward. As the two astronauts settled down in front of the TV in the den with Blanch Lovell, it seemed their job would not be an easy one.

"Apollo 13 is now 37,000 miles from Earth and moving at 7,000 miles per hour," correspondent Bill Ryan said during the opening minutes of the *Today* show, "its course now adjusted so that we should see it land in the Pacific six hours from now. The helicopter carrier Iwo Jima is on station there, and the weather, which had been on again, off again for the past few days, is now on again.

"Some of the spacecraft's most critical maneuvers are still approaching. At 8:23 eastern time, the astronauts are to jettison their service module, and at 11:53 they are to cast off the moon-landing compartment that has been their lifeboat since the power failed aboard the main ship.

"As one astronaut, Alan Bean of Apollo 12, commented, once the lunar module has blown away about an hour before splashdown, the reentry will be about the same as on any other mission, and the emergency will be essentially over."

In front of the television, Armstrong and Aldrin flinched a bit at the words "lifeboat" and "emergency," and looked worriedly at the woman sitting between them. But if Blanch Lovell heard anything amiss, she didn't show it. She turned to the handsome young men on either side of her — both astronauts like her son, but both no doubt just ordinary ones, or else *they* would be flying in space today and he would be the one following things on TV — and smiled at them. Armstrong and Aldrin smiled back.

Out in the family room, Marilyn Lovell watched the same broadcast but responded to it very differently. Alan Bean, who had walked on the moon just last November, might *say* the upcoming reentry would be like any other, but Bean, Marilyn believed, knew better. No other command module had ever taken the beating this one had taken, and no other crew had ever been through so much improvised work on so little rest.

All at once Marilyn heard a small commotion on the front lawn, something that sounded like applause. Hurrying to the window, she was in time to see several of her neighbors moving through the crowd of reporters and crossing the lawn, carrying what appeared to be cases of champagne. Marilyn eyed the cases and took a quick mental inventory of her refrigerator. If she moved a few of the casseroles the same helpful neighbors had brought by earlier and her other guests ate a few more, she should be able to find room for at least some of the bottles. She dearly hoped they wouldn't have to stay on ice too long.

Nobody in Mission Control got very excited when Jim Lovell lit his attitude-control jets for the brief — and what he hoped would be the last — adjustment necessary to place his spacecraft back in the center of its reentry corridor. A short burn from four thrusters that had been working flawlessly for five days was not the stuff of headlines to the controllers, even if that burn was absolutely essential if the crew was going to survive reentry. Practically everything the men at the consoles had to do this morning was absolutely essential if the crew was going to survive reentry. At shortly before 7 A.M. Houston time, as the *Today* show was entering its second hour and Lovell was successfully firing his maneuvering jets, Mission Control was a hive of activity. Three hours earlier, according to the plan Gene Kranz had come up with during the week, Milt Windler's Maroon Team had left the consoles, and for the first time since the PC+2 burn Tuesday night, Kranz's controllers — abandoning their Tiger Team title and resuming their White Team designation — took over. The Maroon Team surrendered their chairs on cue, but not a single member of Windler's group left the room, instead hovering behind their stations or leaning against the walls sipping coffee. Joining them there were most of the members of the Gold Team and most of the Black Team. All of them wanted to stay out of the way of the newly reconstituted White Team, but none of them were willing to be anywhere else but in the auditorium. The White Team controllers plugged in their headsets, swiveled to face their monitors, and went to work on what was to be their first, and perhaps trickiest, maneuver of the day: jettisoning the service module.

"Aquarius, Houston," Joe Kerwin called from the Capcom station.

"Go, Joe," Fred Haise answered.

"I have attitudes and angles for service module separation if you want to copy. You don't need a pad for it, just any old blank sheet of paper will do."

In the spacecraft, Lovell, Haise, and Swigert were in their accustomed positions, all awake and all feeling reasonably alert. Lovell had decided against the Dexedrine tablets Slayton had prescribed for his crew last night, knowing that the lift from the stimulants would be only fleeting, and the subsequent letdown would leave them feeling even worse than they did now. For the time being, the commander had decided, the astronauts would get by on adrenaline alone. Haise, his cheeks still flushed by fever, needed the adrenaline rush more than his crewmates, and at the moment he appeared to be getting it.

"Go ahead, Houston," he said, tearing a piece of paper from a flight plan and producing his pen.

"O.K., the procedure reads as follows: First, maneuver the LEM to the following attitude: roll, 000 degrees; pitch, 91.3 degrees; yaw, 000 degrees." Haise scribbled quickly and did not immediately respond. "Do you want those attitudes repeated, Fred?"

"Negative, Joe."

"The next step is for you or Jim to execute a push of 0.5 feet per second with four jets from the LEM, have Jack perform the separation, then execute a pull at the same 0.5 feet per second in the opposite direction. Got that?"

"Got that. When do you want us to do this?"

"About thirteen minutes from now. But it's not time critical."

Lovell cut into the line. "Can we do it anytime?"

"That's affirmative. You can jettison whenever you're ready."

With clearance from the ground to proceed, Swigert shot up the tunnel into Odyssey and took his position in front of the jettison switches in the center of his instrument panel. Lovell and Haise went to their windows. Near each of their stations, the three men had already left cameras floating, in the hope of photographing the service module's presumably blast-damaged exterior. Swigert had already taken the precaution of wiping Odyssey's five windows clear of condensation, to provide an unobscured view to the outside.

"Houston, Aquarius," Lovell called. "Jack's in the command module now."

"Real fine, real fine," Kerwin said. "Proceed at any time."

"Jack!" the commander shouted up the tunnel. "You ready?"

"All set when you guys are," the call came back.

"All right, I'll give you a five count, and on zero I'll hit the thrusters. When you feel the motion, let 'er go."

Swigert shouted a "roger," reached over with his left hand and picked up his big Hasselblad, then positioned the index finger of his right hand over the SM JETT switch. His paper "NO" flapped to the left of it. Lovell, in the LEM, took his camera in his left hand and his thruster control in his right. Haise picked up his camera as well.

"Five," Lovell called up the tunnel, "four, three, two, one, zero."

The commander eased his control upward, activating the jets and nudging the two-spacecraft stack into motion. In the command module, Swigert responded immediately, snapping the service module switch.

"Jettison!" he sang out.

All three crewmen heard a dull explosive pop and felt a simultaneous jolt. Lovell then pulled down on the controller, activating an opposite set of nozzles and reversing course.

"Maneuver complete," he called.

At their separate windows, Lovell, Swigert, and Haise leaned anxiously forward, raised their cameras, and flicked their eyes about their patches of sky. Swigert had chosen the big, round hatch window in the center of the spacecraft, but pressing his nose against it now he saw . . . nothing. Jumping to his left, he peered out Lovell's window and there too saw nothing at all. Scrambling across to the other side of the spacecraft, he banged into Haise's porthole, scanned as far as the limited frame would allow him, and there, too, came up empty.

"Nothing, damnit!" he yelled down the tunnel. "Nothing!"

Lovell, at his triangular window, swiveled his head from side to side, also saw nothing, and looked over to Haise, who was searching as frantically as he was and finding just as little. Cursing under his breath, Lovell turned back to his glass and all at once saw it: gliding into the upper left-hand corner of the pane was a mammoth silver mass, moving as silently and smoothly and hugely as a battleship.

He opened his mouth to say something, but nothing came out. The service module moved directly in front of his window, filling it completely; receding ever so slightly it began to roll, displaying one of the riveted panels that made up its curved flank. Drifting away a little more, it rolled a little more, revealing another panel. Then, after another second, Lovell saw something that made his eyes widen. Just as the mammoth silver cylinder caught an especially bright slash of sun, it rolled a few more degrees and revealed the spot where panel four was — or should have been.

In its place was a wound, a raw, gaping wound running from one end of the service module to the other. Panel four, which made up about a sixth of the ship's external skin, was designed to operate like a door, swinging open to provide technicians access to its mechanical entrails, and sealing shut when it came time for launch. Now, it appeared, that entire door was gone, ripped free and blasted away from the ship. Trailing from the gash left behind were sparkling shreds of Mylar insulation, waving tangles of torn wires, tendrils of rubber liner. Inside the wound were the ship's vitals — its fuel cells, its hydrogen tanks, the arterial array of pipes that connected them. And on the second shelf of the compartment, where oxygen tank two was supposed to be, Lovell saw, to his astonishment, a large charred space and absolutely nothing else.

The commander grabbed Haise's arm, shook it, and pointed. Haise followed Lovell's finger, saw what his senior pilot saw, and his eyes, too, went wide. From behind Lovell and Haise, Swigert swam frantically down the tunnel holding his Hasselblad.

"And there's one whole side of that spacecraft missing!" Lovell radioed to Houston.

"Is that right," Kerwin said.

"Right by the — look out there, would you? Right by the high-gain antenna. The whole panel is blown out, almost from the base to the engine."

"Copy that," said Kerwin.

"It looks like it got the engine bell too," Haise said, shaking Lovell's arm and pointing to the big funnel protruding from the back of the module. Lovell saw a long, brown burn mark on the conical exhaust port.

"Think it zinged the bell, huh?" Kerwin asked.

"That's the way it looks. It's really a mess."

"O.K., Jim," Kerwin said. "We'd like you to get some pictures, but we want you to conserve fuel. So don't make any unnecessary maneuvers."

With that transmission, Lovell shook himself and realized that pictures were, after all, part of the purpose of this exercise, and so far his crew had gotten none. Already, the blast-damaged hulk was drifting away. Moving to his left, Lovell grabbed Swigert by the arm and pulled him to the window. The command module pilot immediately began shooting frame after frame through his telephoto lens. In the small patch of window left to him, Lovell set up his camera and began snapping just as frantically. On the right side of the ship, Haise also snapped away. The crew followed the module until it had faded to little more than a tumbling star hundreds of yards from the ship. More than twenty minutes after Swigert had thrown the SM JETT switch, the three crewmates fell away from their windows.

"Man," Haise mumbled to no one in particular, "that's unbelievable."

"Well, James," Kerwin called up, "if you can't take better care of a spacecraft than that, we might not give you another one."

"This is Apollo Control, Houston, at 138 hours, 15 minutes into the mission. Apollo 13 presently 34,350 nautical miles out from Earth, traveling at a speed of 7,212 miles per hour. Meanwhile, in the Mission Control viewing room the crowd is beginning to increase. Already here are Dr. Thomas Paine, NASA administrator; Mr. George Low, a NASA deputy administrator; Representative George Miller from California and chairman of the House space committee; Representative Olin Teague of Texas; and Representative Jerry Pettis of California. Dave Scott and Rusty Schweickart of Apollo 9 are among the astronauts in the viewing room, as well as Lew Evans, the president of Grumman. Needless to say, all of the distinguished visitors in the control center were most interested in the report from Apollo 13 of the service module condition as the crew moved away following the jettison. This is Apollo Control, Houston."

The group around the EECOM station was a big one when the time came to power up Odyssey. John Aaron, of course, had been there since 4 A.M., when the Tiger Team came out of room 210 and re-claimed their consoles. But as the morning wore on and 10 A.M. approached — the time when the spacecraft would be less than three hours from splashdown — the number of men at the EECOM console in the second row grew. Sy Liebergot showed up first, pulled up a chair, and sat down at Aaron's left. Clint Burton, the Black Team EECOM, joined them, forgoing a chair and standing behind Aaron. Charlie Dumis of the Maroon Team arrived and stood behind Lieber-got. At most of the other consoles, at least one other member of another team hovered next to the White Team controller on duty, but only at the EECOM console was the full complement of engineers present.

"Flight, EECOM," Aaron called into the loop, glancing over his shoulder at the troika of controllers around him.

"Go, EECOM," Kranz answered.

"Ready for power-up anytime the crew is."

"Roger, EECOM," Kranz said. "Capcom, Flight."

"Go, Flight," Kerwin answered.

"EECOM says the command module can come on line anytime."

"Roger, Flight," Kerwin said. "Aquarius, Houston."

"Go, Houston," said Lovell.

"You're go to start powering up Odyssey."

In the cockpit of Aquarius, Lovell looked at Swigert and motioned him to the tunnel. Unlike the reading of the power-up checklist four-teen hours earlier, the execution of the list would be a simple matter, requiring less than half an hour of work by the command module pilot.

As the first switch was thrown, sending a surge of power through the long-cold wires, Lovell braced for the sickening pop and sizzle indicating that the condensation soaking the instrument panel had indeed found an unprotected switch or junction and shorted the ship right back out. It was a sound he had first heard over the Sea of Japan and one he dearly hoped he would never hear again. But as the power-up in this cockpit proceeded, as Swigert threw his first breaker, and his second, and his third, and so on, all the crewmen heard was the reassuring hum and gurgle indicating that their spacecraft was coming back to life.

To the extent that there was any other drama in the exercise, it would take place not in the spacecraft but at John Aaron's station.

The way Aaron had ciphered things out, the ship could afford to pull no more than 43 amps of juice if it hoped to stay alive for the full two hours of reentry. But, having won the argument in room 210 over when to turn the telemetry on, he wouldn't know if he was actually staying within this power budget until the command module was completely powered up and the data started streaming back from the ship. If it turned out that Odyssey was consuming juice above the 43-amp level, even for a short while, there was a real chance its batteries would be exhausted before it ever hit the ocean.

When Lovell sent Swigert up into Odyssey, Aaron, Liebergot, Dumis, and Burton leaned over their EECOM console expectantly. For twenty minutes, almost no communication came from the ship. Finally Lovell passed the news to the ground that the last of the switches had been thrown, including the telemetry switches. Slowly, the screen at the EECOM station began to strobe to life. When the amp readout materialized, the four EECOMs recoiled as if burned. The number that appeared was 45.

"Shit," Aaron spat out. "What the hell are those 2 amps doing there!"

"I have no idea," Liebergot said.

"I'll be damned if I know," Burton echoed.

"Well, they sure as hell don't belong there. We're blowing half our margin!" Aaron now signaled to his backroom. "Electronics, EECOM."

"Go, EECOM," the voice came back.

"We're pulling 2 amps we shouldn't be."

"I see 'em, EECOM."

"Run through the checklist and see what we left on."

"Roger."

Aaron clicked off the air and leaned to his right, toward the guidance and navigation console. "You got anything on over there that shouldn't be on?"

"Not that I can see, John."

"Well, scan. We've got 2 rogue amps."

As Aaron was talking to his GNC, Liebergot, Dumis, and Burton fanned out through the first three rows, to see if any of the other controllers might have left any instrument on that would be pulling amps it shouldn't. Even before those men could respond, however, Aaron's backroom came back on the loop.

"EECOM," the controller said.

"Go ahead."

"Got it. It's the B-MAGs, the backup gyros. Tell the GNC to have the crew shut 'em off."

Aaron instantly leaned back to the GNC at his left. "Check your B-MAGs. Are they on?"

The guidance and navigation officer looked at his screen and slumped. "Aw, hell," he groaned.

"Flight, EECOM," Aaron quickly called. "Tell Capcom to tell the crew to shut off their backup gyros."

Joe Kerwin relayed Aaron's message up to Odyssey, Swigert threw the appropriate switch, and the amp readout on the EECOM screen dropped back down to 43. But, as Aaron had said, a few of Odyssey's precious amps were gone for good.

With the power-up finally, if imperfectly, completed, the lunar module Aquarius became an expendable ship. At 140 hours and 52 minutes into the mission — less than two hours before splashdown — Apollo 13 was 16,000 miles above the earthly cloud tops and closing at better than 10,000 miles per hour. No longer a discrete and distant disk surrounded on all sides by stars and space, Earth was now a vast blue mass looming hugely ahead, spilling out of all three sides of the LEM's triangular window frames.

Staring through his porthole at the panorama outside, Lovell said, "Freddo, it's about time we bailed out of this ship."

From behind him, Haise said nothing.

"Freddo?"

Lovell turned to face his crewmate and was brought up short by what he saw. Braced against the bulkhead, Haise was a paler shade of gray than Lovell had seen him the entire trip. With his eyes closed and his arms folded tightly against his chest for warmth, he had begun to shake violently with chills.

"Fred!" Lovell said, allowing more alarm than he intended to creep into his voice. "You look awful."

"Forget it," Haise said with an unconvincing wave. "Forget it. I'm fine."

"Yeah," Lovell said, drifting over to him. "You look just terrific. Can you hold out two more hours?"

"I can hold out as long as I have to."

"Two hours, that's all you have to hang on for. After that, we're floating in the South Pacific, we open the hatch, and it's 80 degrees outside."

"Eighty degrees," Haise repeated a little dreamily, and began to shiver again.

"Man," Lovell muttered, "you are a mess." Moving up behind Haise, the commander wrapped him in a bear hug, to share his body heat. At first the gesture seemed to accomplish nothing, but gradually the trembling subsided.

"Fred, why don't you get upstairs and help Jack out," Lovell said. "I'll finish up here,"

Haise nodded and prepared to jump up the tunnel. But before he did, he stopped and took a long look around Aquarius's cockpit. Impulsively, he pushed back toward his station. Attached to the wall was a large screen of fabric netting used to prevent small items from floating behind the instrument panel. Haise grabbed hold of the netting and gave a sharp pull; it tore free with a ripping sound.

"Souvenir," he said with a shrug, wadding the netting into a ball, stuffing it into his pocket, and vanishing up the tunnel.

Alone in the lunar module, Lovell too glanced slowly around it. The debris of four days of close-quarters living was collected in the cluttered cockpit, and Aquarius now looked less the intrepid moonship it had been on Monday than a sort of galactic garbage scow. Lovell waded through the scraps of paper and rubbish and moved back toward his window. Before jumping ship himself, he had one more job: steering the twin vehicles to the attitude Jerry Bostick had specified, so the LEM would drop into the deep water off New Zealand.

Lovell took the attitude control for the last time and pushed it to the side. The ship yawed slightly, jostling some of the floating paper. Without the inert mass of the service module skewing the center of gravity so badly, Aquarius was far more maneuverable, much closer to the nimble ship the simulators in Houston and Florida had conditioned Lovell to expect before this mission began. With a few practiced adjustments, he moved the lander to the proper position, then called the ground.

"O.K. Houston, Aquarius. I'm at the LEM separation attitude and I'm planning on bailing out."

"I can't think of a better idea, Jim," Kerwin replied.

Lovell finished configuring the LEM's switches and systems and then, like Haise, decided that a souvenir might be in order. Reaching to the top of his window, he grabbed the optical sight and gave it a twist. It unscrewed easily and Lovell pocketed it. Looking toward the stowage area at the back of the cockpit, he found the helmet he would have worn on the surface of the moon, picked it up, and tucked it under his arm. Finally, he turned to another cabinet and retrieved the plaque he and Haise would have clamped to the LEM's front leg once they had emerged from the lander and begun to explore. None of the workers in NASA's metal shop who had manufactured the plaque had ever expected to see it again. Now, Lovell reflected, they could stop by his office or den and take a look whenever they chose.

Holding his collected booty, Lovell sprang up the tunnel into Odyssey's lower equipment bay, stashed his souvenirs in a storage cabinet, and moved in the direction of the couches. Instinctively, he moved toward the left-hand station, but when he shimmied out of the equipment bay, he discovered that while Haise was buckled into his familiar right-hand seat, Swigert had claimed Lovell's left-hand spot. It was customary during the descent and reentry phase of a lunar mission for a commander to relinquish his seat to his command module pilot. During a flight in which so many of the critical moments belonged to the commander and the LEM pilot, the man in the center couch was oftentimes overlooked. Reentry, however, when the LEM that had taken his shipmates to the surface of the moon was nothing but a jettisoned memory, was essentially a command module pilot's operation, and as a gesture of respect for both his competence as a flier and the thankless job he had performed so far, he was usually allowed to bring the ship in for its landing. Now, as reentry approached, and the commander of this mission approached his familiar station, he had to switch course and move back to a less familiar one.

"Reporting aboard, skipper," Lovell said to Swigert.

"Aye-aye," Swigert answered, a bit self-consciously.

Lovell donned his headset and nodded, then Swigert signed on the air.

"O.K., Houston, we're ready to proceed with hatch close-up."

"O.K., Jack. Did Jim get all of the film out of Aquarius?"

Lovell looked at Swigert and nodded yes.

"Yes," Swigert said. "That's affirmative. And we remembered to get Jim out too."

"Good deal, Jack," Kerwin said. "Then what we want you to do is seal the hatch and vent the tunnel until you get down to about 3 pounds per square inch. If the hatch holds pressure for a minute or so, you're O.K. and you can feel free to release Aquarius."

"O.K.," Swigert said. "Copy that."

Lovell, indicating to Swigert that he should stay where he was, wriggled back out of his couch and glided toward the lower equipment bay. Swimming into the tunnel, he slammed the LEM's hatch and sealed it with a turn of its lever. Then he backed into Odyssey, retrieved its hatch from the spot where he had tied it down on that Monday night so long ago, and fitted it into place.

If this hatch evidenced the same balkiness it had four days ago, the LEM could not be jettisoned and the reentry could not proceed as planned. Even if the hatch did seal, it would be a few minutes before the onboard pressure sensors would confirm that the seal was tight and the spacecraft wasn't leaking air. Naturally, without this confirmation, a safe reentry would be impossible. Lovell regarded the hatch suspiciously and then threw its locking mechanism. The latches closed with a satisfying snap. Reaching for the tunnel vent switch, he bled the air out of the passageway and into space until the pressure read 2.8 pounds per square inch. Flipping the vent switch shut, he swam back to his seat.

"Sealed?" Swigert asked.

"I hope so," Lovell said.

With this tepid reassurance, the command module pilot flipped several switches on his instrument panel and brought the oxygen system to life, feeding fresh O_2 into the cockpit. For several taut seconds he stared at his flow indicator.

"Oh, no," Swigert groaned.

"What's wrong?" Lovell and Haise asked, practically in unison.

"Flow is high. It looks like we've got a leak."

On the ground, John Aaron hunched over his EECOM screen and spotted the oxygen rate at the same time Swigert did.

"Oh, no," he groaned.

"What's wrong?" Liebergot, Burton, and Dumis asked, practically in unison.

"Flow is high. It looks like we've got a leak."

On the air-to-ground loop, Swigert's voice called out, "O.K. Houston, we've got an O_2 flow high."

"Roger, Jack," Kerwin answered. "Let us check it."

As Swigert kept his eyes on his instruments, Aaron hailed his backroom. He and his engineers muttered on the line about the source of the potential leak while the three other EECOMs in the second row fretted aloud among themselves.

Within minutes, Aaron believed he had the problem sorted out. The

LEM operated at a slightly lower pressure than the command module. Over the past four days, with the hatches opened up and Odyssey shut off, it was Aquarius that determined the pressure in both ships. When the command module was powered up and its door was closed, its pressure sensors spotted that difference and immediately tried to pump the internal atmosphere up to what they thought it should be. In a few moments, Aaron figured, the necessary air should have been added to the cockpit and the high flow rate would stop.

"Sit tight for another minute," he said to the people around him. "I think we'll be all right."

Forty seconds later, the numbers in the spacecraft and on the EECOM's screen indeed began to stabilize.

"O.K.," Swigert said with audible relief, "it's dropping now, Joe."

"Roger," Kerwin called. "In that case, when you are comfortably ready to release the LEM, you can go ahead and do it."

Lovell and Swigert looked at the mission timer on their instrument panel. It was 141 hours and 26 minutes into the flight.

"Do it in four minutes?" Swigert asked.

"Seems like a nice round figure," Lovell answered.

"O.K., Houston," Swigert announced. "We'll punch off at 141 plus 30."

Outside the cockpit's five windows, the astronauts could see nothing of Aquarius but its reflective silver roof plates, just a few feet away from the glass of their portholes. Three and a half minutes elapsed.

"Thirty seconds to LEM jettison," Swigert said.

"Ten seconds."

"Five."

Swigert reached up to the instrument panel, ripped away his "NO" note, and balled it up in his palm.

"Four, three, two, one, zero."

The command module pilot flipped the toggle switch and all three crewmen heard a dull, almost comical pop. In their windows, the silver roof of the lunar lander began to recede. As it did, its docking tunnel became visible, then its high-gain antenna, then the array of other antennas that bristled from its top like metal weeds. Slowly, the unbound Aquarius began a graceful forward somersault.

Lovell stared as the face of the ship — its windows, its attitude-control quads — rolled into view. He could see the forward hatch from which he and Haise would have emerged after settling down in the dust of Fra Mauro. He could see the ledge on which he would have stood while opening his equipment bay before climbing down to the lunar surface. He could see the reflective, almost taunting, nine-rung ladder he would have used to make that final descent. The LEM rolled some more and was now upside down, its four splayed legs pointing up to the stars, the crinkly gold skin of its descent stage shining back at Odyssey.

"Houston, LEM jettison complete," Swigert announced.

"O.K., copy that," Kerwin said softly. "Farewell, Aquarius, and we thank you."

With the loss of the lander, Apollo 13 was at last reduced to its irreducible essence. Shorn of the 36-story Saturn 5 booster that had lifted it off the pad, the 59-foot third-stage booster that blasted it toward the moon, the 25-foot service module that was to have provided it with air and power, and finally the 23-foot LEM that was to have carried Lovell and Haise into history, the spacecraft was now nothing more than an 11-foot-tall, wingless pod, heading inexorably toward a free fall through the fast-approaching atmosphere and a collision with the fast-growing ocean. Before that could happen, however, there was still one more job for the crew to perform.

"How do we stand on that moonset check?" Haise asked Lovell from the right side of the spacecraft.

"You ready for it?" Lovell asked Swigert from the center.

"As soon as we hit nighttime," Swigert answered.

The earthly nighttime Swigert spoke of was still some minutes away, but though the planet below was brightly illuminated, Lovell, Swigert, and Haise could see none of it. Like Apollo 8's approach to the moon sixteen months earlier, Apollo 13's approach to the Earth was rump-forward. In order for the spacecraft to survive, it would have to enter the atmosphere heat-shield-first, its ablative bottom absorbing all of the punishment of the sizzling plunge through the air. As the final hour of the mission ticked down, the astronauts thus found themselves backing blindly toward their planet in an approach that their instruments, but not their eyes, told them was bringing them closer and closer to the ocean below.

For several long minutes the spacecraft continued this way, until gradually it began to arc around the globe, passing over the twilight of western Africa and western Europe, and then into the darkness of the Middle East. When the ship had dropped low enough and flown backward far enough, the lightless landmass began to spread out in front of it. Through their windows, the astronauts could at last see the great curved shadow that they knew was both destination and home. Hanging over it, tiny as a tablet, was the bright white gibbous moon.

"Houston," Swigert called, "proceeding with moonset check."

The command module pilot glanced at his 8 ball to confirm Odyssey's attitude and then shifted his gaze out the window, watching the moon slowly descend toward the dark horizon. As the spacecraft fell farther and farther and the horizon climbed higher and higher, the moon began to drop lower and lower.

"It's coming down, Joe," Swigert called to Kerwin. "We're down to about 45 degrees, and it's coming on down."

"Roger that."

"Down to about 38 degrees now."

"O.K., Jack. Sounds real good."

In the center and right-hand seats, Lovell and Haise watched the timer on the instrument panel as Swigert kept looking out the window. The moon dropped from 38 degrees to 35 to the 20s to the high teens. The seconds until the predicted time of moonset that Jerry Bostick had calculated melted away, until there were just fifteen seconds to go.

"Got anything, Jack?" Lovell asked.

"Nothing yet."

"Now?"

"Negative."

"Now? Just three seconds left."

"Not yet," Swigert answered. Then, at precisely the instant the FIDO in Houston had predicted, the moon dropped a fraction of a degree more and a tiny black nick appeared in its lower edge. Swigert turned to Lovell with a giant grin.

"Moonset," he said, and clicked on the air. "Houston, attitude checked out O.K."

"Good deal," said Joe Kerwin.

From the center seat, Jim Lovell turned to look at the men on either side of him and smiled. "Gentlemen," he said, "we're about to reenter. I suggest you get ready for a ride."

Unconsciously, the commander touched his shoulder belts and lap belts, tightening them slightly. Unconsciously, Swigert and Haise copied him.

"Joe, how far out do you show us now?" Swigert asked his Capcom.

"You're moving at 25,000 miles per hour, and on our plot map board, the ship is so close to Earth we can't hardly tell you're out there at all."

"I know all of us here want to thank all you guys for the very fine job you did," Swigert said.

"That's affirm, Joe," Lovell agreed.

"I'll tell you," Kerwin said, "we all had a good time doing it."

In the spacecraft, the crew fell silent, and on the ground in Houston, a similar stillness fell over the control room. In four minutes, the leading edge of the command module would bite into the upper layer of the atmosphere, and as the accelerating ship encountered the thickening air, friction would begin to build, generating temperatures of 5,000 degrees or more across the face of the heat shield. If the energy generated by this infernal descent were converted to electricity, it would equal 86,000 kilowatt-hours, enough to light up Los Angeles for a minute and a half. If it were converted to kinetic energy, it could lift every man, woman, and child in the United States ten inches off the ground. Aboard the spacecraft, however, the heat would have just one effect: as temperatures rose, a dense ionization cloud would surround the ship, reducing communications to a hash of static lasting about four minutes. If radio contact was restored at the end of this time, the controllers on the ground would know that the heat shield was intact and the spacecraft had survived; if it wasn't, they would know that the crew had been consumed by the flames. At the flight director's station, Gene Kranz stood, lit a cigarette, and clicked on to his controllers' loop.

"Let's go around the horn once more before reentry," he announced. "EECOM, you go?"

"Go, Flight," Aaron answered.

"RETRO?"

"Go."

"Guidance?"

"Go."

"GNC?"

"Go, Flight."

"Capcom?"

"Go."

"INCO?"

"Go."

"FAO?"

"We're go, Flight."

"Capcom, you can tell the crew they're go for entry."

"Roger, Flight," Kerwin said. "Odyssey, Houston. We just had one last time around the room, and everyone says you're looking great. We'll have loss of signal in about a minute. Welcome home."

"Thank you," Swigert said.

In the sixty seconds that followed, Jack Swigert fixed his eyes out the left-hand window of the spacecraft, Fred Haise fixed his out the right, and Jim Lovell peered through the center. Outside, a faint, faint shimmer of pink became visible, and as it did, Lovell could feel an equally faint ghost of gravity beginning to appear. The pink outside gave way to an orange, and the suggestion of gravity gave way to a full g. Slowly the orange turned to red — a red filled with tiny, fiery flakes from the heat shield — and the g forces climbed to two, three, five, and peaked briefly at a suffocating six. In Lovell's headset, there was only static.

In Mission Control, the same steady electronic hiss also streamed into the ears of the men at the consoles. When it did, all conversation on the flight controllers' loop, the backroom loops, and in the auditorium itself stopped. At the front of the room, the digital mission clock read 142 hours, 38 minutes. When it reached 142 hours, 42 minutes, Joe Kerwin would hail the ship. As the first two minutes went by, there was almost no motion in either the main room or the viewing gallery. As the third minute elapsed, several of the controllers shifted uneasily in their seats. When the fourth minute ticked away, a number of men in the control room craned their necks, casting glances toward Kranz.

"All right, Capcom," the flight director said, grinding out the cigarette he had lit four minutes ago. "Advise the crew we're standing by."

"Odyssey, Houston standing by, over," Kerwin called.

Nothing but static came back from the spacecraft. Fifteen seconds elapsed.

"Try again," Kranz instructed.

"Odyssey, Houston standing by, over." Fifteen more seconds.

"Odyssey, Houston standing by, over." Thirty more seconds.

The men at the consoles stared fixedly at their screens. The guests in the VIP gallery looked at one another. Three more seconds ticked slowly by with nothing but noise on the communications loop, and then, in the controllers' headsets, there was a change in the frequency of the static from the ship. Nothing more than a flutter, really, but a definitely noticeable one. Immediately afterward, an unmistakable voice appeared.

"O.K., Joe," Jack Swigert called.

Joe Kerwin closed his eyes and drew a long breath, Gene Kranz pumped a fist in the air, the people in the VIP gallery embraced and applauded.

"O.K.," Kerwin answered without ceremony, "we read you, Jack."

Up in the no longer incommunicado spacecraft, the astronauts were enjoying a smooth ride. As the ion storm surrounding their ship subsided, the steadily thickening layers of atmosphere had slowed their 25,000-mile-per-hour plunge to a comparatively gentle 300-mile-per-hour free fall. Outside the windows, the angry red had given way to a paler orange, then a pastel pink, and finally a familiar blue. During the long minutes of the blackout, the ship had crossed beyond the nighttime side of the Earth and back into the day. Lovell looked at his g meter: it read 1.0. He looked at his altimeter: it read 35,000 feet.

"Stand by for drogue chutes," Lovell said to his crewmates, "and let's hope our pyros are good." The altimeter ticked from 28,000 feet to 26,000. At the stroke of 24,000, the astronauts heard a pop. Looking through their windows, they saw two bright streams of fabric. Then the streams billowed open.

"We got two good drogues," Swigert shouted to the ground.

"Roger that," Kerwin said.

Lovell's instrument panel could no longer measure the snail-like speed of his ship or its all but insignificant altitude, but the commander knew, from the flight plan profile, that at the moment he should be barely 20,000 feet above the water and falling at just 175 miles per hour. Less than a minute later, the two drogues jettisoned themselves and three others appeared, followed by the three main chutes. These tents of fabric streamed for an instant and then, with a jolt that rocked the astronauts in their couches, flew open. Lovell instinctively looked at his dashboard, but the velocity indicator registered nothing. He knew, however, that he was now moving at just over 20 miles per hour.

On the deck of the USS Iwo Jima, Mel Richmond squinted into the blue-white sky and saw nothing *but* blue and white. The man to his left scanned silently too, and then muttered a soft imprecation, suggesting that he saw nothing either; the man to his right did the same. The sailors arrayed on the decks and catwalks behind them looked in all directions.

Suddenly, from over Richmond's shoulder, someone shouted, "There it is!"

Richmond turned. A tiny black pod suspended under three mammoth clouds of fabric was dropping toward the water just a few hundred yards away. He whooped. The men on either side of him did the same, as did the sailors on the rails and decks. Nearby, the network cameramen followed where the spectators were looking, and trained their lenses in the same direction. Back in Mission Control, the giant main viewing screen in the front of the room flashed on, and a picture of the descending spacecraft appeared. The men in that room cheered as well.

"Odyssey, Houston. We show you on the mains," Joe Kerwin shouted, covering his free ear with his hand. "It really looks great." Kerwin listened for a response but could hear nothing above the noise around him. He repeated the essence of the message: "Got you on television, babe!"

Inside the spacecraft that the men in Mission Control and the men on the Iwo Jima were applauding, Jack Swigert radioed back a "roger," but his attention was focused not on the man in his headset but on the man to his right. In the center seat, Jim Lovell, the only person in the falling pod who had been through this experience before, took a final look at his altimeter and then, unconsciously, took hold of the edges of his couch. Swigert and Haise unconsciously copied him.

"Hang on," the commander said. "If this is anything like Apollo 8, it could be rough."

Thirty seconds later, the astronauts felt a sudden but surprisingly painless deceleration, as their ship — behaving nothing like Apollo 8 — sliced smoothly into the water. Instantly, the crewmates looked up toward their portholes. There was water running down the outside of all five panes.

"Fellows," Lovell said, "we're home."

Marilyn Lovell laughed out loud as Jeffrey cried out once and started to squirm. Through a film of tears and a swarm of people, she watched the television screen in her family room as Odyssey struck water and the three parachutes that had carried it down flattened themselves on the surface of the ocean. Throughout the slow descent, she had been holding her son in her lap, and as the spacecraft fell, Marilyn had unknowingly clutched him tighter and tighter. At the moment of splashdown, Jeffrey shouted in protest.

"I'm sorry," Marilyn said, laughing and crying and kissing him on top of his head. "I'm sorry." She hugged him once more and set him down on the floor. When she did, Betty Benware appeared from nowhere and embraced her tightly. Then Adeline Hammack materialized, then Susan Borman. At the edge of the room, Pete Conrad opened the first bottle of champagne, followed by Buzz Aldrin and Neil Armstrong, followed by who knew who else. Marilyn got up, found her other children, and, ducking the spray from the popping bottles, hugged them as well. Somebody put a drink in her hand. She took a long, fizzy swallow, and more tears — this time from the bubbles — came to her eyes. In the distance, Marilyn faintly heard the phone ring in the master bedroom. It rang a second time and Betty disappeared to answer it. A moment later she reappeared.

"Marilyn, it's the White House again."

Marilyn handed her drink to someone nearby, ran to the bedroom, and picked up the dangling receiver.

"Mrs. Lovell?" a woman's voice said. "Hold for the president."

Several seconds went by, and then Marilyn once again heard the deep, familiar voice.

"Marilyn, this is the president. I wanted to know if you'd care to accompany me to Hawaii to pick up your husband."

Marilyn Lovell paused absently and smiled into the middle distance, picturing the spacecraft she had just seen bobbing on the waters of the South Pacific. The line from Washington crackled slightly.

"Mr. President," she said at last, "I'd love to."

EPILOGUE

Christmas 1993

IF JIM LOVELL had turned just a second later, his granddaughter might have cracked Odyssey's heat shield. Actually, it wasn't Odyssey's *entire* heat shield that ten-month-old Allie Lovell would have damaged as she pulled herself up to look at the top of the credenza in her grandfather's den, just a cork-sized plug of it sealed in a Plexiglas paperweight.

Lovell rather liked this modest trophy, and in the months following the splashdown of Apollo 13, when NASA had had a dozen or so of the mementos made up, he had hoped to snag one for himself. The little relics were not really intended for any of the crewmen, but for the heads of state the three astronauts would meet during their hastily arranged five-nation tour following their return from space. When their overseas trip was complete, however, and one of the keepsakes was left over, the man who had commanded the ship from which the scorched sample was taken had packed up the remaining souvenir and brought it home.

"Whoa!" Lovell now said as Allie swept an exploratory hand over the credenza, threatening to knock the twenty-three-year-old artifact to the floor. "You don't want to touch that."

Crossing the room in two quick steps, Lovell scooped up his granddaughter, hoisted her to his shoulder like a flour sack, and kissed her on the forehead. "Maybe we'd better go find your dad," he said.

The day was just getting under way, and Lovell had the feeling it would be a frantic one, filled with such near calamities. It wasn't only his youngest son, Jeffrey, who would be here, offspring in tow, for Christmas dinner, but his other children as well. In all, the second-generation Lovells would be bringing seven third-generation Lovells with them — ranging in age from ten months to sixteen years — and a lot more mementos displayed in the paneled room might be placed in jeopardy.

There were the rows of plaques, the wall of proclamations, the framed letters of congratulation from presidents and vice presidents, governors and senators, all of which had flowed in following the successful flights of Gemini 7, Gemini 12, and Apollo 8. There were the small cloth flags and uniform patches Lovell had worn during those missions, now in their own sealed frames. There was the Emmy given — utterly in earnest — to Lovell, Frank Borman, and Bill Anders following their lunar-orbit TV broadcast twenty-five Christmases ago. There were the other trophies and medals that flanked the Emmy — the Collier Trophy, the Harmon Trophy, the Hubbard Medal, the deLavaulx Medal — all presented as tips of the cap to the first three trips Lovell had made into space. Most cherished were the relics of the vehicles in which those missions had been flown: the systems books, flight plans, pens, utensils, even toothbrushes, all of which had once floated easily in the zero-g, 5-pounds-per-square-inch atmosphere of a spacecraft, and all of which now sat stationary on a shelf, pulled from below by one g and squashed from above by 15 pounds per square inch of sea-level air.

Almost entirely missing from this quiet room, this harbor of memories, were the mementos from Lovell's fourth and last flight, his one unsuccessful flight. There are no Harmon Trophies for missions that did not achieve their primary objectives; no Collier Awards for spaceships that blew up before they got where they were going. Besides the heat shield plug, all that was here to commemorate the flight of Apollo 13 was a framed letter of congratulation from Charles Lindbergh and, on a nearby windowsill, two of the last artifacts collected from the long-ago incinerated lunar module Aquarius: the crew's optical sight and the commemorative plaque intended for its forward leg.

Leaving these mementos, Lovell carried Allie to the kitchen of his comfortable home in Horseshoe Bay, Texas, where he found Marilyn talking to Jeffrey and his wife, Annie.

"I believe this is yours," Lovell said to Jeffrey as he handed his granddaughter over.

"Is she getting into things?" Jeffrey asked.

"Just starting to."

"Well, brace yourself," Marilyn said. "There are six more on the way."

Lovell smiled at the warning, but he did not especially need it. In the sixteen years he and Marilyn had lived with four children of their own in a small house in Timber Cove, they had gotten used to tumultuous holidays. Of course, the Timber Cove years had long since passed, and were now, like so much else from the Apollo days, becoming an increasingly distant memory.

It had been in the mid-1970s that the families that had settled in the suburbs surrounding the Manned Spacecraft Center had begun to pack up, decamp, and scatter. The emigration had been slow at first — Neil Armstrong announced he was going back to Ohio to become a college professor and an industry consultant, Michael Collins left for Washington to work for the State Department, Frank Borman accepted a position at Eastern Airlines — but it had also been inevitable. When Apollo 11 landed on the moon in 1969, heady NASA planners had envisioned sending at least nine more LEMs to nine more spots on the lunar surface through the early 1970s. By the eighties, so the rosiest scenarios went, the first stakes of the first permanent lunar base would be driven into the moon, built on one of the sites that the previous crews had scouted out.

That, of course, was not to be. By the time Apollo 13 flew, Apollo 20 had already been canceled, the victim of a newly parsimonious administration and a public that had begun to question why the country was going back to the moon when it had already proven it could go there once. After Apollo 13, when three astronauts had nearly died in this exercise in cosmic redundancy, Apollos 19 and 18 were quickly scratched too. Washington did concede that Apollos 14 through 17, which were practically bought and paid for already, could proceed as planned, and over the next two and a half years, those final four missions — and the lucky twelve astronauts assigned to them — were sent off to the moon.

In December of 1972, when the last lunar crew splashed down in the Pacific Ocean, a few members of the test pilot community that had grown up around the Apollo program did choose to stay around it. Fred Haise, who through circumstance, poor luck, and a bum service module had been denied the chance to set foot on the moon, was tentatively promised the command of Apollo 19. When that mission was scrubbed, the one-time LEM pilot lent a hand in a few early glide tests of the prototype space shuttle before giving up and going to work for Grumman in the late 1970s. Ken Mattingly, who through circumstance, good luck, and the absence of rubella antibodies had been denied a seat on the disastrous Apollo 13, eventually flew aboard the successful Apollo 16, and also volunteered his piloting skills for the upcoming space shuttle program. Deke Slayton, who had been promised a flight into space in 1959 and then saw that promise broken in 1961 when he was diagnosed with a heart fibrillation, stayed stubbornly at the edges of the astronaut corps until 1975, when he was finally chosen to fly aboard a dusted-off Apollo ship during a politically priceless, if scientifically useless, rendezvous with a Soviet Soyuz spacecraft in Earth orbit.

"I want to warn you," Chris Kraft had said in a call to his NASA superior, George Low, when he drew up his crew manifest for the mission, "that I'm about to recommend Deke for this flight. If you've got any problems with that, you'd better tell me, because that's what I'm going to do."

"Why Deke, Chris?" Low asked wearily, having had this discussion with Kraft before. "You don't have anyone else who can fly this mission?"

"Why?" Kraft said. "Because we've screwed that guy long enough, George. That's why. And that's plenty reason."

Later that summer, Slayton — along with Tom Stafford and Vance Brand — climbed into the cockpit of NASA's final Apollo spacecraft and at last took the ride atop a rocket he'd been waiting more than a decade and a half to take.

With the exception of these pilots and a few others, most of the men who had enlisted in NASA in the earliest days of the lunar program retired when the Agency turned its attention elsewhere. Jim Lovell left the astronaut corps in 1973, working first for a marine company and later in telecommunications. Harrison Schmitt, the Apollo 17 LEM pilot, returned to New Mexico and ran successfully for the U.S. Senate. Even Jack Swigert, who had distinguished himself so well in so impossible a space flight and could no doubt have charted whatever career course he chose at the Agency, decided not to press his celestial luck any further and returned to Colorado, where he too entered politics.

Like Schmitt, Swigert ran first for the Senate, but unlike Schmitt, he lost. In 1982, the retired astronaut declared his candidacy again, this time running for the House of Representatives, and this time winning. A month before his election in November, however, Swigert was diagnosed with an especially aggressive case of lymphoma. Three days before his planned inauguration in January, he was dead. Poor Jack, Lovell often thought, things always started out so bright for him — and then they always turned so dark.

Of course, in the spring of 1970, when Swigert, Lovell, and Haise returned safely from the moon, the luck of all three men seemed good indeed. It was at 12:07 P.M. Houston time on April 17 that the command module Odyssey hit the Pacific, and the national sigh of relief that followed the news of the splashdown was the loudest and longest since John Glenn's return from America's first manned orbital mission eight years earlier. "Astronauts Land Gently on Target, Unharmed by Their Four-Day Ordeal," the *New York Times* exclaimed. "Applause, Cigars, and Champagne Toasts Greet Capsule's Landing."

Moments after the spacecraft sliced into the water, Lovell, Swigert, and Haise were helped into a life raft — the LEM pilot first, then the command module pilot, then the commander — and hoisted into a hovering helicopter. Landing on the deck of the Iwo Jima, they stepped from the chopper, acknowledged the cheers of the sailors with bleary smiles and weak waves, and were whisked below. The men underwent post-flight physicals, which, while revealing them to be less than spectacularly healthy, turned up no surprises. In addition to Haise's infection and fever, all three suffered from dehydration, all three displayed the fuzzy headedness and disorientation characteristic of fatigue, and all three had lost a considerable amount of weight. Lovell, who weighed 170 pounds before the mission, had lost the most, shedding 14 pounds in six days.

Following the physicals, Lovell and Swigert were installed in visitors' quarters, and Haise was placed in the infirmary. That evening, the two ambulatory astronauts joined the Iwo Jima's officers for dinner, a shrimp and salad, prime rib and lobster, alcohol-free champagne affair that, according to the hastily mimeographed menu, would also include such dessert treats as "Moonfruit Melba" and "Apollo Cookies." Overall, the fare, while perhaps unmemorable by the standards applied in the civilian world, was downright ambrosial to two men who had spent the better part of a week sipping cold rations out of plastic pouches.

The next day, all three astronauts, now dressed in freshly laundered, blue flight suits with the Apollo 13 patch stitched to the left side of the chest, were flown by helicopter to American Samoa, where they boarded a C-141 transport for the short flight to Hawaii. Waiting for them there, they had been told, would be Air Force One.

True to his word, President Nixon had flown to Houston earlier that day, picked up Marilyn Lovell, Mary Haise, and Dr. and Mrs. Leonard Swigert, Jack's parents, and brought them out to Honolulu to welcome the returning space crew. According to protocol in such ceremonial greetings, the presidential entourage should land on the tarmac first, so the chief executive could welcome the honorees personally. When the C-141 approached Hawaii, however, Air Force One was nowhere in sight, and the men who had intended to spend part of last week orbiting the moon spent part of Sunday orbiting Honolulu, waiting for the president to fly into view. Only when Nixon's plane was on the runway and the members of his party were positioned on the tarmac did the C-141 land. When it did, Nixon unexpectedly put protocol aside.

"Why don't you go over there first?" he said to the families. "This should be a private hello." Marilyn Lovell, Mary Haise, and the Swigerts ran across the runway to the somewhat dazed crew.

Nixon's concession to sentiment notwithstanding, there was little else about that day or the next that was even remotely private. For the forty-eight hours the Apollo 13 crewmen were in the South Pacific, the media shadowed them, beaming coverage of the welcome-home rites around the world. The stories and pictures were uniformly positive, indeed almost fawning. It was only when the astronauts arrived back in Houston that the press coverage began to get a bit barbed. At 6:30 on Monday evening, one week to the night after the accident, NASA scheduled a press conference in which the astronauts would face the media for the first time since before liftoff. Soon after the Public Affairs officer's opening statement, a reporter asked the question that Lovell — and NASA — had been hoping no one would.

"Captain Lovell," the reporter called out from the crowd, "what did you have in mind during the mission when you made the remark, 'I think this is going to be the last moon flight for a long time'?"

Lovell stalled for a few moments. On the flight from Hawaii, he had tried composing an answer to this all but inevitable question — and it was an answer that required some advance composing. The direct response was that he meant just what he said. Hurtling toward the shadowed side of the moon in a spacecraft with little air, little power, and little chance of returning you safely to Earth does not inspire confidence in the prospects for the next men who would fly out there, and when Lovell questioned whether anyone else would ever try, his doubts were deep and sincere. But while that might be the answer you'd confide to your friends or your wife or your crew, it was not the answer you'd give to a roomful of reporters. That kind of answer took a good deal more thought, and Lovell began his response haltingly.

"That's a good question," the astronaut flattered the reporter. "First of all, you must realize our position at that time. We were going around the moon, we didn't know what had happened to our space-craft, and we were looking out the window, trying to get as many photographs as we could before we whipped around and came home. At that time, I had *perhaps* thought that we should take so many pictures because maybe this was going to be the last moon flight for a long time. But looking back on it now, and looking back on the way that NASA had responded in helping us get home, I don't believe that anymore. I think it's going to be a situation of just analyzing our problems, and I foresee that we can get this incident over with and charge ahead. I wouldn't be scared to fly with the fix."

Lovell paused and looked around the room. It was not a perfect answer; it was not the answer he would give again if he had a little more time to think it through. But it was, he realized, essentially true. He only hoped that someone else would ask the next question fast, so he could let it go at that.

Another reporter now obliged. "Jim, speaking of that subject, flying again, you told us this was going to be your last flight, but that you did want to walk on the moon before giving up flying. How do you feel now? Would you want to go back and take a crack at Apollo 14, 15, 16, or is Marilyn . . ."

The reporter trailed off, letting the word "Marilyn" hang in the air, and when he did, an appreciative chuckle rippled through the room. Lovell laughed along with the group and waited for silence before beginning his response.

"Well," he said, "I'm very much disappointed, just as Fred is and Jack is, that we couldn't complete the mission. We certainly wanted to make a lunar landing. Fra Mauro has so much to offer, we thought.

But this was my fourth space flight, and there are many people in our organization who have not flown, and who deserve to fly, and who are talented enough to fly. And they deserve a mission. If NASA feels that this team should go back to Fra Mauro, I'm certainly willing to go back. But otherwise, I think other people ought to do it."

Unlike the answer to the previous question, Lovell had not given this one much thought. But even as he spoke the words aloud, he knew that he meant them, too. Four flights *were* enough; more than twenty other pilots *were* waiting; and as the reporter had implied, there *was* the question of Marilyn. After Pax River and Oceana, Gemini 7 and Gemini 12, Apollo 8 and Apollo 13, the wife of the man with more hours in space than any other American had a right to expect that he wouldn't be adding any more time to that tally. Jim Lovell, though a test pilot by nature, training, and long experience, was inclined to respect that expectation.

If the Apollo 13 commander's personal exploration of the moon had come to an end, however, NASA's hadn't. In the factories at Grumman and North American Rockwell and the assembly buildings on the Space Center grounds, there was still a quiver of Saturn 5 boosters and a fleet of Apollo spacecraft ready to be launched. But before Agency flight planners could even discuss sending another crew back into the void, the cause of the accident that nearly killed the last one would have to be determined.

So far, there were few clues. After examining the pictures the Apollo 13 crew had brought back to Earth, NASA concluded that it had not been a meteor or other rogue projectile that had damaged their ship. The injury to Odyssey's hull was a clean one, consistent not with an errant rock striking the ship broadside and destroying an oxygen tank on the way in, but with an explosion of some sort in the tank itself, which released a burst of energy into the body of the craft and blasted away the hull on the way out. On April 17, only hours after the command module hissed into the ocean, NASA administrator Thomas Paine formed a commission to determine just what that event was.

The board Paine appointed was headed by Edgar Cortright, the director of the Agency's Langley Research Center in Virginia. Serving with Cortright were fourteen others, including the still-lionized Neil Armstrong, a dozen engineers and administrators from NASA, and, significantly, an independent observer from outside the Agency. Still chafing from the old-boy, in-house inquiry that had followed the Apollo 1 fire, Congress, NASA knew, would want such an observer to monitor the group's proceedings; still chastened from the outcry in Washington that had followed that secretive inquest, NASA readily cooperated.

The Cortright Commission quickly fell to work, and while none of the men on the panel knew what they would find when they began to look for the cause of the Apollo 13 explosion, they pretty much knew what they wouldn't find: a single smoking gun. As aviators and test pilots had discovered since the days of cloth and wood biplanes, cataclysmic accidents in any kind of craft are almost never caused by one catastrophic equipment failure; rather, they are inevitably the result of a series of separate, far smaller failures, none of which could do any real harm by themselves, but all of which, taken together, can be more than enough to slap even the most experienced pilot out of the sky. Apollo 13, the panel members guessed, was almost certainly the victim of such a string of mini-breakdowns.

The first step the Cortright Commission took in following the space-craft's hardware trail was to examine the long manufacturing history of oxygen tank two. Every major component in an Apollo spacecraft, from gyros to radios to computers to cryogenic tanks, was routinely tracked by quality control inspectors from the moment its first blueprints were drawn to the moment it left the pad on launch day; any anomaly in manufacturing or testing was noted and filed away. Generally, the thicker the file any part amassed by the time it was ready to fly, the more headaches it had caused. Oxygen tank two, it turned out, had quite a dossier.

The problems with the tank began in 1965, around the time Jim Lovell and Frank Borman were deep in training for the flight of Gemini 7, and North American Aviation was building the Apollo command-service module that would ultimately replace the two-man ship. Like any contractor tackling such an enormous engineering job, North American did not attempt to complete all of the design and engineering work by itself, but farmed out individual parts of the project to subcontractors. One of the most delicate of the delegated tasks was the construction of the spacecraft's cryogenic tanks, a job assigned to Beech Aircraft in Boulder, Colorado.

Beech and North American knew that the tanks the new ship needed would have to be more than just insulated bottles. To handle contents as temperamental as liquid oxygen and hydrogen, the spherical vessels would require all manner of safeguards, including fans, thermometers, pressure sensors, and heaters, all of which would have to be immersed directly in the supercold slush that the tanks were designed to hold, and all of which would have to be powered by electricity.

The Apollo spacecraft's electrical system was designed to operate on 28 volts of current — the amount of juice provided by the service module's three fuel cells. Of all of the systems inside the cryogenic tanks that would be driven by this relatively modest power system, none required more rigorous monitoring than the heaters. Ordinarily, cryogenic hydrogen and oxygen were maintained at a constant temperature of minus 340 degrees. This was cold enough to keep the frigid gases in a slushy, non-gaseous state, but warm enough to allow some of the slush to vaporize and flow through the lines that fed both the fuel cells and the atmospheric system of the cockpit. Occasionally, however, the pressure in the tanks dropped too low, preventing the gas from moving into the feed lines and endangering both the fuel cells and the crew. To prevent this, the heaters would occasionally be switched on, boiling off some of the liquid and raising the internal pressure to a safer level.

Of course, immersing a heating element in a pressurized tank of oxygen was, on its face, a risky business, and in order to minimize the danger of fire or explosions, the heaters were supplied with thermostat switches that would cut the power to the coils if the temperature in the tank climbed too far. By most standards, that upper temperature limit was not very high: 80 degrees was about as hot as the engineers ever wanted their supercold tanks to get. But in insulated vessels in which the prevailing temperature was usually 420 degrees lower, that was a considerable warm-up. When the heaters were switched on and functioning normally, the thermostat switches remained closed — or engaged — completing the heating system's electrical circuit and allowing it to continue operating. If the temperature in the tank rose above the 80-degree mark, two tiny contacts on the thermostat would separate, breaking the circuit and shutting the system down.

When North American first awarded the tank contract to Beech Aircraft, the contractor told the subcontractor that the thermostat switches — like most of the other switches and systems aboard the ship — should be made compatible with the spacecraft's 28-volt power grid, and Beech complied. This voltage, however, was not the only current the spacecraft would ever be required to accept. During the weeks and months preceding a launch, the ship spent much of its time connected to launch-pad generators at Cape Canaveral, so that pre-flight equipment tests could be run. The Cape's generators were dynamos compared to the service module's puny fuel cells, regularly churning out current at a full 65 volts.

North American eventually became concerned that such a relative lightning bolt would cook the delicate heating system in the cryogenic tanks before the ship ever left the pad, and decided to change its specs, alerting Beech that it should scrap the original heater plans and replace them with ones that could handle the higher launch-pad voltage. Beech noted the change and modified the entire heating system — or *almost* the entire heating system. Inexplicably, the engineers neglected to change the specifications on the thermostat switches, leaving the old 28-volt switches in the new 65-volt heaters. Beech technicians, North American technicians, and NASA technicians all reviewed Beech's work, but nobody discovered the discrepancy.

Although 28-volt switches in a 65-volt tank would not necessarily be enough to cause damage to a tank — any more than, say, bad wiring in a house would necessarily cause a fire the very first time a light switch was thrown — the mistake was still considerable. What was necessary to turn it into a catastrophe were other, equally mundane oversights. The Cortright Committee soon found them.

The tanks that eventually flew aboard Apollo 13 were shipped on March 11, 1968 — complete with their 28-volt switches — to the North American plant in Downey, California. There, they were attached to a metal frame, or shelf, and installed in service module 106. Module 106 was scheduled to fly during 1969's Apollo 10 mission, when Tom Stafford, John Young, and Gene Cernan would conduct the first test of a lunar module in orbit around the moon. But over the following months, additional technical improvements were made in the design of the oxygen tanks, and the engineers decided to remove the existing tanks from the Apollo 10 service module and replace them with newer ones. The tanks that had been installed on the ship would be upgraded and placed in another service module, for use on another flight.

Removing cryogenic tanks from an Apollo spacecraft was a delicate job. Since it was nearly impossible to separate any one tank from the tangle of pipes and cables that ran from it, the entire shelf — along with all of its associated hardware — would have to be removed. In order to do this, technicians would attach a crane to the edge of the shelf, remove the four bolts that held it in place, and pull the assembly out. On October 21, 1968, the day Wally Schirra, Donn Eisele, and Walt Cunningham splashed down after the eleven-day flight of Apollo 7, Rockwell engineers unbolted the tank shelf in spacecraft 106 and began to lift it carefully from the ship.

Unknown to the crane operators, one of the four bolts had been left in place. When the winch motor was activated, the shelf rose only two inches before the bolt caught, the crane slipped, and the shelf dropped back into place. The jolt caused by the drop was a small one, but the procedure for dealing with it was clear. Any accident on the factory floor, no matter how minor, required that the spacecraft components involved be inspected to ensure that they hadn't suffered any damage. The tanks on the dropped shelf were examined and found to be unharmed. Shortly afterward, they were removed, upgraded, and reinstalled in service module 109, which was to become part of the spacecraft more commonly known as Apollo 13. In early 1970, the Saturn 5 booster with Apollo 13 mounted at its tip was taken out to the launch pad and readied for an April liftoff. It was here, the Cortright Commission discovered, that the final piece of the disaster puzzle fell into place.

One of the most important milestones in the weeks leading up to an Apollo launch was the exercise known as the countdown demonstration test. It was during this hours-long drill that the men in the spacecraft and the men on the ground would first rehearse all of the steps leading up to the actual ignition of the booster on launch day. To make the dress rehearsal as complete as possible, the cryogenic tanks would be fully pressurized, the astronauts would be fully suited, and the cabin would be filled with circulating air at the same pressure used at liftoff.

During Apollo 13's countdown demonstration test, with Jim Lovell, Ken Mattingly, and Fred Haise strapped into their seats, no significant problems occurred. At the end of the long dress rehearsal, however, the ground crew did report a small anomaly. The cryogenic system, which had to be emptied of its supercold liquids before the spacecraft was shut down, was behaving balkily. The draining procedure for the cryogenic tanks was not ordinarily complicated; it required engineers simply to pump gaseous oxygen into the tank through one line, forcing the liquids out through another line. Both hydrogen tanks, as well as oxygen tank one, emptied easily. But oxygen tank two seemed jammed, venting only about 8 percent of its 320 pounds of supercold slush and then releasing no more.

Examining the schematics of the tank and its manufacturing history, the engineers at the Cape and at Beech Aircraft believed they knew what the problem was. When the shelf was dropped eighteen months earlier, they now suspected, the tank had suffered more damage than the factory technicians at first realized, knocking one of the drain tubes in the neck of the vessel out of alignment. This would cause the gaseous oxygen pumped through the line leading into the tank to leak directly into the line leading out of the tank, disturbing almost none of the liquid oxygen it was supposed to be pumping away.

For a spacecraft in which engineers maintained a near-zero tolerance for errors, such a glaring malfunction would ordinarily have set off alarms. In this case, it didn't. The detanking method would be used only during pad tests. During the flight itself, the liquid oxygen contained in the vessel would be channeled out not through the venting tube but through an entirely different set of tubes leading either to the fuel cells or to the atmospheric system that pressurized the cockpit with breathable air. If the engineers could figure out some way to get the tank emptied today, therefore, they could fill it up again on launch day and never have to worry about the fill lines and drain lines again. The technique they came up with was an elegantly simple one.

At its present supercold temperature and relatively low pressure, the liquid in the tank wasn't going anywhere. But what would happen, one of the technicians wondered, if the heaters were used? Why not just flip the warming coils on now, cook the slush up, and force the entire load of O_2 out of the vent line?

"Is this the best solution we've got?" Jim Lovell asked the pad technicians when he was called to a meeting at the Cape's operations building, where the procedure was explained to him.

"The best one we can come up with," he was told.

"The tank's doing everything else it's supposed to do?"

"It is."

"There are no other glitches that you can see?"

"There aren't."

"And the drain tube will have no role during the flight itself?"

"None."

Lovell thought for a moment. "How long would it take just to remove the tank altogether and replace it with a new one?"

"Only forty-five hours, but we'd still have to test it and check it out. If we miss the launch window, the entire flight will be delayed by at least a month."

"Well," Lovell said after another thoughtful pause, "if you're all comfortable with this, then I am too."

Months later, during the Cortright hearings at the Cape, Lovell stood by this decision. "I agreed with the solution," he said. "If it worked, we could launch on time. If it didn't, we would probably have to replace the tank, and the launch date would slip. None of the launch-pad test crew knew that the wrong thermostat was in the tank, or thought what would happen if the heaters stayed on for too long."

But the wrong thermostat switch — the 28-volt switch — *was* in the tank, and, as it turned out, the heaters stayed on for a long, long time. On the evening of March 27, fifteen days before Apollo 13's scheduled liftoff, the warming coils in spacecraft 109's second oxygen tank were flipped on. Given the huge load of O_2 trapped in the tank, the engineers figured it would take up to eight hours before the last few wisps of gas would vent away. Eight hours was more than enough time for the temperature in the tank to climb above the 80-degree mark, but the technicians knew they could rely on the thermostat to take care of any problem. When *this* thermostat reached the critical temperature, however, and tried to open up, the 65 volts surging through it fused it instantly shut.

The technicians on the Cape launch pad had no way of knowing that the tiny component that was supposed to protect the oxygen tank had welded closed. A single engineer was assigned to oversee the detanking procedure, but all his instruments told him about the cryogenic heater was that the contacts on the thermostat remained shut as they should be, indicating that the tank had not heated up too much. The only possible clue that the system was not functioning properly was provided by a gauge on the launch pad's instrument panel that constantly monitored the temperature inside the oxygen tanks. If the readout climbed above 80 degrees, the technician would know that the thermostat had failed, and he would shut the heater off manually.

Unfortunately, the readout on the instrument panel wasn't *able* to climb above 80 degrees. With so little chance that the temperature inside the tank would ever rise that far, and with 80 degrees representing the bottom of the danger zone, the men who designed the instrument panel saw no reason to peg the gauge any higher, designating 80 as its upper limit. What the engineer on duty that night didn't know — couldn't know — was that with the thermostat fused shut, the temperature inside this particular tank was climbing indeed, up to a kiln-like 1,000 degrees.

For most of the evening the heater was left running, all the while the temperature needle registering a warm but safe 80. At the end of eight hours, the last of the troublesome liquid oxygen had cooked away as the engineers had hoped it would — but so too had most of the Teflon insulation that protected the tank's internal wiring. Coursing through the now empty tank was a web of raw, spark-prone copper, soon to be reimmersed in the one liquid likelier than any other to propagate a fire: pure oxygen.

Seventeen days later and nearly 200,000 miles out in space, Jack Swigert, responding to a routine daily request from the ground, switched on the cryogenic fan to stir up the contents of the oxygen tanks. The first two times Swigert had complied with this instruction, the fan had operated normally. This time, however, a spark flew from a naked wire, igniting the remains of the Teflon. The sudden buildup of heat and pressure in the pure-oxygen environment blew off the neck of the tank, the weakest part of the vessel. The 300 pounds of oxygen inside the tank flashed instantly into gas and filled bay four of the service module, blowing out the ship's external panel and causing the bang that so startled the crew. As the curved piece of hull flew past, it collided with the orbiter's high-gain antenna, causing the mysterious channel switching that the communications officer on the ground reported at the same moment the astronauts were reporting their bang and jolt.

Though tank one was not directly damaged by the blast, it did share some common plumbing with tank two; as the explosion ripped these delicate pipes away, the undamaged tank found a leak path through the lines and bled its contents away into space. Making matters worse, when the explosion shook the ship, it caused the valves that fed several of the attitude-control thrusters to slam shut, permanently disabling those jets. As the ship rocked from both the tank one venting and the explosion itself, the autopilot began firing the thrusters to try to stabilize the spacecraft's attitude. But with only some of the jets working, the ship could not regain its footing. When Lovell took over manual control of the half-crippled attitude system, his luck was little better. Within two hours, the spacecraft was drifting and dead.

These were the theories, at least, that the members of the Cortright Commission came up with, but it was only when their engineering hunches were put to the test that they were confirmed. In vacuum chambers at the Space Center in Houston, technicians switched on a heater in a sample tank precisely as Apollo 13's heater had been switched on and found that the thermostat did in fact fuse shut; they then left the heater on just as Apollo 13's heater had been left on and found that the Teflon on its wires indeed burned away; finally, they stirred up its cryogenics exactly as Apollo 13's cryos had been stirred and found that a spark indeed flew from a wire, causing the sample tank to rupture at the neck and blow off the side panel of a sample service module with it.

The only other mystery that had yet to be solved was what had caused the shallowing of the trajectory on the way home, and it was left to the TELMUs to dope this one out. Aquarius, so these flight controllers concluded, had been pushing itself steadily off course, not with some undetected leak from a damaged tank or pipe, but from wisps of steam wafting from its cooling system. The tendrils of vapor that the water-based sublimator emitted as it carried excess heat out into space had never disturbed a LEM's trajectory before, but only because the lander was typically not powered up until it was already in lunar orbit, ready to separate from the mother ship and descend to the surface. For such a short-haul trip, the invisible plume of steam would not be strong enough to nudge the lander in any one direction. Over the course of a slow, 240,000-mile glide back to Earth, however, the almost unmeasurable thrust would be more than enough to alter the spacecraft's flight path, pushing it out of its reentry corridor altogether.

It was in late spring that the Cortright Commission released its findings, contritely acknowledging that none of these technical problems should have happened in the first place, but implying that the problems were *merely* technical — that NASA had at least avoided the specter of three dead astronauts perpetually orbiting the Earth in an equally dead spacecraft.

Most of the Houston space community pounced on the report when it was published, but Jim Lovell, Jack Swigert, and Fred Haise were not among them. By that time, the men whose lives had been most directly affected by the fused thermostat and the low-pegged thermometer and the blasted tank and the steaming sublimator were out of the country altogether, busy with the five-nation tour the Agency had planned for them as one of the final chores associated with their mission.

Eight months after the Apollo 13 crew returned from their goodwill trip, Apollo 14 — equipped with upgraded thermostat switches, shielded wires, and a third oxygen tank installed on a separate shelf in its service module — took off for Fra Mauro. Jim Lovell spent much of that flight in Mission Control, watching expressionlessly as Al Shepard and Ed Mitchell left footprints in the soil of the foothills where he and Fred Haise would never get to tread. Shortly afterward, Lovell, permanently out of the lunar flight rotation, left the Apollo program and transferred to the shuttle program, which was just gearing up. There he worked with contractors as they drew up their proposals for the new ship's massive instrument panel.

One afternoon when Lovell was out at the McDonnell Aircraft plant in St. Louis, studying blueprints and switch placements, and examining mockup dashboards, he glanced up and took a slow look around himself. It dawned on him that it was in this very room, at this very plant, that he had worked fifteen years before, a young Navy man just in from Pax River helping to design the instrument panel for the new F4H Phantom. After nearly a generation of flying that included two fiery rides into Earth orbit and two more out to the neighborhood of the moon, he suddenly realized he had completed the circle. That night, Jim Lovell climbed into his T-38 and returned home — this time for good — to his family in Timber Cove.

The rest of the clan showed up at Jim and Marilyn Lovell's house in Horseshoe Bay just before noon on Christmas Eve. Like all such arrivals since the fifth and sixth and seventh grandchildren were born, this one was clamorous. Coming through the door first were sixteen-year-old Lauren, fourteen-year-old Scott, and nine-year-old Caroline. Following them in a more boisterous swirl were Thomas, twelve, Jimmy, eight, and John, four. Following *them* were their already frazzled parents. Allie, the baby, only just settling down from her breathless exploration of the breakables in the house, perked up again in the presence of so many faces and crawled over to join the crush. Hellos were exchanged and bundles were dropped, and then, as Lovell could have predicted, one of his grandchildren, John, sped toward his den. There had not been a visit that Lovell could recall when John had not gravitated toward the woody room with all the toy-like trophies, and not a time when Lovell hadn't wondered if his grandson saw all the knickknacks as *more* than toys.

Today, Lovell allowed John to play alone for a few minutes, and then wandered in behind him. Like so many times before, John had stationed himself in front of a lunar globe standing in a corner of the room. The globe was a big one, three and a half feet in diameter, and had been hand painted to capture even the smallest features on the surface of the mottled moon. Scattered across the face of the sphere were also fifteen tiny paper arrows, glued to the spots where manned and unmanned vehicles had touched down over the years. There were the sites of the American Ranger probes and the Russian Luna probes, the American Surveyors and the Russian Lunokhods. And, of course, there were the American Apollos.

On Lovell's globe at this moment, none of those arrows, or any of the other surface details, were visible. John, as was his custom, had set the big ball spinning and was staring at it intently, speeding it up with a push from his right hand every time it threatened to slow. Lovell stood nearby, watching the craters and mares, hills and rills, spin by in one great monochrome blur, and then walked up behind his grandson. Reaching out, he slowed the globe with the flat of one hand, and with the other led the boy away, toward the windowsill where the optical sight from Aquarius sat.

"John," the former commander said, "let me show you something you might like."

Behind Lovell, the lunar globe came squeakily to a stop, one of its little paper arrows pointing perpetually to Fra Mauro.

Appendix A

APOLLO 13 MISSION TIMELINE

GET *(ground elapsed time)*	*Houston Time and Date*	*Mission Event*
00:00:00	Sat. April 11 1:13 P.M.	Liftoff
2:35:46	Sat. April 11 3:48 P.M.	Translunar injection
30:40:50	Sun. April 12 7:53 P.M.	Mid-course correction burn to leave free-return trajectory
55:11:00	Mon. April 13, 8:24 P.M.	Beginning of last TV transmission
55:54:53	Mon. April 13, 9:07 P.M.	Oxygen tank two explodes
57:37:00	Mon. April 13 10:50 P.M.	Crew abandons Odyssey
61:29:43	Tues. April 14 2:43 A.M.	Aquarius's engine fired for return to free-return trajectory
77:02:39	Tues. April 14 6:15 p.m.	Spacecraft disappears around far side of moon

GET (ground elapsed time)	Houston Time and Date	Mission Event
79:27:39	Tues. April 14 8:40 P.M.	Aquarius's engine fired for PC+2 speed-up burn
86:24:00	Wed. April 15 3:38 A.M.	Crew begins constructing lithium hydroxide adapters
97:10:05	Wed. April 15 2:23 P.M.	Battery two in Aquarius explodes
105:18:28	Wed. April 15 10:31 P.M.	Aquarius's engine fired to correct trajectory
108:46:00	Thurs. April 16 1:59 A.M.	Aquarius's helium disk bursts
137:39:52	Fri. April 17 6:52 A.M.	Aquarius's attitude jets fired to correct trajectory
138:01:48	Fri. April 17 7:14 A.M.	Service module jettisoned
141:30:00	Fri. April 17 10:43 A.M.	Aquarius jettisoned
142:40:46	Fri. April 17 11:53 A.M.	Reentry begins
142:54:41	Fri. April 17 12:07 P.M.	Splashdown

Appendix B

APOLLO 13 DRAMATIS PERSONAE

John Aaron — Electrical and environmental command officer (EECOM), Maroon Team

Arnie Aldrich — Systems chief, Flight Operations Directorate

Don Arabian — Director, Mission Evaluation Room

Stephen Bales — Guidance officer (GUIDO), Maroon Team

Jules Bergman — Science correspondent, ABC News

George Bliss — EECOM backroom engineer, White Team

Bill Boone — Flight dynamics officer (FIDO), Black Team

Jerry Bostick — FIDO, Maroon Team

Vance Brand — Capsule communicator (Capcom) and astronaut, Gold Team

Dick Brown — EECOM backroom engineer, White Team

Clint Burton — EECOM, Black Team

Gary Coen — Guidance, navigation and control officer (GNC), Maroon Team

Edgar Cortright — Director, NASA's Langley Research Center

Chuck Deiterich — Retrofire officer (RETRO), Gold Team

Brian Duff — Director of Public Affairs, Manned Spacecraft Center, Houston

Charlie Duke	Apollo 13 backup LEM pilot; Apollo 16 prime LEM pilot
Charlie Dumis	EECOM, White Team
Max Faget	Director, engineering and development branch, Manned Spacecraft Center
Bill Fenner	GUIDO, White Team
Bob Gilruth	Director, Manned Spacecraft Center
Alan Glines	Instrumentation and communications officer (INCO), White Team
Jay Greene	FIDO, Maroon Team
Gerald Griffin	Flight director, Gold Team
Fred Haise	Apollo 13 lunar module pilot
Jerry Hammack	Chief of spacecraft recovery team
Willard Hawkins	Flight surgeon, White Team
Bob Heselmeyer	Telemetry, electrical, EVA mobility unit officer (TELMU) for the lunar module, White Team
Tom Kelly	Lunar module engineering manager, Grumman Aerospace
Joe Kerwin	Capcom and astronaut, Maroon Team
Jack Knight	TELMU, Maroon Team
Chris Kraft	Deputy director, Manned Spacecraft Center
Gene Kranz	Lead flight director, White Team
Sy Liebergot	EECOM, White Team
Hal Loden	Lunar module flight control officer (CONTROL), Black Team
Jack Lousma	Capcom and astronaut, White Team
Jim Lovell	Apollo 13 commander
George Low	Director of Spacecraft and Flight Missions
Glynn Lunney	Flight director, Black Team
Ken Mattingly	Apollo 13 prime command module pilot; Apollo 16 backup command module pilot

Jim McDivitt	Gemini 4 and Apollo 9 commander; head of Apollo Program office
Bob McMurrey	NASA protocol officer
Merlin Merritt	TELMU, Black Team
Thomas Paine	NASA administrator
Bill Peters	TELMU, Gold Team
Dave Reed	FIDO, Gold Team
Gary Renick	GUIDO, Black Team
Mel Richmond	Recovery officer
Ken Russell	GUIDO, Gold Team
Phil Schaffer	FIDO, Gold Team
Larry Sheaks	EECOM backroom engineer, White Team
Sig Sjoberg	Director of Flight Operations
Deke Slayton	Director of Flight Crew Operations; astronaut
Ed Smylie	Chief of Crew Systems Division; inventor of lithium hydroxide adapter
Bobby Spencer	RETRO, White Team
Bill Stoval	FIDO, White Team
Bill Strable	GNC, White Team
Larry Strimple	CONTROL, White Team
Jack Swigert	Apollo 13 command module pilot
Ray Teague	GUIDO, White Team
Dick Thorson	CONTROL, Gold Team
Glenn Watkins	Propulsion officer, TELMU backroom
John Wegener	CONTROL, Maroon Team
Tom Weichel	RETRO, Black Team
Terry White	NASA Public Affairs officer
Buck Willoughby	GNC, Gold Team
Milt Windler	Flight director, Maroon Team
John Young	Apollo 13 backup commander; Apollo 16 prime commander

Appendix C

THE MANNED APOLLO MISSIONS

APOLLO 7

Crew: Wally Schirra, commander
Donn Eisele, command module pilot
Walt Cunningham, lunar module pilot
Launched: October 11, 1968
Splashdown: October 21, 1968
Mission: First Earth-orbit test of the Apollo command-service module.
No lunar module.

APOLLO 8

Crew: Frank Borman, commander
Jim Lovell, command module pilot
Bill Anders, lunar module pilot
Launched: December 21, 1968
Splashdown: December 27, 1968
Mission: First manned orbit of the moon. Command-service module
only.

APOLLO 9

Crew: James A. McDivitt, commander
Dave Scott, command module pilot
Rusty Schweickart, lunar module pilot

Launched: March 3, 1969
Splashdown: March 13, 1969
Mission: First Earth-orbit test of both command-service module and
  lunar module.

APOLLO 10

Crew: Tom Stafford, commander
        John Young, command module pilot
        Gene Cernan, lunar module pilot
Launched: May 18, 1969
Splashdown: May 26, 1969
Mission: First test of both command-service module and lunar module
  in orbit around the moon. Stafford and Cernan pilot LEM to within
  50,000 feet of the lunar surface.

APOLLO 11

Crew: Neil Armstrong, commander
        Michael Collins, command module pilot
        Buzz Aldrin, lunar module pilot
Launched: July 16, 1969
Splashdown: July 24, 1969
Mission: First lunar landing. Armstrong and Aldrin land in Sea of
  Tranquillity and spend 2 hours and 31 minutes walking on the
  moon. Collins orbits overhead in the command module.

APOLLO 12

Crew: Pete Conrad, commander
        Dick Gordon, command module pilot
        Alan Bean, lunar module pilot
Launched: November 14, 1969
Splashdown: November 24, 1969
Mission: Second lunar landing. Conrad and Bean land in Ocean of
  Storms, collect rocks and retrieve parts from unmanned Surveyor
  spacecraft, which landed nearby in April 1967.

APOLLO 13

Crew: Jim Lovell, commander
Jack Swigert, command module pilot
Fred Haise, lunar module pilot
Launched: April 11, 1970
Splashdown: April 17, 1970
Mission: Third attempted lunar landing. At 55 hours, 54 minutes, and 53 seconds into the mission, a cryogenic tank explodes, causing a loss of breathable oxygen and power in the command-service module. Crew abandon ship and survive in the LEM until just a few hours before splashdown, when they return to the command module, jettison the LEM, and reenter the atmosphere.

APOLLO 14

Crew: Alan Shepard, commander
Stuart Roosa, command module pilot
Ed Mitchell, lunar module pilot
Launched: January 31, 1971
Splashdown: February 9, 1971
Mission: Third lunar landing. Shepard and Mitchell touch down in the Fra Mauro highlands, the intended destination of Apollo 13.

APOLLO 15

Crew: Dave Scott, commander
Al Worden, command module pilot
Jim Irwin, lunar module pilot
Launched: July 26, 1971
Splashdown: August 7, 1971
Mission: Fourth lunar landing. Scott and Irwin touch down at Hadley Rille in the Apennine Mountains. First test of the four-wheel-drive lunar roving vehicle.

APOLLO 16

Crew: John Young, commander
Ken Mattingly, command module pilot
Charlie Duke, lunar module pilot

Launched: April 16, 1972

Splashdown: April 27, 1972

Mission: Fifth lunar landing. Young and Duke landed in the Cayley-Descartes highlands, drive lunar roving vehicle 16.8 miles, and collect 213 pounds of lunar samples.

APOLLO 17

Crew: Gene Cernan, commander
Ron Evans, command module pilot
Harrison Schmitt, lunar module pilot

Launched: December 7, 1972

Splashdown: December 19, 1972

Mission: Sixth and last moon landing. Cernan and Schmitt touch down in the Taurus Mountains near the Littrow crater, collect 243 pounds of samples, and lift off from the lunar surface after seventy-five hours and three moonwalks.

AUTHORS' NOTES

It is one of the ironies of historical journalism that telling the story of a headline-making event can often take longer than the event itself. The crew of Apollo 13 needed about two years to train for their mission to the moon and then flew it in just six days. Researching and writing this book did not exceed that total by much — about two and a half years from inception to completion — but it did exceed it.

Like many nonfiction books of this kind, one of the authors was also one of the participants in the history being recounted; unlike many books of this kind, *Apollo 13* is written in the third person. Had the key events of the mission taken place exclusively in the spacecraft, a first-person account, in the singularly well-informed voice of the commander of the mission, would have made the most narrative sense. But as the men and women who were involved in the flight uniformly agree, the tale of Apollo 13 was one with many venues. For this reason, we have tried to take the reader to as many of those places as possible — newsrooms, conference rooms, homes, hotels, factories, naval vessels, offices, ready rooms, laboratories, and of course Mission Control and the spacecraft themselves. To achieve this kind of omniscient sweep, the third-person voice seemed the only way to go.

Happily, even twenty-three years after the mission of Apollo 13 ended, reconstructing the flight was a relatively straightforward matter. Thousands of pages of documents and hundreds of hours of tapes pertaining to both the mission itself and the investigation that followed have been preserved in NASA's libraries, all of which were graciously made available to us. Most helpful were the recordings and transcripts of the conversations conducted throughout the flight on the flight director's loop, the air-to-ground loop, and the various backroom loops around Mission Control. Often, these exchanges make for compelling listening and reading.

Just as often, however, they lapse necessarily into technical jargon. Therefore, though the in-flight conversations included in the text were taken directly from the tapes and transcripts, in many cases we have edited, compressed, or paraphrased them in the interests of both comprehensibility and pacing. But in no case has the meaning or substance of any of the exchanges been changed. Other dialogue included in the book that was not preserved on tape or paper was reconstructed through interviews with at least one — and usually more than one — of the principals involved. Information about Jack Swigert's thoughts and state of mind were taken from his writings, the recollections of his crewmates, and from a taped interview conducted before his death and generously made available to us by screenwriter and filmmaker Al Reinert.

It goes without saying — though we would be remiss in not saying it — that just as the Apollo 13 astronauts were incalculably indebted to a modest-sized army of people for bringing them home safely, so too do we owe thanks to a somewhat smaller group for giving of their time in order to help make this book possible. Many of these people were the same folks who performed so heroically during that harrowing week in the middle of 1970. Others were people who remember Apollo 13 merely as a historical event, but who had the wisdom to recognize it as one worth recounting.

Among those people in the first group to whom we're the most indebted are Gene Kranz, Chris Kraft, Sy Liebergot, Gerald Griffin, Glynn Lunney, Milt Windler, John Aaron, Fred Haise, Chuck Deiterich, and Jerry Bostick. Also lending indispensable assistance were Don Arabian, Sam Beddingfield, Collins Bird, Clint Burton, Gary Coen, Brian Duff, Bill Fenner, Don Frenk, Chuck Friedlander, Bob Heselmeyer, John Hoover, Walt Kapryan, Tom Kelly, Howard Knight, Russ Larsen, Hal Loden, Owen Morris, George Paige, Bill Peters, Ernie Reyer, Mel Richmond, Ken Russell, Andy Saulietis, Ed Smylie, Dick Snyder, Wayne Stallard, John Strakosch, Jim Thompson, Dick Thorson, Doug Ward, Guenter Wendt, and Terry Williams.

Providing special insight was also a small and decidedly elite corps of men who could understand, perhaps better than anyone else, what the crew of Apollo 13 experienced during their voyage, and who gave their time to share their thoughts. That rarefied group includes Buzz Aldrin, Bill Anders, Neil Armstrong, Frank Borman, Scott Carpenter, Pete Conrad, Gordon Cooper, Charlie Duke, Dick Gordon, Jack Lousma, Jim McDivitt, Wally Schirra, and Deke Slayton.

For helping to open the doors and the archives of NASA we would also like to acknowledge Brian Welch, of the Johnson Space Center's Public Affairs office; Hugh Harris and Ed Harrison, of the Kennedy Space Center's

Public Affairs office; Peter Nubile, of the NASA audio office; and especially Lee Saegesser of the NASA history office in Washington, D.C.

In addition to the people in the space community who lent us their help, innumerable people in publishing and journalism also contributed their time and energy. Without the considerable talents and limitless enthusiasm of Joy Harris, of the Lantz-Harris Literary Agency, and Mel Berger, of the William Morris Agency, there would have been no *Apollo 13*. And without the practiced eye and editorial guidance of John Sterling of Houghton Mifflin Company, the *Apollo 13* we initially envisioned would never have been improved and focused into the *Apollo 13* that ultimately appeared.

While nearly all of our thanks are extended jointly, each of us would also like to acknowledge some folks individually. Jim Lovell could never have made it through Gemini 7, Gemini 12, Apollo 8, and, especially, Apollo 13, without the love and support of Marilyn, Barbara, Jay, Susan, and Jeffrey, and could never have undertaken the effort to tell the story of those flights without the same love and support of the same people. Special thanks go to Marilyn, for reading every page of the manuscript as it was produced, Darice Lovell, for her patience and skill in typing revisions, and Mary Weeks, for her extraordinary secretarial help.

Jeffrey Kluger also extends thanks and love to Splash, Steve, and Garry Kluger, Bruce Kluger and Alene Hokenstad for their unflagging support and for listening, with something closely resembling interest, to endless descriptions of the science of gimbal lock and the physics of descent propulsion burns. Moon-sized gratitude also goes to the folks at *Discover* magazine and Disney Publishing, especially Marc Zabludoff and Rob Kunzig for reading and — on carefully selected occasions — advising; Dave Harmon and Denise Eccleston, for providing a wonderful place to work and play; and most especially Lori (T.C.) Oliwenstein, without whose timely — and succinctly expressed — encouragement this book would surely never have been written. Admiration and appreciation also go to Taj Jackson, as well as to Nancy Finton, Josie Glausiusz, and Theres Luthi, of New York University's Science and Environmental Reporting Program, for transcribing hours upon hours of no doubt incomprehensible interviews. Finally, thanks also go to Evelyn Windhager, for her generous reader's eye; Marnie Cooper, for her generous enthusiasm; and David Paul Jalowsky, for good advice and counsel a long time ago.

INDEX